Springer Tracts in Civil Engineering

Springer Tracts in Civil Engineering (STCE) publishes the latest developments in Civil Engineering - quickly, informally and in top quality. The series scope includes monographs, professional books, graduate textbooks and edited volumes, as well as outstanding PhD theses. Its goal is to cover all the main branches of civil engineering, both theoretical and applied, including:

- Construction and Structural Mechanics
- Building Materials
- Concrete, Steel and Timber Structures
- Geotechnical Engineering
- Earthquake Engineering
- Coastal Engineering; Ocean and Offshore Engineering
- Hydraulics, Hydrology and Water Resources Engineering
- Environmental Engineering and Sustainability
- Structural Health and Monitoring
- Surveying and Geographical Information Systems
- Heating, Ventilation and Air Conditioning (HVAC)
- Transportation and Traffic
- Risk Analysis
- Safety and Security

Indexed by Scopus

To submit a proposal or request further information, please contact:
Pierpaolo Riva at Pierpaolo.Riva@springer.com (Europe and Americas) Wayne Hu at wayne.hu@springer.com (China)

Gang Wu · De-Cheng Feng · Chun-Lin Wang

Novel Precast Concrete Structure Systems

东南大学出版社
SOUTHEAST UNIVERSITY PRESS

Springer

Gang Wu
Southeast University
Nanjing, China

De-Cheng Feng
Southeast University
Nanjing, China

Chun-Lin Wang
Southeast University
Nanjing, China

ISSN 2366-259X ISSN 2366-2603 (electronic)
Springer Tracts in Civil Engineering
ISBN 978-981-19-6820-4 ISBN 978-981-19-6821-1 (eBook)
https://doi.org/10.1007/978-981-19-6821-1

Jointly published with Southeast University Press
The print edition is not for sale in China (Mainland). Customers from China (Mainland) please order the print book from: Southeast University Press.
ISBN of the Co-Publisher's edition: 978-7-5641-8846-7

This Springer imprint is published by the registered company Springer Nature Singapore Pte Ltd.
The registered company address is: 152 Beach Road, #21-01/04 Gateway East, Singapore 189721, Singapore

Preface

The scale of infrastructure construction in China is huge, and the total annual construction volume exceeds that of other countries in the world combined. However, the current traditional construction methods have shortcomings such as high labor intensity, serious environmental pollution, high energy consumption, and backward construction modes. In order to promote the structural adjustment, transformation, and upgrading of China's construction industry, the State Council has clearly put forward a series of development strategies such as "promoting construction industrialization actively" since 2011. The new type of construction industrialization with the development direction of "greening, industrialization, and informatization" has been valued and risen.

The prefabricated building is the basic way to realize the industrialization of construction. It adopts industrialized production and prefabricated construction, which is in line with the transformation of the national construction industry from extensive to refined. At present, the research on prefabricated concrete structures in China is basically focused on the "equivalent cast-in-place" structure. The so-called equivalent cast-in-place refers to that the design and construction of prefabricated structures are aimed at achieving the performance of cast-in-place structures. Since 2010, among the projects funded by the National Natural Science Foundation of China, there have been 13 projects for "equivalent cast-in-place" prefabricated concrete structures; In the national key R&D plan of the 13th five-year plan, hundreds of millions of yuan have been invested in the "equivalent cast-in-place" prefabricated concrete structure project. With the relevant codes and regulations clearly proposing the "equivalent cast-in-place" prefabricated structure design and construction method, a relatively complete "equivalent cast-in-place" structural design theory has been formed.

For another branch of the prefabricated concrete structure system, the "non-equivalent cast-in-place" structure, the research on its basic performance and the application in practical engineering are still in its infancy. Compared with the "equivalent cast-in-place" type, the "non-equivalent cast-in-place" prefabricated concrete structure, which connects all parts together by bolts, welding, or prestressing, has the advantages of convenient installation, rapid construction, superior performance, and so on. There is also no on-site wet operation, the construction efficiency is greatly improved. At the same time, the self-reset of the node and the quick repair after the earthquake can be realized by adding energy-consuming parts. Obviously, the "non-equivalent cast-in-place" structure is more in line with the essence of "fabricated" technology. However, due to the wide variety of "non-equivalent cast-in-place" structures, the complex stress mechanism, the lack of a unified and reliable design theory, and the lack of a complete design–production–construction system, this situation greatly limits the promotion and practical engineering application of the "non-equivalent cast-in-place" connection structure.

Faced with the development background of the "equivalent cast-in-place" structure and the technical status of the "non-equivalent cast-in-place" structure, a group of colleges and universities, mainly Southeast University, have carried out in-depth research under the support of the National Key R&D plan (2016YFC0701400) and the National Natural Science Foundation of China (51838004). This book sorts out and summarizes the above research results, systematically expounds seven new prefabricated concrete structural systems, and focuses on the structural characteristics, stress modes, performance advantages, and design methods of each new prefabricated concrete structural system. And the combination of the new structural system and construction industrialization technology and the advantages compared with the traditional "equivalent cast-in-place" structural system are analyzed. Relevant achievements can promote the basic theoretical research and technical development and application of "non-equivalent cast-in-place" prefabricated concrete structural systems, complementing the existing "equivalent cast-in-place" type each other, and build a complete prefabricated concrete structure system, which provides theoretical and technical support for large-scale application of fabricated concrete structures in china. This book has prominent emphasis and rich contents, which can be used as a reference book for the personnel in R&D, design and construction of prefabricated concrete structures in China.

The book is divided into eight chapters, the main contents include: In Chapter 1 Introduction, systematically combing the characteristics and advantages of China's prefabricated concrete structures, combined with the classification and application status of typical systems, analyzes its development limitations and future trends. In Chapter 2 Ductile Precast Concrete Frame with Dry-Connections, based on the component characteristics of the ductile connection device, focuses on its influence on the seismic performance of the prefabricated concrete frame, as well as the shear force transfer mechanism of the system and the design method of the member capacity. In Chapter 3 Prestressed Precast Concrete Frame with External Dissipaters, based on three types of bamboo-shaped energy-dissipating rods of all-steel, aluminum alloy, and partially restrained, from the external replaceable and prestressed assembly levels, the damage control and performance effects of the prefabricated concrete frame are emphatically described. In Chapter 4 Friction Damped Self-Centering Precast Concrete Frame system, focuses on the friction energy dissipation from five aspects: the construction mechanism of the prefabricated structure, the test of the nodal energy dissipation device, the numerical modeling of the structure, the seismic performance of the structure and the analysis of the structural fragility. In Chapter 5 Cast-In-Place Frame-Prefabricated Sub-Frame System, combined with the design and construction problems of prefabricated reinforced concrete structures in high-intensity areas and the energy dissipation and shock absorption technology, focuses on the the energy consumption of the primary and secondary frame connections is higher than the connection performance of the nodes and the seismic requirements between the primary and secondary structures. In Chapter 6 Prefabricated Rocking Wall Structural System, combined with theoretical analysis, experimental research, numerical simulation, and other methods, focuses on describing the seismic performance of the new rocking wall. Damage control mechanism and residual displacement control ability. In Chapter 7 Prefabricated Concrete Cassette Structure system, from the perspective of two types of substructures of hollow sandwich panel and grid frame wall, focuses on the static and dynamic performance of the system, combined with the actual Engineering comparison and analysis of its economy. In Chapter 8 Modularized Suspended Building Structure system, based on the concept of substructure modularization of suspension structure and shaking table test, mainly describes the vibration reduction mechanism, force transmission characteristics, and optimization method of the system.

The writing of this book was jointly completed by Prof. Wu Gang, Prof. Wu Jing, Prof. Zhou Zhen, Prof. Wang Chunlin, Prof. Feng De-Cheng, and Prof. Wang Jian of Harbin Institute of Technology (Wu Gang, Feng De-Cheng: Chaps. 1, 6, 7, and 8; Wu Jing: Chap. 2; Wang Chunlin: Chaps. 3 and 8; Zhou Zhen: Chap. 4; Wang Jian: Chap. 5), the whole book was drafted by Wu Gang and Feng De-Cheng. During the writing process, doctoral candidates Wang Zhun, Cao Xuyang, Ye Zhihang, Chen Zhipeng, Cui Haoran, Liu Ye, Xie Luqi, etc., made great contributions, and I would like to express my heartfelt thanks here.

This book strives to present the latest research and practical results of prefabricated structural systems to readers. Due to the rush of time, the current development of prefabricated buildings (especially "non-equivalent cast-in-place" structures) is in its infancy, and the limitation of engineering practice and technical accumulation, there must be omissions and deficiencies in the book. Your suggestions would be appreciated.

Nanjing, China

Gang Wu
De-Cheng Feng
Chun-Lin Wang

Contents

Chapter 1
Introduction

Abstract Building industrialization is widely comprehended as a new mode of production integrating design, production, construction, and other sustainable developments of the entire construction industry chain. It is the major direction to upgrade the construction industry in China. The development of precast/prefabricated building structures is the key part of building industrialization. In this chapter, firstly, the essential advantages of prefabricated constructions are summarized. Secondly, prefabricated concrete structures are categorized into several major types (including frame, shear wall, and frame-shear wall system) and then minor types within each major type, with comprehensive illustration and discussion. Thirdly, several well-applied systems, as well as their key technologies and representative cases, are introduced. Lastly, the limitations of the current prefabricated concrete structure systems are discussed, mainly from the perspective of the currently prevailing wet connections and the gradually emerging dry connections. The necessity to deeply study the "nonequivalent cast-in-place" prefabricated concrete structure system is pointed out.

Keywords Building industrialization · Prefabricated building · Essential advantages · Categorization · Wet connection and dry connection · Nonequivalent cast-in-place concrete structure

1.1 Background

As one of the pillar industries of China's national economy, the construction business has been booming in the last 20 years with the sustained and rapid development of the national economy. However, its extensive decentralized, low-level and inefficient traditional handicraft production mode still occupies a dominant position, which does not match current large-scale economic construction and is far from the requirements for the development of new urbanization, industrialization, and information technology. The existing approach cannot meet the requirements of the entire construction industry and social progress.

With the improvement of energy conservation and environmental protection requirements and the fading out of China's demographic dividend, the phenomena of "recruitment difficulty" and "labor shortage" in the construction industry have

G. Wu et al., *Novel Precast Concrete Structure Systems*, Springer Tracts in Civil Engineering, https://doi.org/10.1007/978-981-19-6821-1_1

appeared and are still worsening. The traditional construction mode has been unsustainable. At present, the construction industry has become the largest single energy-consumptive industry in our country. According to statistics, in 1993, China's building energy consumption accounted for only 16% of the total energy consumption of the entire society, rising to 28% in 2012. The energy consumption per unit of building area is 2–3 times that of developed countries. If effective measures are not taken, the consumption will increase to more than 3 times that of the present by 2020. On the other hand, construction in China mainly adopts the traditional production mode dominated by on-site construction, with the problems of a low degree of industrialization, poor working environment, high labor intensity, serious environmental pollution, backward construction mode, loss of building materials such as cement, steel and wood and a large amount of construction waste. These are not only one of the main sources of $PM_{2.5}$ and urban noise, but also one of the biggest obstacles to energy conservation and emission reduction. Environmental protection, energy savings, low carbon emissions, and emission reduction in the construction sector have become the main contradictions affecting the transformation of China's national economic growth mode and the sustainable development of the national economy. Therefore, to change the current situation of China's construction business, we must eliminate the dependence on and bondage to the traditional mode path and strive to seek a new direction for the industrial development of new buildings.

The industrialization of new construction is a new mode of production integrating design, production, construction, and other sustainable developments of the entire construction industry chain. This is the direction of development of the construction industry. At present, China is in a critical period of economic transformation. The construction business is facing the important tasks of updating the development concept, reforming the mode of production, and transforming the production results. In this special historical period, it is of great significance to promote the industrialization of new construction. First, it is necessary to realize both the leap of construction from "construction" to "manufacturing", and the transformation of the production mode of the construction industry to one with high efficiency, low carbon footprint, and environmental protection requirements. Second, it effectively improves the quality of science and technology in the construction industry, reduces resource consumption and environmental pollution, promotes the optimization and upgrading of the construction industry structure, and promotes the transformation of the development model of the construction industry from extensive to intensive, efficient and scientific. Third, through modular design, factory manufacturing, integrated construction, building factory production and construction capacity, it significantly improves the labor production efficiency of the construction industry while more effectively ensuring safety and quality.

The universally recognized comprehensive definition of "building industrialization" is proposed in the Policy and Measure Guidance of The Government to gradually realize the industrialization of the construction industry (1974) issued by the United Nations, that is, to build the construction industry by the mode of large-scale industrial production. Its core is design standardization, factory processing and production, on-site installation and assembly, and scientific organization management. Its main purpose is to change the traditional construction industry production model through the use of new technological achievements, improve construction efficiency, speed up construction speed, improve the quality of the project, reduce the construction cost, and optimize production safety and environmental effects. In the past 10 years, China's central government has issued several policies to encourage building industrialization, which brightens the prospect of building industrialization. Zhejiang, Jiangsu, Shandong, and other provinces have also issued relevant documents related to the industrialization of the construction industry, vigorously promoting its industrialization.

In June 2006, the National Housing Industrialization Base trial method was issued. This indicates that we should rely on technological innovation to promote the transformation of the extensive housing construction mode and to improve the standardization and industrialization of the housing industry. It emphasizes developing a new housing system of land and energy savings and enhancing the sustainable development ability of the housing industry. The paper put forward the development route of promoting housing industrialization: by establishing the demonstration base of housing industrialization, cultivating and developing a batch of enterprises that are closely related to the modernization of the housing industry and have a strong driving ability, and further guiding and driving the comprehensive and healthy development of housing industrialization through the demonstration function of these enterprises. The paper also pointed out the direction of development of housing industrialization and pointed out that the development of a complete set of housing industrialization technologies and construction systems should meet the requirements of environmental protection, energy savings, land savings, water savings, and material savings to meet the needs of urban and rural residents to improve the quality, performance, and quality of housing.

On March 5, 2016, Premier Li Keqiang proposed in his government work report that in the process of further promoting new urbanization, we should "actively promote green buildings and building materials, and vigorously develop prefabricated constructions." The State Council issued Several Opinions on Deepening the Construction of New-type Urbanization and put forward the following: "New-type urbanization is the only way to modernization, the biggest potential of domestic demand, an important driving force for economic development, and an important project for people's livelihood. It includes: promoting the extension of infrastructure and public services to rural areas, promoting the integrated development of primary, secondary, and tertiary industries in rural areas, and so on."

In January 2016, The Ministry of Housing and Urban–Rural Development of the People's Republic of China and the General Administration of Quality Supervision, Inspection, and Quarantine of the People's Republic of China jointly promulgated and implemented the Evaluation Standard for Industrial Buildings (GB/T 51129-2015). The standard clearly defined professional terms such as "precast rate", "assembly rate" and "prefabricated components". "Precast rate" refers to the volume ratio of the amount of precast concrete in the main structure and envelope above the outdoor floor of an industrial building to the total amount of concrete in the corresponding component. "Assembly rate" is the ratio of the number (or area) of prefabricated components and building parts in industrial buildings to the total number (or area) of similar components or parts. In addition, the standard also made it clear that industrial buildings should conform to the basic characteristics of standardized design, factory production, construction and assembly, integration of decoration and management information, prefabrication rate should not be less than 20%, and assembly rate should not be less than 50%. The implementation of this standard has important guiding significance for strengthening the construction plan, construction technology, quality control, material supply, and responsibility division of industrial construction projects.

Facing the urgent need for transformation and upgrading the construction industry to modern practices across the country, under the support of national policies, more than 30 provinces and cities in China have successively issued policies to support the development of the relevant construction industry and promote the construction of industrialization bases and pilot demonstrations. Local provinces and cities, including research and development units, real estate development enterprises, general contracting enterprises, and colleges and universities, are actively developing and exploring construction industrialization. Domestic scientific research institutes, colleges, and universities have cooperated with relevant enterprises to establish several construction industrialization innovation strategic alliances to jointly develop and establish new industrialized building structure systems and related technologies, striving to make the proportion of prefabricated constructions in new buildings reach 30% in approximately 10 years. In recent years, the prefabricated construction system represented by Heilongjiang Yuhui Group, Changsha Broad Group, Nanjing Dadi Group, and Zoina Construction Group has been continuously improved. In 2016, there were 118 prefabricated demonstration projects nationwide, including 41 concrete projects. Prefabricated construction has shown great potential.

1.2 Advantages of Prefabricated Constructions

China's construction industrialization started late. The large prefabricated slab buildings of the Soviet Union were mainly introduced in the 1950s and 1960s. In the late 1970s, the prefabricated construction represented by fully assembled concrete large slab buildings prospered and developed, and the relevant policies and standards began to be matched and improved. After that, the disadvantages of poor seismic performance and insufficient waterproofing of large prefabricated slab buildings hindered the further development of prefabricated buildings. Until 2005, industrial buildings rose again and developed rapidly, and various new prefabricated steel structures, wood structures, hybrid structure systems, concrete structures, and related technologies were vigorously developed and applied. Zhou Fulin, the academician of the Department of Civil, Hydraulic, and Architectural Engineering of the Chinese Academy of Engineering, said at the green construction and sustainable development forum hosted by the first construction bureau that at present, the degree of construction industrialization in China is only 3–5%, while that in Europe and America is 75%, that in Sweden is 80%, and that in Japan is 70%.

Prefabricated buildings are an important position and the main way to realize construction industrialization. In February 2016, several opinions of the CPC Central Committee and the State Council on Further Strengthening the Management of Urban Planning and Construction took the development of new construction methods as an important future direction for urban planning and construction management. The paper proposed to formulate and improve the design, construction, and acceptance standards and specifications of prefabricated buildings as soon as possible by vigorously promoting prefabricated buildings. It also indicated that we should improve the standardization of parts and components, promote the industrialized production of building parts and components, reduce construction waste, control dust pollution, shorten the construction period, and improve the project quality. Third, it puts forward the measurement scheme of building a national prefabricated building production base and encouraging construction enterprises to implement industrialized production and on-site assembly construction. In addition, it also drew a blueprint for the development of new construction methods, that is, to increase policy support, striving to cause the proportion of prefabricated buildings in new buildings to be 30% in approximately 10 years. It can be seen that prefabricated buildings are the inevitable trend and the method for realizing the development of construction industrialization.

Compared with the traditional extensive manual construction industry, prefabricated buildings have many outstanding characteristics:

1. They can promote each other with urbanization. At present, China has entered the late stage of industrialization, and the process of industrialization and urbanization has accelerated. In the interactive development process of construction industrialization and urbanization, on the one hand, the rapid development of urbanization and the continuous expansion of the construction scale provide a good material foundation and market conditions for the rapid development of construction industrialization. On the other hand, construction industrialization has brought new industrial support to urbanization. Industrialized production can also effectively solve the employment problem and promote the transformation of migrant workers into industrial workers and skilled workers.
2. The prefabrication of buildings can improve construction quality and reduce construction accidents. The prefabricated building adopts the components prefabricated in advance by the factory and assembled on site. Through the standardized design module and manufacturing process, the impact on project safety caused by the low professional quality of personnel and uncertain factors in the construction process is reduced. It also standardizes the rationality and safety of building structure design and layout. It also simplifies the previous traditional construction mode, which is conducive to the supervision of the government and the division of safety responsibilities, to better ensure the construction quality and ensure the safety of human life and property from the legal perspective.
3. It can greatly shorten the construction time. The production model of factory prefabrication with local assembly can generally shorten the construction period by approximately 20%, which is convenient for the rapid completion and occupancy of prefabricated buildings. At the same time, it is also conducive to the

construction of relevant infrastructure, building a modern new residence with a beautiful environment, supporting facilities, and perfect functions in combination with urban planning.

4. It is conducive to improving the service life of buildings. Prefabricated components of prefabricated buildings can be reused, dismantled, protected, and maintained conveniently. At the same time, with continuous improvement in people's material standard of living, it is more convenient to adjust the building and structural layout to meet the increasing building needs.
5. It can meet the various construction needs of people. With the gradual improvement in both the material standard of living and cultural needs, residents' requirements for their living environment and aesthetics are also gradually increasing, and their requirements for residential buildings are becoming more stringent. Compared with previous forms of construction, prefabricated components with fine characteristics can not only be customized according to demand but can also simulate various details, such as imitation stone facades and column carving, which not only reduces the time, cost, and inconvenience of later decoration but also reduces pollution and damage to the ecological environment. It can also be used to build prefabricated buildings with cultural characteristics.
6. It is conducive to the realization of green buildings. Xiao Xuwen, the academician of the Chinese Academy of Engineering and chief consultant engineer of the technical center of CSCEC, pointed out that green construction can be realized only by realizing industrialization. "At present, the real estate industry accounts for 7% of China's total economy, drives more than one-third of related industries, and its energy consumption accounts for more than 40%. Therefore, reducing its energy consumption is an important channel for the realization of green construction, energy conservation and environmental protection in China." The use of standardized components can avoid environmental hazards such as construction waste, noise pollution, and environmental pollution caused by on-site construction. Second, prefabricated constructions can be designed and constructed according to local conditions and the local climate environment to reduce energy consumption, achieve energy conservation, and provide a comfortable and agreeable living environment for residents.

1.3 Classification of Prefabricated Concrete Structures

The construction method, construction efficiency, and regional applicability of prefabricated buildings all depend on the difference in building structure systems. Different building structure systems have different ways of splitting and connecting prefabricated components, so the concept of the system is inseparable when discussing prefabricated buildings. In the development of prefabricated buildings at home and abroad, scholars and engineers have explored beneficial engineering practices for the industrialization development of different building structure systems.

Summarizing these development experiences and engineering achievements will significantly promote the future development of prefabricated buildings in China.

At present, there are many types of prefabricated buildings, including frame structure systems, shear wall structure systems, frame shear wall structure systems, etc. The most widely used prefabricated construction structure systems in China are the shear wall structure system and frame structure system. Next, the structural characteristics, development history, and application status of the frame, shear wall, and other structural systems will be described in detail.

1.3.1 Prefabricated Frame Structure

A frame structure refers to the structural system form using the connection of beams and columns. It has the advantages of flexible space division, light weight, and flexible coordination with the building plane layout, which is conducive to the arrangement of building structures requiring large space. At the same time, the beams and columns of the frame structure can resist the vertical and horizontal loads caused by wind and earthquakes. It has good seismic performance and has been widely used all over the world. The beam and column members of the frame structure can be easily qualitatively standardized, and are thus very suitable for assembly construction. The prefabricated frame structure system includes a prefabricated concrete frame structure system, a prefabricated steel frame structure system, and a prefabricated bamboo wood frame structure system. The use of prefabricated frame structures can both improve construction efficiency and reduce environmental pollution, and ensure the quality of the building structure. Therefore, the prefabricated frame structure is one of the most widely used structural systems worldwide.

The prefabricated concrete frame structure is generally composed of precast columns (cast-in-place columns), precast beams, precast floors, precast stairs, external wall panels, and other components. Force transmission path within the structure is clear, the assembly efficiency is high, and there are fewer wet-process operations. These are mainly used in factories, warehouses, shopping malls, parking lots, office buildings, teaching buildings, medical buildings, business buildings, and other buildings that require a large space. In recent years, these have also gradually been used in residential buildings and other city buildings.

Prefabricated concrete frames can be divided into (1) wet connection frames and (2) dry connection frames. A wet connection frame refers to the whole frame structure being formed by pouring concrete or cement slurry between the precast members of the frame structure. This connection mode realizes the strength and ductility of the assembled frame structure and the cast-in-place frame, so it is also called the imitation cast-in-place connection. This connection requires cast-in-place concrete, whose formwork support and maintenance greatly reduce the construction speed and relatively high cost of prefabricated frame structures. A dry connection frame refers to the dry connection between the prefabricated components of the frame. By implanting steel plates or other steel components in the connected components, the overall frame is formed using through-bolt connection or welding. Another obvious difference between a dry connection and a wet connection is that in a wet connection frame, the plastic deformation allowed by design is usually set outside the connection area, and the connection area remains elastic. In contrast, the dry connection frame contains prefabricated components that are maintained in the elastic range, and the design requirements of plastic deformation are often limited to the connection area itself. This can result in a concentrated crack in the beam-column joint surface. Therefore, compared with similar cast-in-place structures, the damage degree of prefabricated concrete structural members can be expected to be much less, and it is easy to achieve post-earthquake repair.

The prefabricated concrete frame structure can be divided into two categories based on the use of prestressed technology. One is a prestressed prefabricated frame structure, mainly including a prefabricated monolithic prestressed slab-column frame structure (IMS system), Scope system (Fig. 1.1), prepressing prefabricated prestressed frame structure, etc. Among them, the scope system is most widely used in China. The other type is a non-prestressed prefabricated frame structure. The Taiwan Ruentex system is the most widely used non-prestressed prefabricated frame structure in China.

A prefabricated steel frame structure refers to a steel beam and column structure with prefabricated construction. Compared with other building structures, the prefabricated steel frame structure with the advantages of industrialization, modularization and standardization is closest to the concept of a "green building" (Fig. 1.2). It also has the characteristics of a short construction period, recyclable steel, and a good comprehensive technical and economic index. This system has produced large-scale residential industrialization in other countries, and has been widely studied and used in China.

The beam-to-column joint is the key to the prefabricated steel frame structure, which is used to transmit the internal force of structural members and coordinate deformation. At present, different countries adopt different classification methods for steel frame beam-column joints. The American Institute of Steel Construction (AISC) classifies joints into pinned connections or bending connections according to the secant stiffness in the serviceability state [1]. The joint is divided into rigid,

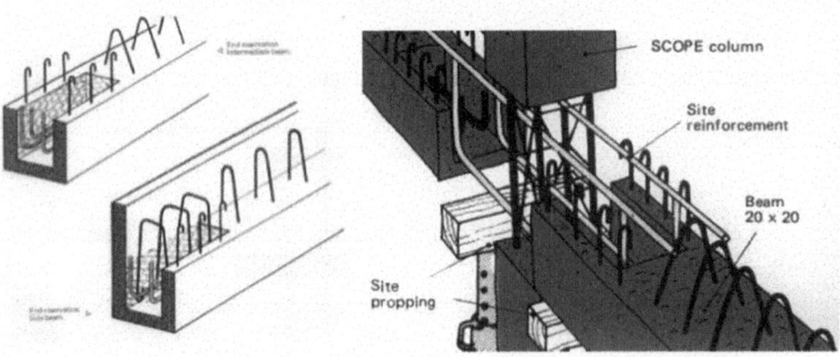

Fig. 1.1 Scope system

Fig. 1.2 Steel frame structure

semi-rigid, or pinned connections according to the rotational stiffness of joints in EC3 [2], while full-strength, partial strength or pinned connections are categorized by comparing the joint design moment resistance with the design moment resistance of the members that it connects. In China, a joint is classified as a rigid, semi-rigid, or pinned connection.

To further improve the seismic performance of steel frame structures, prefabricated braced frames and self-centering energy dissipation frames have been proposed and studied. A concentrated braced frame (CBF) is an economical seismic-resistant frame with large stiffness. After using a buckling restrained brace, the ductility of the CBF can be improved, but the residual deformation is still large. Prestressed reinforcements and energy dissipation devices are used in the self-centering energy dissipation frame. The prestressing enables the frame to return to its original position when subjected to seismic action, and the energy dissipation devices enhance the seismic performance of the frame. However, due to the difficulty of construction, practical applications are not common.

A prefabricated bamboo-wood frame structure is a new prefabricated structure, which uses natural bamboo and wood materials to make basic glued wood, wood-based composites, and engineering bamboo. And it is an integral building structure formed by these factory-made bamboo and wood profiles.

The steel-bamboo composite columns, beams, and floor slabs are connected by metal connectors to form the prefabricated bamboo frame structure. The composite beams directly carried the transverse and vertical loads from the composite floor slabs, and then the loads were transmitted to the composite columns. Finally, the loads are transmitted to the foundation by the columns. The steel-bamboo composite wall only plays the role of maintenance and separation, and the wall and floor slab can be filled with thermal insulation and waterproof materials (Fig. 1.3). The design of beam-to-column joints is the key issue in the beam-to-column bamboo frame structure. A commonly used joint is an energy dissipation composite joint (Fig. 1.4), which connects bamboo beams and columns to form a prefabricated bamboo frame structure with large stiffness and ductility. Under seismic action, the steel plates in joint areas

Fig. 1.3 Prefabricated bamboo frame

Fig. 1.4 Steel-bamboo/wood composite joint

dissipate the energy through nonlinear deformation. Therefore, adopting this kind of joint promotes the ductility and energy dissipation capacity of the prefabricated bamboo frame.

The prefabricated wood-frame structure can be divided into continuous and platform wood frame structures according to the characteristics of the internal structure of the building. Continuous wood frame structures appeared in the light wood frame structure buildings of the United States in the 1930s. This structure is composed of floor beams, walls, ceiling beams, and roof rafters. The wood is 38 mm thick and is used as the building material, and the wood skeleton is connected with long nails. The continuous wood frame structure became the main form of residential, catering, and other buildings at that time due to its safety, fast construction, and comfort. Platform wood frame structure was the dominant form of wood structure in North America in the late 1940s. The most obvious difference between the platform and continuous wood frame structure is that when the height of the first floor is the same, the wall of the second floor can be directly carried out on the floor slab that has been built. The walls of the platform wood frame structure can be prefabricated directly, which meets the development of construction and installation. Therefore, the platform wood-frame structure in North America has replaced the continuous wood-frame structure (Fig. 1.5).

Fig. 1.5 Dormitories at the University of British Columbia, Canada (https://bbs.feng.com/mobile-news-read-0-657331.html)

1.3.2 Prefabricated Shear Wall Structure

Shear wall structures have been widely used in high-rise buildings in China. A prefabricated shear wall structure is an industrial building structure suitable for national conditions in China with the advantages of high quality, fast construction, environmental protection, resource conservation, and sustainable development. It is a concrete structure with prefabricated or semi-prefabricated walls as the main components, which are precast or partially cast in place. Prefabricated large panel shear wall structures first appeared in the development of prefabricated shear wall structures. Based on this structure, Japan has developed a prefabricated reinforced concrete frame-shear wall structure (WR-RC). An unbonded post-tensioned prefabricated shear wall structure was proposed by the precast seismic structure systems (PRESS) project, which was carried out by the United States and Japan in the 1990s. To date, prefabricated superimposed slab shear wall structures, monolithic prefabricated concrete shear wall structures, and prefabricated prestressed shear wall structures have been applied in China.

Prefabricated shear wall structures mainly consist of prefabricated large panel shear wall structures, prefabricated superimposed slab shear wall structures, monolithic prefabricated concrete shear wall structures, and prefabricated prestressed shear wall structures.

The prefabricated large panel shear wall structure is a highly industrialized building structure prefabricated with precast reinforced concrete walls and floor slabs. The main advantages of this structure are commercialized manufacturing and high construction efficiency. However, stress centralization often occurs in the joint seams of this structure, resulting in reduced overall performance and deformation capacity.

The prefabricated superimposed slab shear wall structure is a prefabricated monolithic shear wall structure consisting of superimposed wall slabs, cast-in-place concrete shear walls, boundary members, beams, and floor slabs (Fig. 1.6). A single-side superimposed slab shear wall and double-side superimposed slab shear wall can be adopted in the construction of the superimposed slab shear wall structure.

Fig. 1.6 Superimposed slab shear wall structure

A double-sided superimposed slab shear wall is a kind of vertical wall member composed of prefabricated wall panels on both sides and cast-in-place concrete in the middle. In this structure, the distribution of prefabricated wall reinforcement is designed to meet the requirements of shear wall force and lateral pressure of concrete. After connecting the prefabricated wall slabs by truss reinforcement, the double-sided superimposed slab shear wall structure is formed by casting concrete in the middle part. The seismic performance and design method of superimposed slab shear wall structures are quite different from those of cast-in-place shear wall structures, and the applicable height is relatively low. According to the "Specification for construction and acceptance of superimposed slab concrete wall structure" in Anhui Province (DB34/T 810-2008), this structure is suitable for multistory and high-rise residential buildings with seismic fortification intensity equal to or less than 7° and non-seismic areas. The height of the buildings is lower than 60 m, and the number of floors is less than 18. If applied to higher buildings, special research and demonstration is needed.

The emergence of fabricated superimposed shear wall structures mainly comes from two aspects of demand. First, with the increasing requirements for building energy efficiency, a prefabricated superimposed shear wall structure with a thermal insulation sandwich has rapidly become popular. Second, this structure can effectively reduce the use of formwork and improve construction efficiency. In other countries, no connectors were previously used in the superimposed wall before, which was only composed of structural wall slabs and nonstructural wall slabs. With the emergence of wall connectors, the mechanical performance became more reasonable.

The fabricated superimposed shear wall structure embodies a high level of industrialization. Due to the existence of a cast-in-place concrete sandwich, the connection of precast components is strengthened, and the seismic performance of the structure is improved. However, the main shortcomings of this structure are that it uses a large amount of on-site wet concrete construction, and the effective connection between components is insufficient.

The prefabricated superimposed shear wall structure is also widely used in China. Similar to the PCF technology of the Vanke Group, using a precast concrete slab as the wall formwork of the building can effectively reduce cost and scaffold engineering. However, this technology is still mainly limited to the external wall and cannot exploit the advantage of this structure. Hefei Sievert Company and Jiangsu Suqian Yuanda Construction Company introduced the German "prefabricated concrete superimposed shear wall structure" technology, which can quickly and automatically produce components including precast concrete superimposed walls, precast concrete superimposed floor slabs, and precast staircases.

The monolithic prefabricated concrete shear wall structure is a reinforced concrete shear wall structure formed by prefabricated concrete shear wall slabs and cast-in-place concrete shear walls as vertical and lateral resistance members of the structure (Fig. 1.7). In this system, part or all of the shear walls are prefabricated. The joint seams between them are connected using wet connections, thus allowing sleeve grouting connection, slurry anchor lap connection, and bottom reserved reinforcement lap connection all to be adopted in the post-pouring area. The reinforcement at the horizontal joints can be in the form of sleeve grouting connection, slurry anchor lap connection, and lap connection of reinforcement in the bottom reserved postcast area. This structure is mainly used in high-rise buildings.

The prefabricated shear wall slab is connected with the adjacent vertical cast-in-place slab by an equal-strength connection to form the shear wall section in the horizontal direction. The bottom is filled by pressure grouting or seat grouting to form the filler course. The top is connected to the prefabricated wall slab of adjacent

Fig. 1.7 Monolithic prefabricated shear wall structure

floors by a horizontal cast-in-place segment and ring beam. The seismic performance of this structure is the same as that of the cast-in-place shear wall structure.

According to the prefabricated wall parts used, the prefabricated shear wall structure can be divided into a fully prefabricated shear wall structure or a partially prefabricated shear wall structure. A fully prefabricated shear wall structure refers to all shear wall slabs that are prefabricated components. The prefabricated rate of this structure is high, while too many joint seams increase the construction difficulty. A partially prefabricated shear wall structure refers to a structure with cast-in-place interior wall slabs and prefabricated exterior wall slabs. The seismic performance of this structure is similar to that of a cast-in-place structure, with a relatively large applicable height and good applicability to the full-prefabricated structure. A prefabricated exterior wall slab can be combined with thermal insulation, decoration, waterproofing, doors and windows, and balconies to exploit the advantages of a full prefabricated structure.

According to the steel connection form of the horizontal joint seams, the prefabricated monolithic shear wall structure can be divided into three main technical systems with vertical reinforcement being connected (1) by sleeve grouting, (2) by a slurry anchor lap, and (3) by the bottom reserved postpouring area. The three technical systems have their advantages, disadvantages, and scope of application. At present, they all have practical engineering applications in China.

The traditional joint seam connection method of a prefabricated shear wall structure has strict requirements for structural details and complex construction. It is difficult to guarantee the quality of this method, which has negative effects on the construction and seismic resistance of the structure. However, the above problems can be avoided by using prestressing technology for joint seam connections. Prestressing can effectively connect the adjacent wall slabs, provide restoring force and reduce residual deformation of the shear wall.

In the 1990s, the Precast Seismic Structural Systems (PRESS) project proposed by the United States and Japan promoted the development of prefabricated shear wall structures [3–5]. A new type of prefabricated concrete shear wall structure was first introduced in PRESS—an unbonded prestressed prefabricated self-centering shear wall structure. The structure is composed of concrete wall slabs with unbonded prestressed steel strands through horizontal seams. Under seismic action, the effect of the earthquake on the structure is effectively reduced through the gaps without fixed connections between wall slabs and between the wall and the foundation. The self-weight and prestress of the structure provide a self-centering force to close the gaps. This shear wall structure has insignificant cracking damage and residual deformation after the earthquake, which is convenient for maintenance. The construction process adopts a full dry connection, which is convenient for construction and meets the requirements of industrial production. However, the unbonded prestressed reinforcement of this shear wall structure remained elastic in the loading process. No yield deformation will occur to dissipate energy. Therefore, the energy dissipation capacity of the unbonded prestressed prefabricated self-centering shear wall structure is poor.

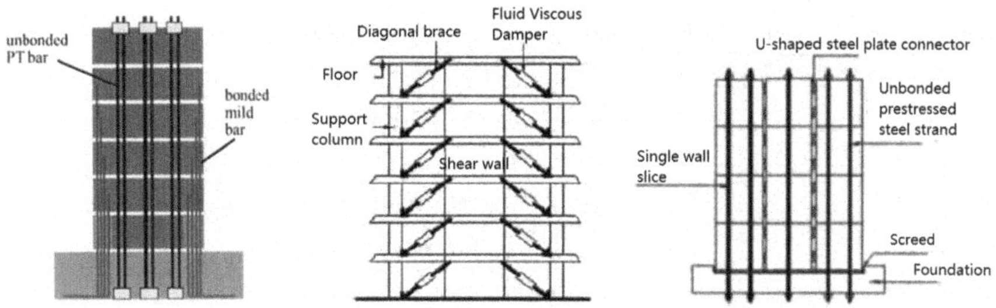

Fig. 1.8 Typical prefabricated hybrid shear wall structure

The prefabricated hybrid prestressed shear wall is a new type of structure that adds energy dissipation devices to improve the seismic performance. Common energy dissipation devices include additional energy dissipation steel bars, viscous dampers, and U-shaped steel plate energy dissipation devices between the walls. Figure 1.8 shows the typical prefabricated hybrid shear wall structures.

The prefabricated hybrid shear wall structure combines the advantages of a post-tensioned unbonded prestressed shear wall structure. Due to its dry connection mode, the construction is convenient and suitable for industrial production. Moreover, the energy is mainly dissipated by the additional energy dissipation devices under seismic action. The damage to the structure can be effectively reduced. This structure shows good seismic performance and is easy to repair after an earthquake.

The concept of the short-pier shear wall structure was first introduced by Bosheng Rong. According to the "Technical Specification for Concrete Structures of Tall Buildings", the range of the ratio of height to the thickness of the short-pier shear wall structure is 5–8.

Because this short-pier shear wall structure is suitable for national conditions, it is widely used in China. By combining the advantages of prestressed connections, some researchers have also proposed the concept of prefabricated prestressed short-pier shear wall structures, including post-tensioned unbonded prefabricated prestressed short-pier shear walls, coupled shear walls and multiple shear walls (Fig. 1.9).

1.3.3 Prefabricated Frame-Shear Wall (Core Tube) Structure

The frame part of the prefabricated frame-shear wall structure is similar to the prefabricated frame structure. The shear wall part can be cast-in-place or prefabricated. If the shear walls are arranged as the core tube, the prefabricated frame-core tube structure is formed. At present, the prefabricated frame-cast-in-place shear wall structure has been applied in China. In Japan, relevant research and engineering practices have been carried out on the prefabricated frame-prefabricated shear wall structure, while the application of this structure in China is limited. Research work is still in progress.

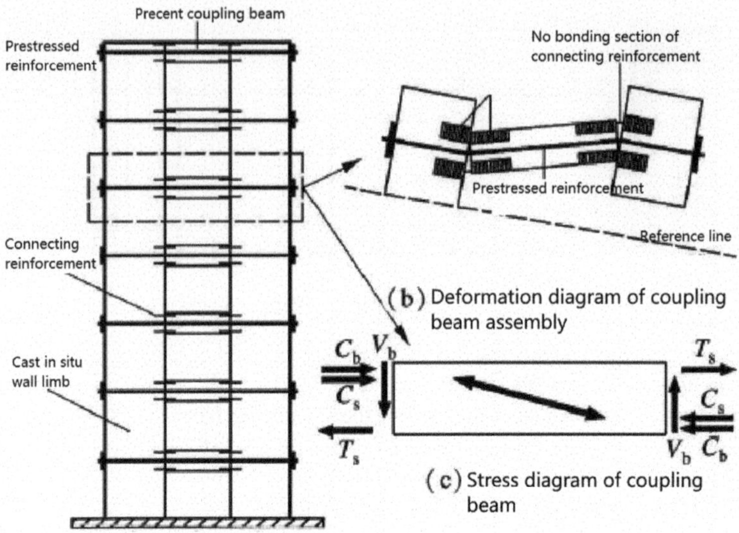

Fig. 1.9 Prestressed prefabricated hybrid coupled wall structure (PPHCW)

1.4 Application Status of Prefabricated Concrete Structure System in China

1.4.1 Prefabricated Monolithic Prestressed Slab-Column Frame Structure

The prefabricated monolithic prestressed slab-column frame structure (IMS system) is a kind of prestressed concrete slab-column structure made of ordinary reinforced concrete materials and prefabricated reinforced concrete floors, columns, and other components made by component factories. After being put in place at the construction site, the IMS system is formed through the assembly of prestressed steel bars and integral tension. It is one of the most widely used industrial building systems in Yugoslavia. The system experienced two major earthquakes in Yugoslavia in 1969 and 1981, showing excellent seismic performance. The prefabricated monolithic prestressed slab-column frame structure is traditionally used in multistory factory buildings. In residential buildings, it is also generally a multistory structure.

Since the Tangshan earthquake, China has introduced a prefabricated monolithic prestressed slab column structure. A large number of experimental studies on components, joints, panels, and machines have been carried out by the National Institute of Architecture, Institute of Earthquake Resistance, and Institute of Design. In Beijing, Chengdu, Tangshan, Chongqing, Shenyang, Guangzhou, Shijiazhuang, and other places, more than ten buildings with 2 to 12 floors, including scientific research buildings, office buildings, residential buildings, workshops, and warehouses, have been built, totaling approximately 40,000 m^2. The precast floor slab of the prefabricated monolithic prestressed slab-column frame structure is square or rectangular. During the construction of the system, the prefabricated reinforced concrete square columns (generally 2–3 floors are a section) are first erected on the site and fixed with temporary supports, and then the prefabricated floor slabs (each span is a whole floor slab) are set aside by connecting brackets. After the first floor slab is fully in place, the full-length prestressed steel bars are laid, and the floor slabs and columns are squeezed against each other by tension, as shown in Figs. 1.10 and 1.11. If necessary, vertical bending force is added to the prestressed reinforcement along the longitudinal and transverse directions to produce sufficient bending force to compensate for the loss of prestress and to provide the lifting force to support the dead weight of the structure. The floor slab is supported and fixed on the column by prestressing and its static friction force, and the prestressed friction joint is formed between the slab and column. Finally, fine aggregate concrete is poured into the side column. Prestressed reinforcement plays the dual roles of structural stress reinforcement and assembly method.

In this structure, a whole large plate between the original columns can be divided into several small plates, and the extrusion stress is transferred between the panels through the cushion blocks, forming the unique cushion block type splicing technology in China, as shown in Fig. 1.12. This not only reduces the size of the plate, making it easy to manufacture, transport and install but also increases the structural span, making its application more flexible. In practical engineering, according to the different column spacing in the vertical and horizontal directions, the division forms

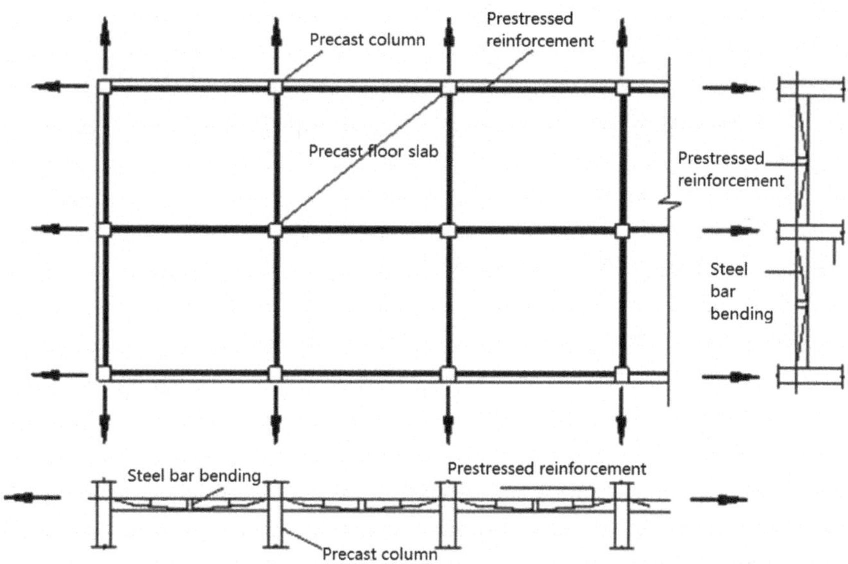

Fig. 1.10 Schematic diagram of the plane layout and prestressing of the slab-column frame system

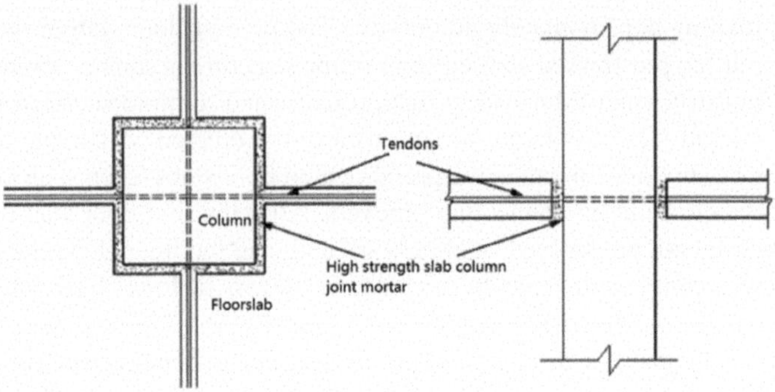

Fig. 1.11 Plane diagram of slab-column joints

of the plates are also different. A whole plate between columns can be divided into two plates, three plates, four plates or six plates.

Compared with the general conventional frame structure, the assembled monolithic prestressed slab-column frame structure has the following characteristics:

(1) The structure has no beam, no post cap, a flat bottom and a large span. The residents can adjust the indoor partition wall according to their needs, which is not restricted by the beam. It is convenient for changes in use and flexible for spatial arrangement.
(2) Different from the basic theory of other structural systems, this structure relies on the friction between plates and columns to support the floor load. Through the application of two-way prestressing, a fully assembled floor without a beam

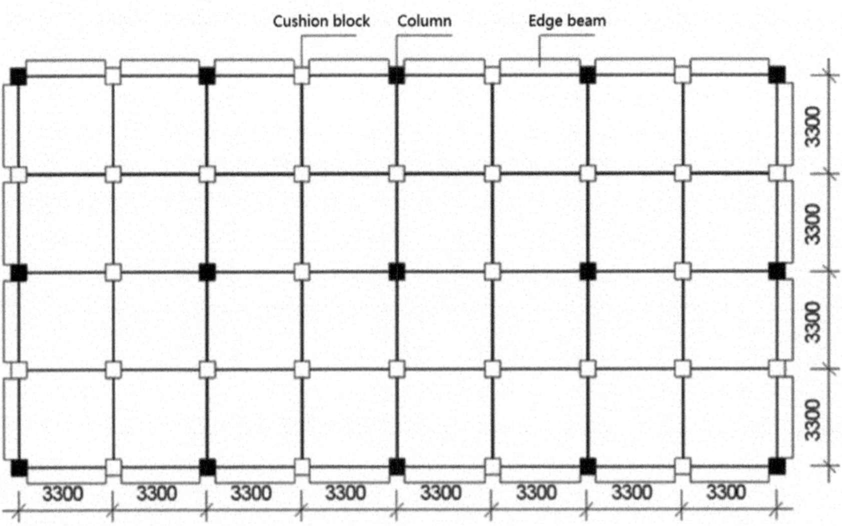

Fig. 1.12 Plane layout of the multi-panel integral prestressed slab-column frame system

and post cap is formed. The two-way prestressed reinforcement makes each axis form a prestressed "ring beam". These ring beams, such as hoops, cause the floor as a whole to have high horizontal stiffness to ensure that horizontal forces such as seismic loads are transmitted to the vertical members.

(3) The connection nodes of the structure are flexible joints with automatic adjustment and active growth. Once the external force is removed, it can be returned to the original position immediately. When the roof is pushed, the deformation of each floor is basically linear. This kind of structure has good integrity and strong seismic capacity.

1.4.2 Scope System

The scope system is a frame structure of a prefabricated prestressed concrete assembly integral frame structure system and prestressed concrete composite plate system based on sleeve pregrouting connection technology. It is the main product of precast prestressed concrete building (PPB) technology in France. Its principle is to use unique key slot beam-column joints to cast-in-place or precast reinforced concrete columns, precast prestressed concrete beams and slabs. Then, the beams, slabs, columns and joints are connected into a whole through postcast concrete.

In practical engineering application, there are mainly three kinds of assembly forms in the scope system: the first is the full assembly of precast columns and precast prestressed concrete composite beams and slabs; the second is to use cast-in-place columns, precast prestressed concrete composite beams and slabs for partial assembly; third, only precast prestressed concrete composite slabs are used, which is suitable for the assembly of various types of structures. The first of these assembly methods saves the most time. Since the main part or all of the houses are manufactured in factories and the piles, columns, beams and plates are all made by special machines, the level of tooling and the degree of standardization are high, so the assembly is convenient. Only by connecting the relevant nodes on site and pouring the concrete compactly can the building structure be formed.

In 2000, Nanjing DADI Construction Group Co., Ltd. introduced the scope system. In the past 10 years, it has completed approximately 1 million square meter of projects in the Nanjing construction market and formulated the recommended technical specification for engineering construction in Jiangsu Province, i.e., technical specification for precast prestressed concrete assembled monolithic frame (Scope system) (JG/T006-2005). The representative buildings include The International Academic Exchange Center of Nanjing Audit University, Nanjing Jeshing International Home Furnishing Plaza Jiangbei Store, Nanjing Red Sun Home Furnishing Plaza Maigaoqiao Store, etc. The International Academic Exchange Center of Nanjing Audit University adopts a fully assembled frame structure of prefabricated columns, prefabricated prestressed concrete composite beams, and composite slabs. The cost of the main project is approximately 10% lower than that of a cast-in-place frame structure. The Nanjing Jeshing International Home Furnishing Plaza Jiangbei

Store and Nanjing Red Sun Home Furnishing Plaza Maigaoqiao Store both adopt a semi-assembled frame structure of cast-in-place column, prefabricated prestressed concrete composite beam, and composite plate. Compared with the cast-in-place structure, the construction period is greatly reduced.

The prefabricated components of the scope system include prefabricated reinforced concrete columns, prefabricated concrete composite beams, and composite slabs. Among them, high-strength prestressed reinforcement (steel strand and stress relief steel wire) is used in the precast part of the composite beam and composite slab, and the pre-tensioning effect is produced.

The precast column bottom and concrete foundation are generally connected by the grouting sleeve. The position of the embedded sleeve in the foundation is shown in Fig. 1.13. Among them, the length of the reserved hole should be longer than the overlapping length of the main reinforcement of the column. The reserved hole should be selected as the bottom-sealed galvanized bellows. The bottom should be dense without slurry leakage. The inner diameter of the pipe should not be less than the diameter of the external tangent circle of the main reinforcement of the column.

Precast beams and columns are connected by key slot joints (Fig. 1.14), which is also the largest feature of the scope system. By reserving grooves at the end of the precast beam, the longitudinal reinforcement of the precast beam overlaps with the U-shaped reinforcement extending into the joint. The U-shaped reinforcement is mainly used to connect the two ends of the joint, and the traditional anchoring method of longitudinal reinforcement in the joint area is changed into the method of prestressed steel bars at the end of the precast beam in the key slot, that is, the overlap connection of the plastic hinge area at the beam end. Finally, high-strength and slight expansion concrete are poured to connect the beam and column joints, respectively.

The connection nodes of the prefabricated prestressed composite slab and prefabricated beam are shown in Fig. 1.15. A typical prefabricated column practice is shown in Fig. 1.16. The cross reinforcement should be added at the connection nodes between layers of prefabricated columns and welded with the longitudinal reinforcement. A cross reinforcement should be set on each side of the precast column. Its diameter should be calculated and determined according to the bearing capacity and deformation requirements in the transportation construction stage and should not be less than 12 mm. In addition, the column is calibrated and fixed with an adjustable

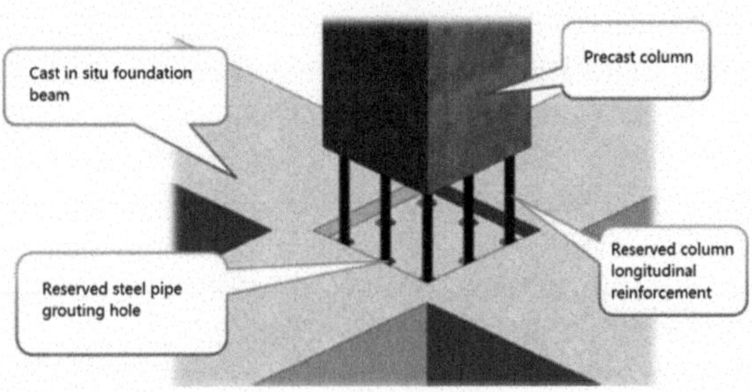

Fig. 1.13 Connection joint between precast column and cast-in-place foundation

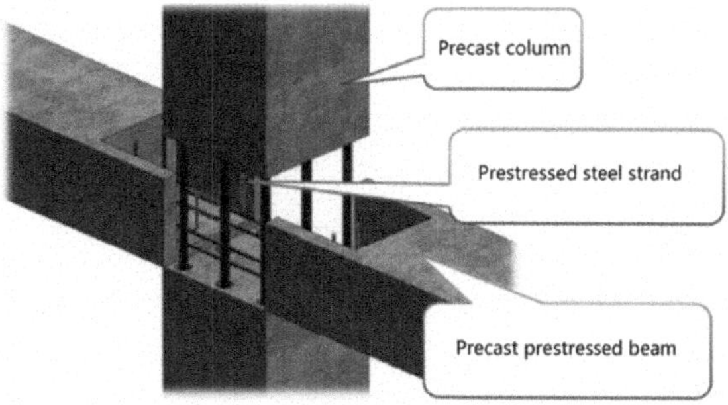

Fig. 1.14 Connection joint between precast beam (key slot) and precast column and composite plate

diagonal brace after it is in position. Due to the limitations on transportation and hoisting of components, the prefabricated column cannot reach the top at one time, so the form of the connecting column must be adopted. The connecting column can be connected by a section steel support or by a sealed steel pipe. The specific connection method depends on the specific project. Figure 1.17 shows on-site construction photos of the precast column and composite slabs.

Compared with the conventional frame structure, the scope system has the following characteristics:

(1) Prestressed high-strength steel bars and high-strength concrete are used in the precast beam and slab. The cross-sections of the beam and slab are reduced, the

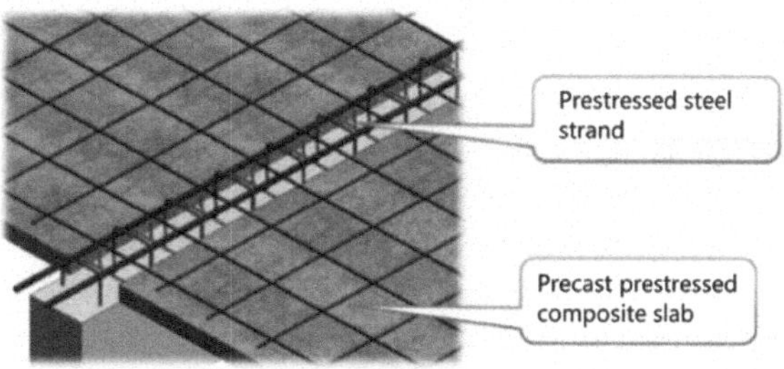

Fig. 1.15 Connection node between prestressed composite slab and precast beam

Fig. 1.16 Interlayer joint of precast column

Fig. 1.17 On-site construction of precast column

amount of reinforcement and concrete is reduced, and the crack resistance of the slabs is improved.

(2) The precast column adopts a segmental column (2–3 layers of column prefabrication), and the field construction of the beam and slab does not require formwork, which reduces the construction period of the main structure.
(3) The bottom of the floor slab is flat, does not need to be painted, reduces the amount of wet work, is conducive to environmental protection, reduces noise pollution, and makes on-site construction more civilized.
(4) The prefabricated part of the composite slab is not limited by the modulus and can be randomly divided at will according to the design requirements, with great flexibility and wide applicability.
(5) Because the arching height of the prestressed composite slab cannot be accurately controlled, obvious assembly cracks may appear after completion.
(6) Generally, the structural system using prestressed composite floor slab is suitable for areas with seismic fortification intensity less than or equal to 8°. Although the special joint structure improves the overall performance and seismic performance of the scope system, as a prefabricated frame structure, its scope of application is limited to areas where the seismic fortification intensity is less than or equal to 7°.

1.4.3 Ruentex System

The Ruentex prefabricated frame structure system is a kind of prefabricated assembled integral frame structure system based on multi-spiral stirrup reinforcement technology. The structure adopts prefabricated reinforced concrete columns, composite beams, and composite slabs. The columns, beams, slabs, and joints are connected through the post-pouring part of reinforced concrete. The core technology of the Ruentex system lies in the development of prefabrication multi-spiral stirrup column, sleeve-type steel connector, ultrahigh early strength shrinkage-free cement mortar, prefabricated isolation construction method, and prefabricated exterior wall surface decoration effort technology.

Since 1995, the Taiwan Ruentex Group has introduced a complete set of prefabrication production technology and dry-mixed mortar production lines from Partex in Finland, as well as Japanese seismic design technology, self-created steel bar processing technology, application of advanced information technology, etc., to give full play to the prefabricated concrete assembly process in Taiwan with constant innovation and development. It has become the pioneer of the composite chemical method in Taiwan's construction industry. With the application of the Ruentex system, more than 5 million square meters of commercial buildings and workshops have been built in Taiwan, and several pilot applications of engineering projects have been completed in Shanghai, Jiangsu, and other places. Recently, technology transfer has been conducted, and Shanghai Urban Construction Group has been guided in implementing the first fully prefabricated integrated structural security housing project in Pujiang.

The prefabricated components of the Ruentex system include precast reinforced concrete columns, precast concrete composite beams, and composite slabs. This system adopts the joint connection method of a traditional assembly monolithic concrete frame, namely, the column and the foundation beam are connected by grouting sleeves, and the precast column and composite beam are connected as a whole through a cast-in-place reinforced concrete joint, as shown in Fig. 1.18. The main characteristics of the connection joint are that the longitudinal reinforcement is extended and bent at the end of the prefabricated beam. The longitudinal reinforcement in the prefabricated column is close to the four corners of the column, and a gap is left in the middle of each side of the column, which allows the longitudinal reinforcement extended from the end of the precast beam to be directly anchored conveniently in the column joint area, and the beam-column joint area and the composite plate are cast-in-place to form a precast assembly overall structure. The construction process of the Ruentex system is to first hoist the precast column in place and grout the precast column with non-shrinking grouting material to realize the connection between the column and the foundation or the upper column and the lower column. Then, the girders are hoisted, the small beams are hoisted, the beam-column joints are sealed, and the large and small beam joints are grouted. Finally, the composite floor slab is hoisted and postcast to form the whole frame. Figures 1.19 and 1.20 is a construction drawing showing the precast column and precast girder.

Figure 1.21 is a schematic diagram of the precast multi-spiral stirrup column. The configuration of the column is composed of a large circular screw hoop at the center and a small circular screw hoop at four corners. This configuration breaks through the traditional limitation that hoops are only suitable for circular section columns. Compared with the square hoop, the structure efficiency and production efficiency of the round screw hoop have been greatly improved.

Figure 1.22 is a schematic diagram of the semi-prefabricated isolation method of the Ruentex system. This construction method was applied in the civil building of Taiwan University in 2008. The traditional isolation construction method is improved

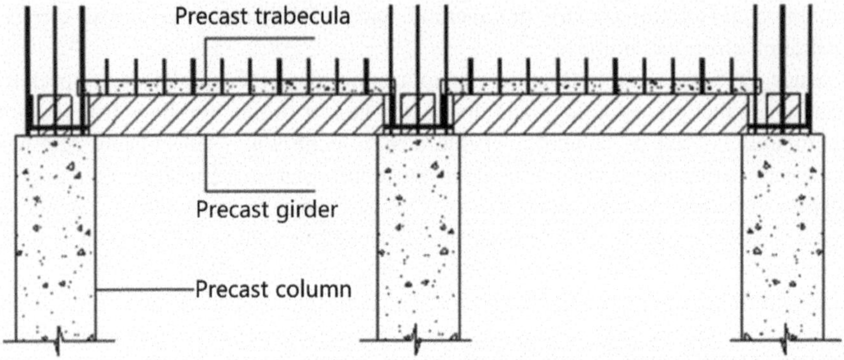

Fig. 1.18 Schematic diagram of frame beam-column joints

Fig. 1.19 Construction drawing of precast column

Fig. 1.20 Construction drawing of precast beam

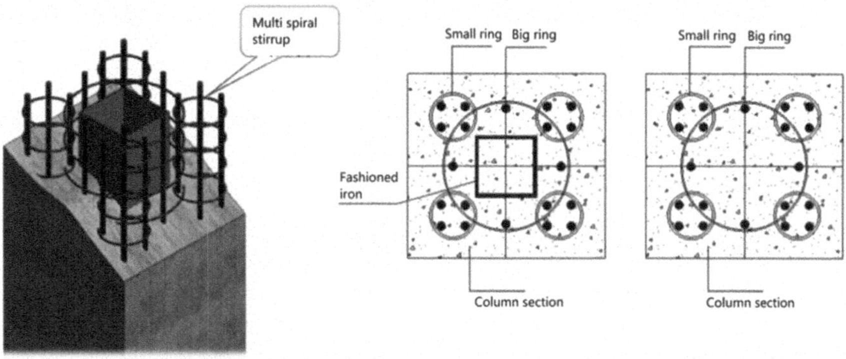

Fig. 1.21 Schematic diagram of precast multispiral stirrup column

by using the concept of prefabrication. The isolation and prefabrication are combined, and the construction speed of the isolation layer is faster than that in Japan.

Compared with the conventional frame structure, the Ruentex system has the following characteristics:

(1) The use of spiral stirrups in the component production stage reduces the amount of stirrup binding in the factory and relatively improves the production cycle of factory components.
(2) The use of prefabricated beams, slabs, and columns reduces the amount of on-site formwork and turnover frame materials;
(3) The cost of the system is higher than that of the cast-in-place frame, the engineering quality is easier to control, and the appearance and durability of the components are good.

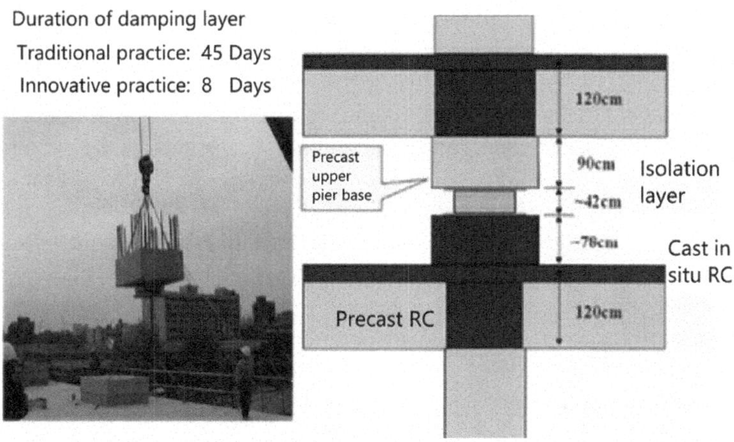

Fig. 1.22 Semi-prefabricated isolation method

(4) The maximum applicable seismic fortification intensity of the assembly frame structure of the Ruentex system is less than or equal to 7°.

1.4.4 Prefabrication and Assembly Technology of Vanke, Sievert, Zoina and Yuhui Groups

At present, the nationwide enterprises that promote the research, development and application of prefabricated technology are mainly the Vanke Group, Hefei Sievert, Zoina Group and Yuhui Group, and they have created their own prefabricated assembly technology systems.

The Vanke Group established an architectural research center in 1999 and a factory center in 2004 to carry out research on prefabrication and assembly technology. It is one of the earliest enterprises to promote industrialization of housing construction in China. At present, PCF technology is mainly produced, namely, precast concrete template technology. This technology takes a prefabricated concrete slab as the external wall template of the building, which can effectively save the template and scaffold engineering. However, the shear wall of the main structure of the system is almost always cast-in-place, and the floor slab is a composite floor slab, which is still cast-in-place with template support. However, the amount of assembly is low, and the amount of wet operation on site is large [6].

Sievert Concrete Prefabrication (Hefei) Co., Ltd. and Jiangsu Suqian Yuanda Architecture Science and Technology Co., Ltd. introduced the German "prefabricated concrete composite shear wall" technology, forming a composite slab concrete shear wall structure system. The structural components are divided into composite floor slabs, composite wall slabs, and prefabricated stairs. The composite slab is composed of the bottom prefabricated slab and lattice reinforcement, which can be used as the template of post-cast concrete. The composite wall slab is made of two layers of precast slabs and lattice reinforcement. After being installed in place on-site, concrete can be poured between the two layers of precast slabs. As a key project of national residential industrialization, the company has industrialized the production of prefabricated residential structure systems with these characteristics [7].

After years of practice, Zoina Group has summarized residential industrialization technology and formed an NPC technology system with the following characteristics. The vertical member shear walls and infilled walls are fully prefabricated, and the horizontal member beams and slabs are superimposed. In the vertical direction, the reinforcement bars (connecting steel bars) are reserved for the lower components, and the metal corrugated slurry anchor pipes are reserved for the upper components to realize a reinforced slurry anchor connection. In the horizontal direction, the cast-in-place connecting belt and cast-in-place concrete connection are set at appropriate points. The horizontal members and vertical members are connected by the reservation of inserted reinforcement into the beam of vertical members, the connection of the slab composite layer, and the cast-in-place concrete of the composite layer. The vertical component and horizontal component are connected to form an integral structure through reinforced mortar anchor joints, cast-in-place connecting belt, and composite cast-in-place [8]. This technology is a leading domestic fully prefabricated assembly technology created by introducing international advanced technology and combining it with the actual national conditions of China. It has the characteristics of new structure, new technology, new support, and new installation [9].

Since 2005, based on the production and construction technology of PC structural components of many enterprises and relying on the scientific research force of Harbin Institute of Technology, the Heilongjiang Yuhui Group has developed the precast assembled monolithic concrete shear wall structure system technology, as well as the corresponding component design, component prefabrication, component assembly, and construction technology, and has edited the provincial local standard of "Technical Specifications for Precast Concrete Shear Wall Structures". The core technology of the system is the "plug-in reserved-hole grouting steel lap connection". The simple explanation is that the vertical connection of the structure adopts the reserved hole plug-in slurry anchor connection, the horizontal connection mode adopts the reinforcement bolt mode, and the composite floor slab and beam joint the cast-in-place mode. The components of this structural system are simple in form and convenient in fabrication. However, due to the need to insert each vertical steel bar into the relevant hole in the construction, the components are large and heavy, which creates higher requirements for construction accuracy and on-site lifting operations.

1.5 The Development Limitations of the Current Prefabricated Concrete Structure System

From the above application status of prefabricated buildings, the most widely used prefabricated building form can be summarized as prefabricated components - node connection - core area post-pouring, as shown in Fig. 1.23a. This form is also known as the "wet" connection prefabricated structure system because there is still wet work on site. Correspondingly, another type of prefabricated building that directly uses prestressing technology, bolts, welding, and other methods without a post pouring section, as shown in Fig. 1.23b, can be called a "dry" connection prefabricated structure system [10].

In the current relevant codes and regulations in China, the "wet" connection system requires its design and construction performance to be "equivalent to cast-in-place" as the goal. The simple explanation is that the design and construction of a prefabricated structure can achieve the performance of the cast-in-place structure, so it can also be called an "equivalent cast-in-place" prefabricated concrete structure system. At present, the most developed prefabricated concrete structure system in China is mainly the "equivalent cast-in-place" prefabricated concrete structure system. Its specific form still follows the traditional structural form, such as frame, frame-shear wall, shear wall, and so on. The main difference is that the connection area adopts a series of structures or measures to ensure that it meets the requirements of "equivalent

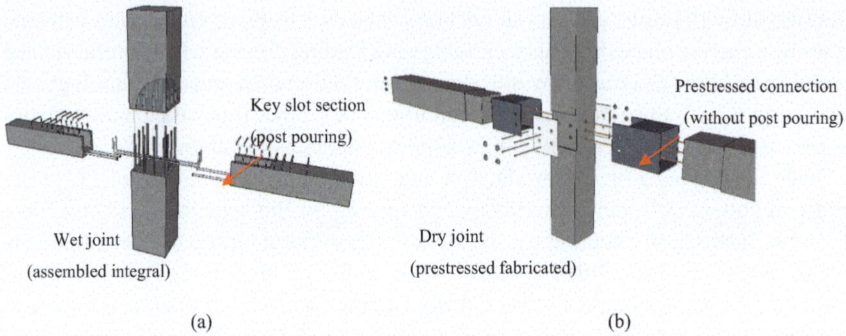

Fig. 1.23 Typical precast concrete connection joint forms: **a** "wet" connection **b** "dry" connection

cast-in-place". After decades of research, a mature "equivalent cast-in-place" structure design theory has been formed here and abroad. At the same time, there are many practical engineering cases of "equivalent cast-in-place" prefabricated concrete structure systems, which verify the feasibility of its structural connection and structural system. The advantages of the rapid connection and high ductility of the prefabricated structure system are fully utilized, and the requirements of seismic fortification are also met. However, the "equivalent cast-in-place" prefabricated concrete structure system still has much wet-work site construction and cannot fully utilize the advantages of prefabricated structures over cast-in-place structures. This greatly limits the development and application of prefabricated buildings.

In contrast, using the form of a "dry" connection has no requirement to pour concrete, but through the embedded connection parts in the components, the parts are connected by bolts, welding, or prestress. There is no wet operation on site, and the construction efficiency can be greatly improved, which is more in line with the characteristics of a prefabricated structure. In addition, the "dry" connection prefabricated concrete structure system can realize the self-reset of the joints through additional energy dissipation components such as angle steel and dampers, and the residual deformation of the joints is small. Moreover, the energy dissipation capacity can meet the design requirements by adjusting the position, type, and bonding length of the prestressed reinforcement and by selecting appropriate and high-quality energy dissipation components or dampers in the design. The damaged local members between beams and columns can be directly replaced after an earthquake. Compared with the "equivalent cast-in-place" prefabricated concrete structure system, it is easier to achieve repair and replacement, which causes the structure to have better recoverability. Therefore, this kind of structural system can also be called a "nonequivalent cast-in-place" prefabricated concrete structure system, which is more in line with both the essence of "assembly" technology and the trend of engineering development. However, this kind of "nonequivalent cast-in-place" connection structure has an unclear force mechanism and calculation method because of its various connection structure forms and local components. Compared with the "equivalent cast-in-place" structure, its structural characteristics, stress mechanism, and calculation method have obvious differences and peculiarities. Although domestic and foreign scholars have carried out many in-depth studies in this area, there is still a lack of mature "nonequivalent cast-in-place" prefabricated concrete structure systems and related theories. Therefore, it is necessary to deeply study the "nonequivalent cast-in-place" prefabricated concrete structure system and establish a reasonable structural form, calculation theory, and design method.

1.6 Main Contents of This Book

To meet the urgent needs of developing the industrialization of prefabricated building systems in China, accelerating the pace of industrialization of the industrial building structure system, and taking into account the structural performance of various types of industrial building systems and the actual complex situation in the construction process, it is necessary to carry out independent innovative research and to make substantial progress in the research on key issues such as new structural systems, joint structures, design theory, construction technology and seismic performance of industrialized buildings. The research and development of prefabricated structural systems with superior structural performance, convenient fabrication, transportation and construction, simple process, and wide adaptability ensures that the structural system is in the lead and that the economic and safety indicators are both excellent. They also provide a scientific basis for the design, construction, and revision of standards and specifications of prefabricated structures. In addition, it is necessary to promote the practical engineering application of research results for the market and accelerate the realization of building industrialization and an ecological civilization.

Given the problems existing in the current "equivalent cast-in-place," prefabricated concrete structure system and the shackles of the development of the "nonequivalent cast-in-place" prefabricated concrete structure system, Southeast University and Harbin Institute of Technology have carried out in-depth research on this and put forward many new types of prefabricated concrete structure systems, such as ductile connection frame systems, prefabricated concrete rocking wall structure systems, prefabricated concrete box structure systems, prefabricated modular suspension structure systems, etc. The research results can promote the development of basic theory and technology of building industrialization, accelerate the scientific and technological innovation for building industrialization in China, and promote the popularization and application of prefabricated building technology. In addition, they also provide theoretical and technical support for the large-scale application of prefabricated concrete structures in China, laying a solid foundation for the further research of prefabricated building technology standard systems and realization of the concept, technology, and industrial upgrading of green building and building industrialization.

This book systematically sorts and summarizes the above research results and presents them to its readers for the first time. The book introduces seven new types of prefabricated concrete structures, including a prefabricated concrete ductile connection frame system (Chap. 2), a prefabricated concrete bamboo-bar energy dissipation frame system (Chap. 3), a prefabricated concrete friction energy dissipation frame system (Chap. 4), a cast-in-place main frame-fabricated secondary frame structure system (Chap. 5), a prefabricated concrete rocking wall structure system (Chap. 6), a prefabricated concrete box structure system (Chap. 7), and a prefabricated modular suspension structure system (Chap. 8). This book focuses on the composition, stress mode, structural characteristics, performance advantages, and calculation methods of each new structural system and analyzes the combination of the new structural system and building industrialization technology and the advantages of the new structure system compared with the traditional structure system.

In the second chapter, a prefabricated concrete ductile connection frame system is described. The ductile connector is embedded in the column and connected with the reinforcement in the beam through bolts and transfer blocks, and the plastic deformation during an earthquake is concentrated on the ductile link, thus protecting other parts of the structure from damage. A ductile connection device is a kind of energy dissipation device that uses plastic deformation of metal during a strong earthquake to dissipate seismic energy. It is generally made of steel and has the characteristics of a full hysteresis curve and stable energy dissipation performance. This chapter mainly introduces the performance and design method of the connecting device and its structural system.

In the third chapter, a prefabricated concrete frame system with external energy dissipation is described. This system combines advanced high-performance energy-dissipating devices with the self-reset mechanism of the structure and has the advantages of rapid assembly, high performance and recovery after an earthquake. This chapter briefly describes the research work that has been carried out from the research and development of metal energy dissipation rods and the performance evaluation of prestressed precast concrete frame joints, focusing on the four aspects of the energy dissipation rod structure, experimental verification, theoretical exploration and joint performance test.

In the third chapter, a prefabricated concrete frame system with external energy dissipation is described. This system combines advanced high-performance energy-dissipating devices with the self-reset mechanism of the structure and has the advantages of rapid assembly, high performance, and recovery after an earthquake. This chapter briefly describes the research work that has been carried out from the research and development of metal energy dissipation rods and the performance evaluation of prestressed precast concrete frame joints, focusing on the four aspects of the energy dissipation rod structure, experimental verification, theoretical exploration, and joint performance test.

The fourth chapter describes a kind of prefabricated concrete friction energy dissipation frame system. The structural system greatly improves the strength and ductility of the prefabricated joints through the stable bearing-energy dissipation dual-function friction energy dissipation, and the post-tensioned prestressed tendons provide a stable elastic restoring force for the structure. In this chapter, the construction mechanism of a prefabricated structure, joint energy dissipation test, structural numerical modeling, structural seismic performance, and structural vulnerability analysis are expounded.

In the sixth chapter, a kind of prefabricated concrete rocking wall structure system is described. The system is an embedded self-resetting rocking structure that uses the self-weight and prestresses of the structure to achieve self-resetting and reduce residual displacement. Replaceable energy dissipation components are added to the middle part of the wall, and the corner of the wall is made of high ductility elastic material so that the damage to the wall can be controlled. This chapter will introduce the prefabricated concrete rocking wall structure system from the two aspects of experimental analysis and finite element simulation, indicating that this type of wall has the characteristics of low damage, high energy consumption capacity, and controllable structural damage under the action of an earthquake.

The seventh chapter describes a prefabricated concrete box structure system, which is a new type of spatial structure system independently developed in China. It is mainly composed of open-web sandwich plates, namely, grid-format frame walls. It has the advantages of light overall weight, strong crossing ability, and high stiffness. Based on existing research and best engineering practice, this chapter studies the application of box structure systems in high-rise structures by analyzing the hysteretic performance, seismic performance, and economic performance of such systems.

In the eighth chapter, a prefabricated modular suspension structure system is described. The system has the advantages of transparent space in each layer of substructure, open space at the bottom layer, and light weight. Using the relative motion of the primary and secondary structures, combined with dampers or energy-dissipating devices, a better damping effect can be achieved. This chapter will introduce the prefabricated modular suspension structure system from the aspects of the vibration damping mechanism, mechanical characteristics, damping performance optimization, and experimental research on the primary and secondary structures.

This book strives to present the latest research and best practices of prefabricated structure systems to the readers, including reference teachers, students, and technicians.

References

1. Aisc A (2010) AISC 341–10, seismic provisions for structural steel Buildings. American Institute of Steel Construction, Chicago, IL
2. Bijlaard FSK (2014) Eurocode 3—design of steel structures. In: Dictionary geotechnical engineering/w rterbuch geotechnik 2014, pp 486
3. Nakaki SD, Stanton JF, Sritharan S (1999) An overview of the PRESSS five-story precast test building. PCI J 44(2):26
4. Priestley M, Sritharan S, Conley JR et al (1999) Preliminary results and conclusions from the PRESSS five-story precast concrete test building. PCI J 44(6):42
5. Priestley M (1991) Overview of PRESSS research program. PCI J
6. Li J (2008) The construction technology of using PC board as exterior wall panel in Shanghai Vanke industrial residence. Shanghai Constr Sci Technol 2008(06):10–11

7. Wu D (2016) Seismic performance evaluation of prefabricated shear wall structure with mortar anchor connection. Southeast University, 2016
8. Guo Z, Dong N, Zhu Z (2011) New progress in construction technology of prefabricated concrete structure in housing construction. Constr Technol 40(11):1–2
9. Haiquan H (2014) NPC technology system to achieve the core competitiveness of enterprises. Architecture 2014(05):29
10. Gang W, Decheng F (2018) Research progress on basic performance of precast concrete frame joints. J Arch Struct 39(2):1–16

Chapter 2
Ductile Precast Concrete Frame with Dry-Connections

Abstract The chapter deals with the dry-connection methods for the precast concrete frame. The emulative cast-in-situ method and dry connection method are compared. The construction and mechanical principle of the replaceable energy dissipation connectors are introduced. Two types of dry-connection beam-column joints and a kind of column base are proposed and experimentally analyzed, respectively. In the end, the design method of dry-connected precast concrete frame based on the replaceable energy dissipation connectors is presented, focusing on the seismic performance and the design procedure. The proposed dry-connected precast concrete frames can be applied in high seismic regions, while the robustness and post-earthquake repair can be conveniently achieved.

Keywords Dry-connection · Precast concrete frame · Replaceable energy dissipation connectors · Design method

2.1 Introduction

At present, the engineering practice of precast concrete buildings in China is still concentrated in the emulative cast-in-situ method, while the "dry connection" is still in the conceptual research stage. Although the research of dry connection precast concrete structure has been more in-depth all over the world, only a few applications are applied in practice. For the emulative cast-in-situ method, the mechanical performance of the frame are similar to that of the cast-in-situ ones, so the design theory, calculation process and structural requirements can generally follow the relevant provisions of the cast-in-situ structure. However, in the process of precast, the exposed connecting reinforcement needs to be reserved at the end of the component, which brings various problems to production, transportation and installation [1, 2]. Because the key parts, such as the joints between the beams and columns, still need the cast-in-situ work, the characteristics of high-efficiency of precast structure are not easy to be achieved completely. For precast concrete structures with dry connection method, the construction is convenient to achieve, which is the mainstream direction of precast concrete buildings in the future [3–5]. Considering that there are obvious differences in the mechanical property between dry connection structures

G. Wu et al., *Novel Precast Concrete Structure Systems*, Springer Tracts in Civil Engineering, https://doi.org/10.1007/978-981-19-6821-1_2

and emulative cast-in-situ structures, this chapter takes a type of precast concrete frame with replaceable energy dissipation connection as an example to introduce the construction, performance and main design methods of dry connected precast concrete frame [6–10]. It can be used as a reference for scientific researchers and engineering practitioners.

2.2 Emulative Cast-In-Situ Method and Dry Connection Method

The reliability of frame depends not only on the design and construction technology of each component, but also on the reliability of the connection between components. The reliability of joints in frame plays a critical role in the seismic performance of frame, which needs to be fully considered in design. The connection joints of precast concrete frame structures can be divided into two types: emulative cast-in-situ connection and dry connection.

The emulative cast-in-situ connection refers to the construction method of connecting precast components through cast-in-situ joints. In order to connect various components of the frame, the precast components manufactured in factory need to reserve extended reinforcement at their ends. After these components are successively assembled on site, the post-cast concrete is poured to connect the components into a whole by the bonding effect between concrete and reinforcement. Emulative cast-in-situ can achieve the similar mechanical performance as that of cast-in-situ frame, in which the design method is also generally the same as that of cast-in-situ frame known to most engineers. Therefore, in the early stage of building industrialization, this construction method has been popularized in a large area.

With the deepening of building industrialization, engineers also began to realize some shortcomings of emulative cast-in-situ structure. Firstly, the anchor reinforcement protruding from the end of precast components brings high requirements on the production and transportation of components. Secondly, in the process of component installation, due to the narrow area and too many rebars for construction, the position conflict of reinforcement in different directions is easy to occur in the joint. In order to avoid the interference of the reinforcement protruding from the end of the former component to the installation of the subsequent component, the installation sequence also needs to be carefully set in the installation process. Additionally, the joints are the internal force intersection area with complex stress distribution, which should have better quality and performance than the components. However, though the component prefabricated in the factory could be guaranteed in high quality, while the quality of the cast-in-situ joint is discrete, and it is difficult to ensure the overall requirements of "strong joints and weak components." In view of this situation, nowadays researchers focus on dry connection.

The dry connection refers to the construction method of precast component by prestress, weld or bolts without post pouring concrete, in which the reinforcement protruding from the end of the member is not needed. Thus, the process of manufacturing, transportation and installation of the precast components can be more convenient, while the mold for pouring concrete can be regular and simple in the manufacturing process. Furthermore, it is conducive to the reuse of the mold without worrying about the collision and deformation of the extended reinforcement in the process of formwork closing, formwork removal, hoisting and transportation. During the installation process, the load can be borne without waiting for the post cast concrete, which effectively compresses the wet operation cycle.

In terms of structural seismic performance, the cast-in-situ or emulative cast-in-situ frames dissipate the inputted energy by the ductility of the components only under large earthquakes. In this case, the damage in the plastic hinge area is not easy to repair after the earthquake, additional reinforcement means are needed to restore the service function of the structure. Additionally, the repair period is also long; Alternatively, in the dry connection frame, the weak area of the component is generally the connection position itself. The disassembly and assembly of the connecting parts is simply constructed, and the repair is also convenient to achieve, which is also one of the reasons why dry connection is paid attention to. However, it is undeniable that due to the great difference between the dry connection and the emulative cast-in-place structure in the section form and stiffness of the components, an independent design theory is needed to be formed, including the mechanical analysis, seismic performance analysis and design method, etc., which is also the key content of this chapter.

2.3 Replaceable Energy Dissipation Connector

2.3.1 Plate Connector

(1) Construction and compose

The replaceable energy dissipation connector (REDC) is a type of metal-yield energy dissipation device, which is similar to a buckling-restraint brace (BRB). The REDCs can be regarded as a part of the reinforcement in the beam end, as well as the damper of the frame. An REDC is composed of the fuse member, restraining member, and connection member, as shown in Fig. 2.1. The fuse member is divided into the yielding segment (YS), transition segment (TS), and connection segment (CS), as shown in Fig. 2.2. Additionally, it is manufactured using steel with high ductility, which could lead to a high energy dissipation capacity. To restrain the buckling amplitude of the fuse member under compression, the restraining member is designed to cover the fuse member. Thus, the fuse member can yield along the whole length of the YS, displaying a plump hysteretic curve under a cyclic load. The restraining member is composed of two restraining plates and three infilling plates, and is assembled using high-strength bolts, as shown in Fig. 2.1. The connection member is composed of two steel blocks that are embedded in the column and beam, and a step should be reserved between the two blocks for assembling the REDCs. Additionally, the blocks are tied to the reinforcement in the beam and anchor rebars in the column using a plug welding method, as shown in Fig. 2.3. The REDCs mainly resist the axial force, while the shear force and bending moment can be neglected based on the results of a mechanical analysis and finite element method, which also confirmed the conclusions mentioned above.

(2) Materials and construction of REDC

Selection of materials: The REDC fuse member should be manufactured from steel with high ductility to give the REDC a high low-cycle fatigue capacity. In this study, a Q235B steel plate defined by the Chinese codes was selected for manufacturing the

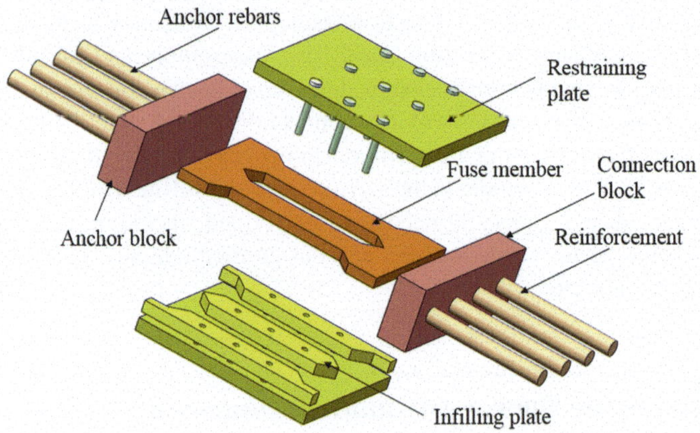

Fig. 2.1 Components of REDC

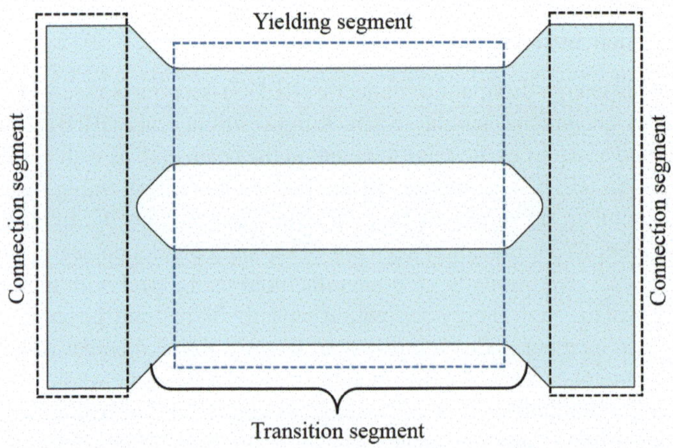

Fig. 2.2 Fuse member

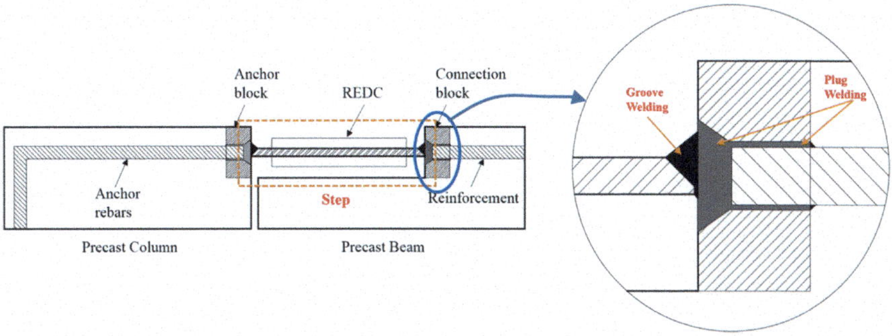

Fig. 2.3 Connection members and steps

fuse member. Its yielding strength was 230–280 MPa, while its elongation capacity under uniaxial tension could reach values greater than 25%.

Construction: As shown in Fig. 2.2, the REDC fuse member is composed of the YS, TS and CS, which are manufactured by wire-electrode cutting steel plates with the same thickness. The YS has the smallest width among the three parts, while that of the CS should be twice as large to restrict the yield concentrated on the YS. Both the CS and TS should remain elastic to ensure the reliability of the connection between the REDC and the structural members. Once the YS yields, the axial stiffness can decrease sharply to 0.6–2.0% of the elastic stiffness, and it tends to buckle under compression. The restraining member is assembled to restrain the buckling of the fuse member. In addition, because the TS should be as short as possible to prevent a stress concentration, the YS was designed as two parallel limbs.

Demands for YS cross-sectional area: The REDC is regarded as part of the rebars in the beam end, in which the YS should remain elastic under vertical loads and frequent earthquakes. Additionally, if a strong earthquake occurs, the YS should yield first to restrain the increase in the internal force in the beams and columns. In the design procedure, the yielding force of the REDC should be lower than those of the rebars in the beam and the anchor bars in the column, as shown below:

$$P_{\max} > \frac{M_{\max}}{d} \tag{2.1}$$

$$0.9 \le \frac{P_{\max}}{\lambda P_{\mathrm{y}}} \le 1.0 \tag{2.2}$$

$$\lambda P_{\mathrm{y}} < P_{\mathrm{y}}^{\mathrm{rebar}}, \tag{2.3}$$

where $P_{\max}$ is the designed maximum axial force of the fuse member; $M_{\max}$ is the maximum bending moment of the beam end; d is the distance between the centroid lines of the two REDCs on the upper and lower sides of the beam end; P_{y} is the yielding force of the fuse member; λ is the overstrength coefficient of the fuse member; and P_y^{rebar} is the minimum yielding force of the longitudinal reinforcement in the beam and anchor bars in the column.

The bending moment in the beam end under vertical loads and frequent earthquakes should be calculated first during the design procedure for the REDC-PCF. Then, the maximum axial force under this condition can be obtained using Eq. (2.1). The yielding force demand of the REDC fuse member can be obtained using Eq. (2.2), which is used to determine the cross-sectional area of the YS. Additionally, the minimum demand for the yielding force and cross-sectional area of the longitudinal reinforcement can be obtained using Eq. (2.3).

TS and CS: The geometrical characteristics of the TS and CS can significantly affect the connection behaviour. In addition, the connection between the CS and connector should also be considered.

Considering the depth of the reversed steps, steel plates without any stiffening ribs were selected for the cross-sections of the TS and CS. Additionally, a smooth shape was selected for the TS to prevent a stress concentration. When the REDC is under compression, the friction between the fuse member and the restraining member can amplify its compression-bearing capacity, which can increase the cross-sectional demand on the TS. The maximum internal force of the TS can be found as follows [11]:

$$F_{yt} = f_y \cdot A_T \leq f_y \cdot (\gamma_m \cdot \zeta_m \cdot A_e) \tag{2.4}$$

where f_y is the yielding strength of the fuse member; A_T is the cross-sectional area of the TS; A_e is the cross-section of the YS; γ_m is the compression-bearing capacity enhancement coefficient, which should be no less than 1.1; and ζ_m is a coefficient that considers the strain hardening effect, which is chosen to be 1.4. According to Eq. (2.4), A_c should be no less than 1.5 times larger than A_e.

The reliability of the connection should be guaranteed by limiting the nominal stress to less than the yielding stress. On the other hand, the actual yielding stress will be larger when considering the overstrength of the material. The cross-sectional area of the CS should satisfy Eq. (2.5):

$$A_c \geq \zeta \cdot \gamma_m \cdot \zeta_m \cdot A_e, \tag{2.5}$$

where A_c is the cross-sectional area of the CS, and ζ is the overstrength ratio of the material, which is selected to be 1.3.

Length of YS: The YSs of the REDCs yield first, while the other members remain elastic under strong earthquakes. The position of the YS on the beam is also a plastic hinge, which is used to concentrate the deformation of the frame under strong earthquakes. Thus, the length of the YS should be determined by the maximum rotation of the beam end. According to GB50011 in China, the maximum relative rotation angle between the beam and the column should be no larger than 2%. The maximum strain amplitude of the YS should be limited to no greater than 3.0% to ensure the low-cycle fatigue performance. Therefore, the length of the YS should meet the requirement that when the relative rotation angle between the beam and the column reaches 2%, the maximum strain amplitude of the YS should be no more than 3%.

(3) Mechanical performance

The mechanical performance of the REDCs are studied through the low-cycle reversed loading test, in which the initial stiffness, post-yield stiffness, hysteretic performance, etc. are tested. It can be seen that the REDC specimens can exhibit a stable hysteretic property and high ductility, which indicated a high low-cycle fatigue capacity. In the tests, most of the failure are observed in the pre-set yielding segments, which can be seen summarized in Fig. 2.4, while the connection segments and the transition segments remains elastic. The reliability of this type of connectors between the precast components can be confirmed.

The hysteretic curves of part of specimens are shown in Fig. 2.5. The key parameters of the test results are listed in Table 2.1. It can be seen in the figures that all of the specimens experienced stable hysteretic properties without stiffness degeneration. The residual deformation was stable without increasing further under the constant amplitude loading tests. Strain strengthening was observed in the specimens tested under the variable amplitude loading tests. These curves revealed that the REDC specimens could exhibit stable energy dissipation performances.

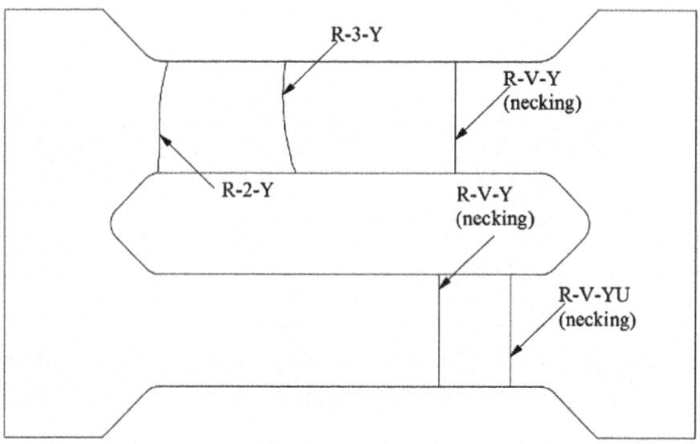

Fig. 2.4 Failure modes and buckling of REDC specimens

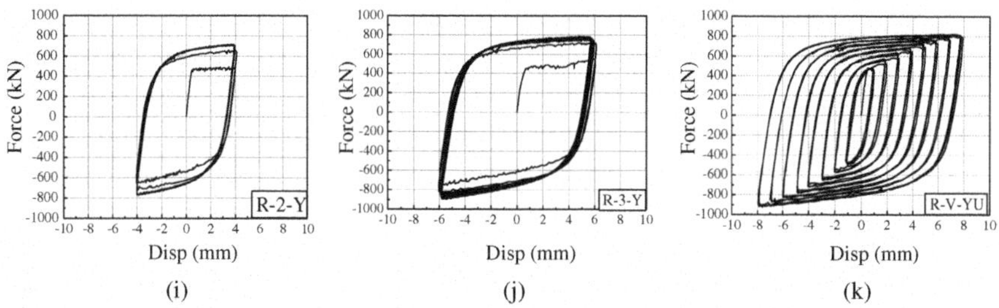

Fig. 2.5 Hysteretic curves of specimens

Table 2.1 Results for specimens

Specimen	ε_{nom} (%)	μ_{max}	β	N_f	*CPD*
R-2-1	2.0	14.98	1.06	54	2897
R-2-2	2.0	14.98	1.17	40	2238
R-2.5-1	2.5	20.20	1.15	25	1974
R-3-1	3.0	22.90	1.21	19	1641
R-3-2U	3.0	23.06	1.07	17	1501
R-V-1U	0.5–4.5	34.59	1.11	25	1698
R-V-2U	0.28–6.0	45.98	1.22 (1.05)	33	1899
R-V-3	0.5–3.5	26.91	1.11	37	1139
R-2-Y	2.0	15.08	1.08	74	3980
R-3-Y	3.0	23.01	1.14	46	4345
R-V-Y	0.5–4.5	35.01	1.15	54	2871
R-V-YU	0.5–4.5	35.07	1.12	45	3072

Note ε_{nom} is the nominal strain amplitude; μ_{max} is the maximum ductility ratio; β is the compression strength adjustment factor; and N_f is the number of failure cycles

2.3.2 *Bars Connector with Non-slipping Threaded Sleeve Assembly*

The replaceable energy dissipation bars (REDB) are composed of energy dissipation bars (ED), connection components, and restraining members, as shown in Fig. 2.6. The ED bars are divided into the connection zone, transition section, and ED segment along the longitudinal axes. Because the area of the ED segment is smaller than that of the threaded connection segment, the ED segment can be designed to yield under an axial load, while the threaded connection segments maintained their elasticity. The bilateral connection segments are provided with threads matching the connection components. Because the outer concrete and stirrups at the bottom of the beam would be unable to restrain the compressive buckling of the ED bars under a large displacement, special restraining members should be installed according to the BRB principle.

The restraining members adopted for the REDB connection had two main forms. The first type of restrainer is shown in detail in Fig. 2.6a. It includes two restraining plates with internal grooves and is assembled as a square box using high-strength bolts. A stopper is settled at the middle of the ED bar to couple the position of the restraining plate, prevent interference between the mutual compressible layers at both ends, and prevent the restraining plate from axial loading. The second type of restraining member consisted of an outer restraining plate coupled to the beam by screws and threaded sleeves embedded in the beam, with grout filling the gap between the restraining plate and beam bottom, where the outer restraining plate and filling material provided the lateral stiffness, as shown in Fig. 2.6b. In contrast with the first type, the deformation of the ED bars in the second type is concentrated on

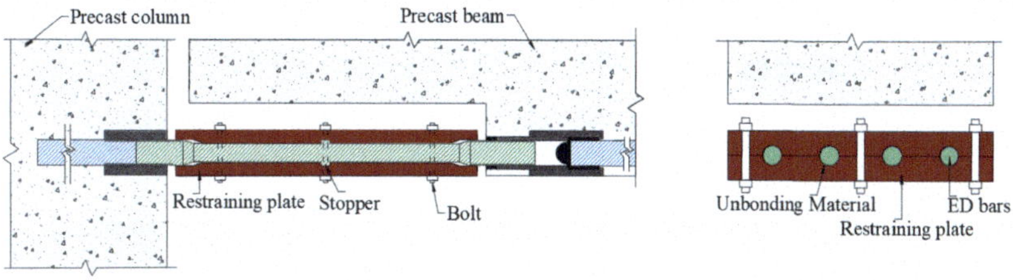

(a) Schematic diagram of constraint system (first type)

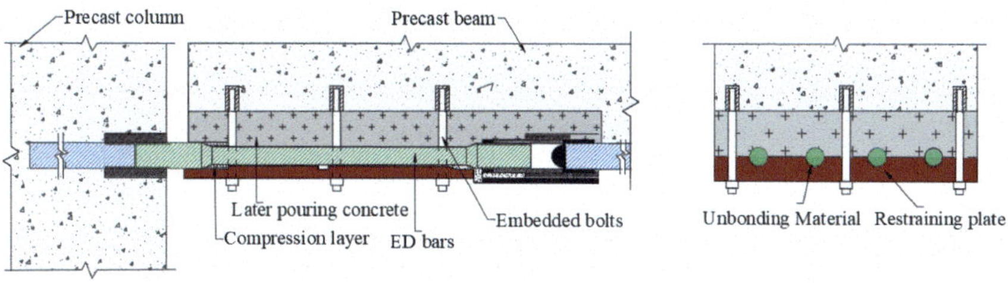

(b) Schematic diagram of constraint system (second type)

Fig. 2.6 Schematic diagram of constraint systems

one side near the column without using a stopper. Therefore, the slot could be utilized to reserve sufficient space for the deformation of the ED bars in rare earthquakes. The second type is mainly adopted in practical projects for REDB-PCF connections.

Two common techniques are implemented for dry connections: welding and threaded couplers [12]. Unfortunately, a welding seam is prone to brittle failure under low cyclic loads. Moreover, there are some problems with the use of threaded couplers due to the high size and position requirements and exacerbation of thread slippage by the tolerance gaps needed for construction. Therefore, as shown in Fig. 2.7a, a new type of the non-slipping threaded sleeve assembly (NSTSA) was employed for the REDB-SYBC connection.

The NSTSA can tightly connect the two segments of concentric rebars without sliding when transferring tension and pressure, and meet the installation precision requirements. It is composed of the outer sleeve, the first inner sleeve built into the outer sleeve, the second inner sleeve, and lock nuts. More specifically, one side of the outer sleeve is designed with equal diameter necking, where the inner diameter is larger than that of the longitudinal reinforcement and smaller than the outer diameter of the first inner sleeve to jam the first inner sleeve. Internal threads are cut into the other side to connect with the second internal sleeve. In addition, one side of the first inner sleeve has a central countersunk hole provided with internal threads to connect the longitudinal reinforcement in the beam. Meanwhile, the guide head on the other side is semi-spherical, which forced the rebar to produce a small bending action through the guiding action between the guide head and the second inner sleeve, adjusting the installation error in the axis direction. For the second inner sleeve, one

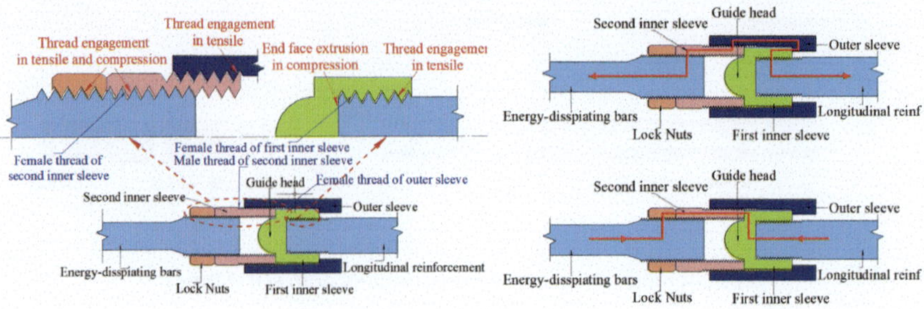

(a) Components of non-slipping threaded sleeve assembly

(b) Load path in tensile and compression

Fig. 2.7 Details of non-slipping threaded sleeve assembly

side is provided with internal threads to connect with the ED bars, and the other side has external threads to couple with the outer sleeve. Locknuts are set on one side of the second inner sleeve to limit the relative movement between the ED bars and the second inner sleeve.

Figure 2.7a shows a magnified view of the critical threads to demonstrate the mutual engagement of threads among the various components of the threaded sleeve assembly in detail. It is found that the tension and pressure could be transmitted without slipping based on the reasonable structural details of the components. Accordingly, the load transfer path of the NSTSA connection is shown in Fig. 2.7b. The reliability of the NSTSA as a connection component for the ED bars was verified in tests.

2.4 Dry-Connected Beam-Column Joint

2.4.1 Double-Side-Yield REDC-PCF Beam-Column Joint

(1) Construction and compose

The REDC-PCF is a kind of precast concrete frame fabricated by a "dry connection" method. The connection is positioned in the beam end. The shear force is borne by the pin shaft system, while the bending moment is borne by the REDCs, as shown in Fig. 2.8. The REDCs are welded to the anchor/connection blocks embedded in the precast components, thus, a stable load-transmission mechanism is formed in the REDC-PCF. The REDCs can be regarded as a part of the longitudinal reinforcement to bear the moment in the beam end under gravity and frequent earthquakes. Once a strong earthquake occurs, the REDCs yield and dissipate the inputted energy, while the beams and the columns remain elastic. The seismic rehabilitation of the REDC-PCF is achieved by replacing the damaged REDCs, which is very convenient.

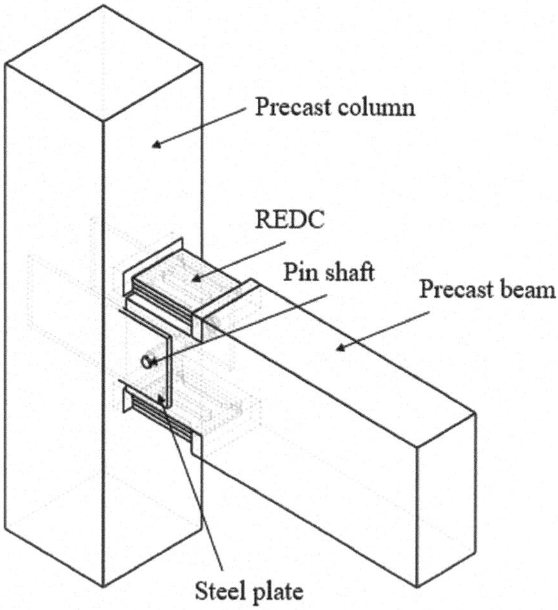

Fig. 2.8 Beam-column joint of the REDC-PCF

The main properties of the beam-column joint are listed below:

1. The beams and the columns are precast in the factory. The connections between the beams and the columns are realized by the REDCs and the pin shaft system.
2. The yield axial force of the REDCs is smaller than that of the longitudinal reinforcement in the beam and the anchor rebars in the column, concentrating damage in the plastic hinge on the beam end.
3. The beam and the column are separated with a given clearance between them; thus, the beam can rotate under the load without having contact with the concrete. The concrete in the beam end and the joint can remain elastic without crushing under strong earthquakes, thus, the ductility of the REDC can be fully developed.
4. The connection method for the REDCs and the rebars embedded in the precast components is welding through the blocks rather than threaded connections, which can largely lighten the demand for precision during construction. Additionally, the weld can be accomplished by a mechanized method, such as robots, while the construction quality can be guaranteed.
5. The cross-section centroid of the REDC is aligned with that of the longitudinal reinforcement and the anchor rebars, which can guarantee that the load-transfer path is direct.

(2) Mechanical performance

The mechanical performance of REDC-PCF beam-column joint is obtained by the low-cycle fatigue tests. The test apparatus and specimens are shown in Fig. 2.9.

Fig. 2.9 Test setup of the joint

According to Fig. 2.10, it can be seen in the initial test that the deformation of the joint was mainly concentrated in the fuse members of the REDCs, and the deformation of the concrete beam is small according to the slight self-closed cracks. These phenomena indicate that the elastoplastic deformation was mainly concentrated in the YSs of the REDC fuse members, while the longitudinal reinforcement and the stirrups in the beam and the column remained elastic. As shown in Fig. 2.10a, the cracks in the beam were mainly caused by the elongation of the longitudinal reinforcement due to the flexural moment. All the cracks developed during the loading process closed when the load was released. The fracture occurred after the rotational drift reached 5.0%, which was far more than the 2.0% threshold for the reinforced concrete frame structures stipulated by code GB50010-2011. The fuse member fractured in the middle zone, in which a necking phenomenon was observed, as shown in Fig. 2.10b. The fuse member tended to buckle when yielding, while the buckling amplitude was restrained by the restraining member, resulting in a multiwave buckling phenomenon, as shown in Fig. 2.10c. It is concluded that the REDC-PCF can exhibit high ductility, while the beams and the columns remain elastic.

Then, the damaged REDCs were replaced by new ones through gas cutting and butt welding. The rehabilitated specimen can be seen in Fig. 2.11a. Then, the low-cycle fatigue test was conducted with the same test protocol as that in the initial test. The test phenomena were similar to that of the initial test, in which the plastic deformation was concentrated in the REDCs with an elastic beam and column. The fracture occurred in the second cycle of the 5% rotational drift, and the failure also happened at the middle section of the YS of one REDC, as shown in Fig. 2.11b. The other REDC did not fracture; however, multiwave buckling was observed, as shown in Fig. 2.11c.

The load–displacement curves of the test specimens are listed in Fig. 2.12, in which the stable energy-dissipation capacities were obtained. The seismic capacity could be confirmed to be better than that of the common concrete frame due to

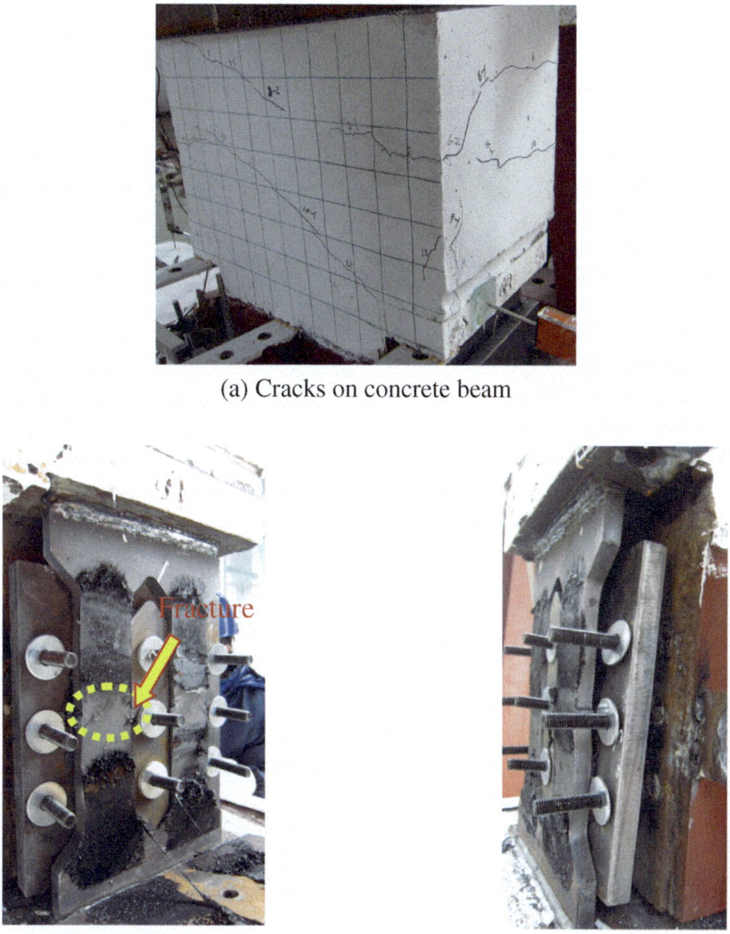

(a) Cracks on concrete beam

(b) Fracture of REDC

(c) Buckling of REDC

Fig. 2.10 Failure modes of the initial test

the plump hysteretic property without pinch as well as its high ductility. The initial stiffness values of the specimens in both the initial test and the rehabilitated test were identical, which further confirmed the damage concentration and the seismic rehabilitation property of the REDC-PCF. Compared with common concrete beam-column joints, the neutral axis of the beam was fixed in the cross section centroid. The fuse member of the REDC could yield under both tension and compression due to the restraint of deformation by the restrainers. As the clearance reserved between the beam end and the column was 15 mm, which is larger than the demand value of the rotation of the beam end, compression of the concrete was prevented; thus, the pinch of the hysteretic curves was eliminated.

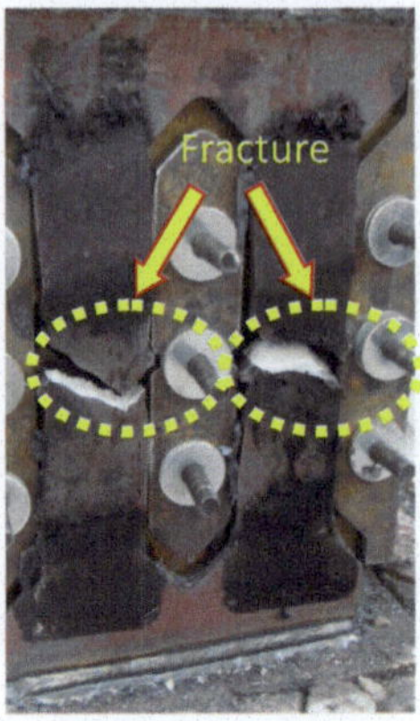

(a) Rehabilitation of the specimen (b) Fracture of the REDC (c) Restrained buckling of the REDC

Fig. 2.11 Failure mode of the rehabitated test

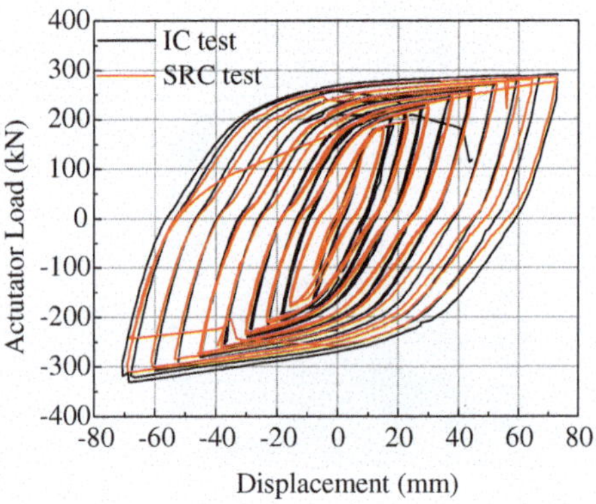

Fig. 2.12 Load–displacement curves for the tests

Comparing the IC and SRC tests, it can be observed that the initial stiffness, post-yield stiffness and energy-dissipation capacity of the two tests were almost the same. The two curves confirm that the seismic rehabilitation of the joint was effective since the mechanical properties and seismic performance could be rehabilitated.

Taking the IC test as an example, the axial force–axial deformation curve of the REDC and the reinforcement strain–load curve in the beam end near the joint are shown in Fig. 2.13a, b, respectively. The axial force of the REDC was calculated as follows:

$$F_A = F_P L/d \tag{2.6}$$

where F_A is the calculated axial force of the REDC, F_P is the actuator load, L is the distance between the centroid of the actuator and the pin shaft, and d is the distance

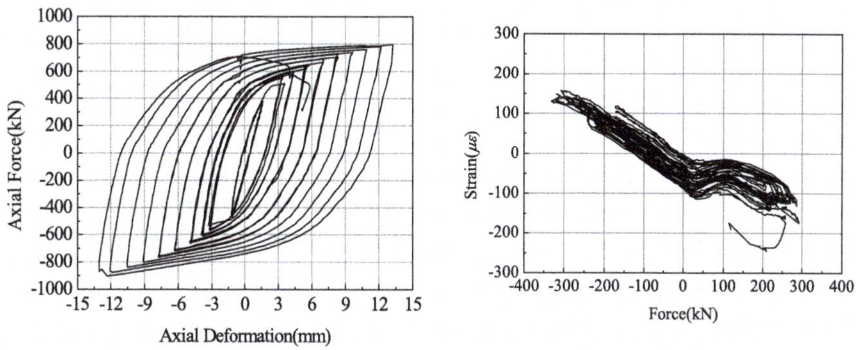

(a) Force-deformation curve for the REDC (b) Strain of the reinforcement bars

Fig. 2.13 Mechanical properties of the REDCs and reinforcement bars

between the two REDCs. Figure 2.13b shows that rebars near the beam end remained elastic, with a maximum strain value of no more than 300 $\mu\varepsilon$, which is far less than the yield strain (1260 $\mu\varepsilon$). Additionally, the cracks in the beam could close once the load on the beam end was relieved, which could further confirm that the beam remained elastic during the whole experimental process.

The skeleton curves of the tests, including the load–displacement curves and the moment-drift curves, are listed in Fig. 2.14a, b, respectively. The moment and the drift ratio on the beam end (shown in Fig. 2.14b) are obtained by the following equations:

$$M = F_{\mathrm{A}} L \tag{2.7}$$

$$\theta = \Delta\varepsilon l_{\mathrm{e}}/d \tag{2.8}$$

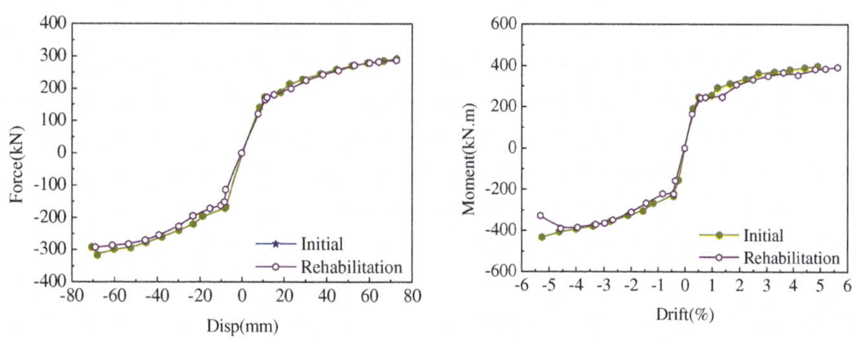

(a) Load-displacement curves (b) Moment-rotation curves

Fig. 2.14 Skeleton curves of the specimens

where M is the moment of the beam end, θ is the drift ratio of the joint, $\Delta\varepsilon$ is the axial strain of the REDC, l_e is the equivalent length of the REDC. The other symbols are the same as Eq. (2.4).

The conclusions are listed below:

(1) Upon comparing the skeleton curves between the two tests, an approximately equal initial stiffness is observed. Once the REDC yielded, the rotational stiffness significantly decreased. The skeleton curves can be simplified as a bilinear constitutive model in finite element analysis. The post-yield stiffness of the joint is determined by the REDC since the beam and the column remained elastic at all times.
(2) The skeleton curves of the tests are symmetrical in the stages of push and pull, which reveals that the mechanical properties of the REDCs on the upper and lower sides were approximately the same.
(3) No stiffness degeneration was observed in the skeleton curves, which is fundamentally different from the case for common concrete frame joints. This property further confirmed that the damage was mainly concentrated in the REDCs.
(4) The ratio of the post-yield stiffness to the initial stiffness is 13.6% in the load–displacement curves, while a value of 8.47% is obtained in the moment-rotation curves. The difference between the two values was caused by the stiffness of the concrete beam. The moment-rotation curves only reflected the rotational stiffness of the joint; the effect of the beam was neglected.

The mechanical and deformation parameters of the tests are listed in Table 2.2.

Table 2.2 shows that the ratios of the maximum loads to the yield loads on the beam end in the two tests are 1.67 and 1.64, respectively. This result revealed that the load increased to a relatively large amplitude after the REDC yielded, which was caused by the post-yield hardening of the Q235 steel.

As shown in Table 2.2, the ratio of the maximum displacement to the yield displacement (Δ_m/Δ_y) for the specimen varied between 5.65 and 6.50, while the ratio of the maximum rotational drift to the yield rotational drift (θ_m/θ_y) varied between 14.63 and 15.60. The values revealed that high ductility was seen in the REDC-PCF beam-column joint.

Table 2.2 Mechanical properties of the specimens

Test	F_y (kN)	F_m (kN)	Δ_y (mm)	Δ_m (mm)	θ_y (%)	θ_m (%)	F_m/F_y	Δ_m/Δ_y	θ_m/θ_y
IC	175.50	292.45	11.22	72.93	0.352	5.49	1.67	6.50	15.60
SRC	178.62	292.71	12.91	72.89	0.328	4.80	1.64	5.65	14.63

Note F_y is the yield force of the joint; Δ_y is the yield displacement on the beam end; θ_y is the yield rotational angle of the joint; F_m is the maximum force of the joint; Δ_m is the maximum displacement at the beam end; and θ_m is the maximum rotational angle of the joint

2.4.2 Single-Yield REDC-PCF Beam-Column Joint

(1) Construction and compose

The Single-yield REDC-PCF (SYPC) joint proposed comprises a precast beam, precast column, shear transfer, and REDC, as is shown in Fig. 2.15.

The upper part of the precast beam and adjacent part between the column and beam in the precast column are used to arrange the top reinforcements; this area also represents the closure pour region. The closure pour region is located at the upper part of the vertical slot along the entire beam length direction, and extends into the column. In this region, the top longitudinal reinforcements in the upper part of the beam are connected with the reinforcements embedded in the column through steel

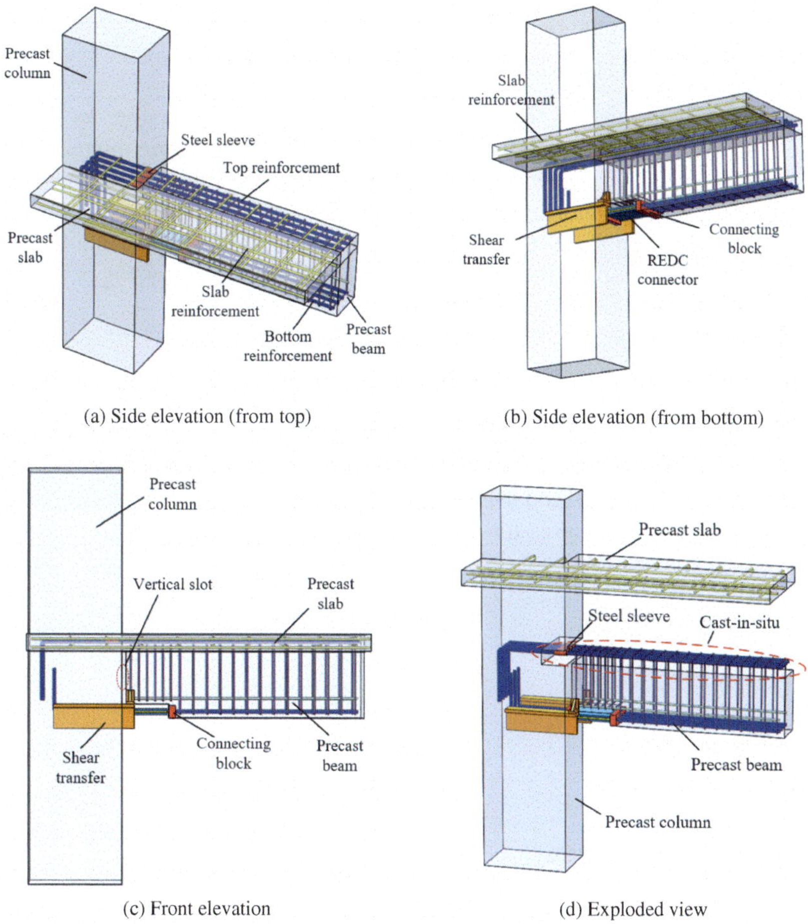

Fig. 2.15 SYPC beam–column joint with REDC

sleeves (Fig. 2.15a), forming a complete connection. As shown in Fig. 2.15b, two shear transfer elements are embedded in the column, and are overlapped with the two shear transfer elements welded on both sides of the beam. Through this construction, the negative shear force (downward shear force) of the beam end can be transferred to the precast column. A positive shear force (upward shear force) is transferred through the compression concrete zone. The REDC is arranged at the bottom of the beam. The energy-dissipation capacity of the SYPC connection is provided by the tension and compression plastic deformation of the REDC. As shown in Fig. 2.15c, a vertical slot is installed at the bottom of the beam, immediately adjacent to the column face. The width of the vertical slot needs to be wide enough to ensure that no contact force is generated owing to the gap close to the maximum drift. The height must be sufficiently high to minimize the pressure provided by the concrete compression zone. Figure 2.15d shows an exploded view of the connection formed by moving the superimposed concrete slab upward. In Fig. 2.15d, the red dashed line is the cast-in-situ part of the beam and column.

(2) Single-yielding mechanism

Figure 2.16 shows the internal force distribution of the SYPC connection. As shown in Fig. 2.16a, when the connection is subjected to a positive bending moment, the REDC provides a tension force (T_s), the top reinforcements provide a compression force (C_s'), and the concrete compression zone above the vertical slot (top hinge) provides a compression force ($C_{c,top}$). Therefore, the force balance formula for the connection under a positive bending moment can be expressed by Fig. 2.16, as follows:

$$T_s = C_s' + C_{c,top} \tag{2.9}$$

As shown in Fig. 2.16b, when the connection is subjected to a negative bending moment, the REDC provides a compression force (C_s), the top reinforcements provide a tension force (T_s'), and the concrete compression zone below the top reinforcements provides a compression force ($C_{c,bottom}$). Therefore, the force balance formula of the connection under a negative bending moment can be expressed by Fig. 2.16, as follows:

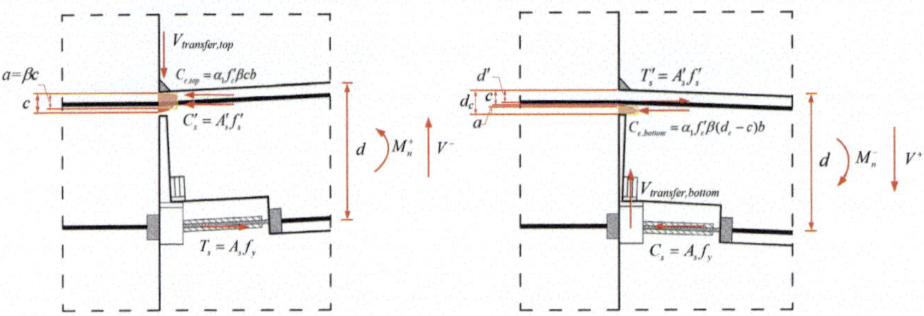

(a) Positive moment (gap opening) (b) Negative moment (gap closing)

Fig. 2.16 Force distribution in the connection under positive and negative bending

$$T_s' = C_s + C_{c,bottom} \tag{2.10}$$

It can be seen from Eq. (2.15) that, under a positive bending moment, because $C_{c,top}$ is large enough to balance T_s independently, C_s' can be controlled to be very small (top reinforcement in elastic), and the single-yielding mechanism can be realized easily. Under the negative bending moment, T_s' must be used not only to balance C_s, but also to balance $C_{c,bottom}$ (Eq. (2.16)). Therefore, to realize the single-yielding mechanism under a negative bending moment, the vertical slot height should be increased by as much as possible. Simultaneously, the total yield force of the top reinforcements in the beam must be greater than the maximum tension/compression force (yield force × strain-hardening coefficient γ) of the REDC. When the vertical slot is sufficiently high, $C_{c,bottom}$ can be ignored. At this time, to realize the single-yielding mechanism, the top reinforcements must satisfy Eq. (2.11), as follows:

$$A_s' f_y' / A_s f_y \geq \gamma \tag{2.11}$$

In the above, A_s' and A_s represent the equivalent cross-sectional areas of the top reinforcements and REDC core plate, respectively; f_y' and f_y represent the yield strengths of the top reinforcements and REDC core plate, respectively; and γ is the strain-hardening coefficient of the REDC.

For the same reason, to avoid the yield of the bottom reinforcements, the total yield force of the bottom reinforcement should be greater than the maximum tension/compression force (yield force × γ) of the REDC, shown in Eq. (2.12), as follows:

$$A_{s,bottom} f_{y,bottom} / A_s f_y \geq \gamma \tag{2.12}$$

Here, $A_{s,bottom}$ represents the equivalent cross-sectional area of the bottom reinforcements, and $f_{y,bottom}$ represents the yield strength of the bottom reinforcements. Based on this structure, the connection can realize a single-yielding mechanism under both positive and negative bending moments.

In addition, because the vertical slot is sufficiently high to ignore the compression force provided by the concrete compression zone, both the positive and negative bending moments of the SYPC connection can be controlled based on the tension/compression force of the REDC. This realizes the design purpose, i.e., the controllable bending moment of the SYPC connection.

(3) Mechanics of the SYPC connection

As shown in Fig. 2.16a, the positive nominal flexural strength ($M + n$) can be calculated similarly to that in the case of a conventional reinforced concrete connection. Because the positive bending moment of the SYPC is governed by the tension/compression force of the REDC core plate, $M + n$ can be calculated by multiplying the REDC yield force of the lever arm based on Eq. (2.13), as follows:

$$M_n^+ = A_s f_y (d - a/2) \tag{2.13}$$

In the above, A_s represents the equivalent cross-sectional area of the REDC core plate, f_y is the yield strength of the REDC core plate, d represents the beam effective depth, and a represents the Whitney equivalent rectangular concrete stress block depth.

As shown in Fig. 2.16b, the negative nominal flexural strength (M –n) cannot be determined by referring to a conventional reinforced concrete connection. This is attributed to the vertical slot, which makes the compression force provided by the concrete compression zone very small, or even close to zero. However, when the slot height is relatively small, the compression force provided by the concrete compression zone cannot be ignored, and must be included in the calculation. In this case, the negative moment of the SYPC connection can be calculated using Eq. (2.14), as follows:

$$M_n^- = A_s f_y (d - d') + \alpha_1 f_c' \beta (d_c - c) b (d_c - a/2 - d') \tag{2.14}$$

Here, d and d' represent the beam effective depth and depth to the top reinforcement, respectively; α_1 and β are concrete stress block factors; f_c' represents the specified 28-d concrete compressive strength; d_c represents the depth of the concrete top-hinge in this novel precast connection; c represents the neutral-axis depth (measured from the top surface of the beam); and b represents the section width of the beam.

However, when the vertical slot is sufficiently high, i.e., such that the height of the concrete compression zone is very small, the neutral axis is very close to the position of the compression resultant of the concrete compression zone. Therefore, to simplify the calculation, $C_{c,bottom}$ can be ignored. In such a case, the negative moment of the SYPC connection can be calculated using Eq. (2.15), as follows:

$$M_n^- = A_s f_y (d - d') \tag{2.15}$$

When the height of the vertical slot is small (d_c is sufficient for transferring the upward shear force), the upward shear force (negative shear force) can be transmitted by the concrete compression zone, and is calculated using Eq. (2.16), as follows:

$$V^- \leq 0.25 \beta_c f_c' b h_0 \tag{2.16}$$

In the above, β_c is a coefficient of the concrete strength (= 1.0 for C40 concrete), and h_0 represents the effective height of the cross-section (h_0 represents the d_c in the SYPC connection).

When the vertical slot is too high (i.e., d_c too small), it cannot transfer the upward shear force. In this case, a top shear transfer element must be arranged at the top of the connection to transfer the shear force. To simplify the design, the upward shear force of the connection is transmitted by the top shear transfer element, as shown in Eq. (2.17).

$$V^- \leq V_{transfer,top} \tag{2.17}$$

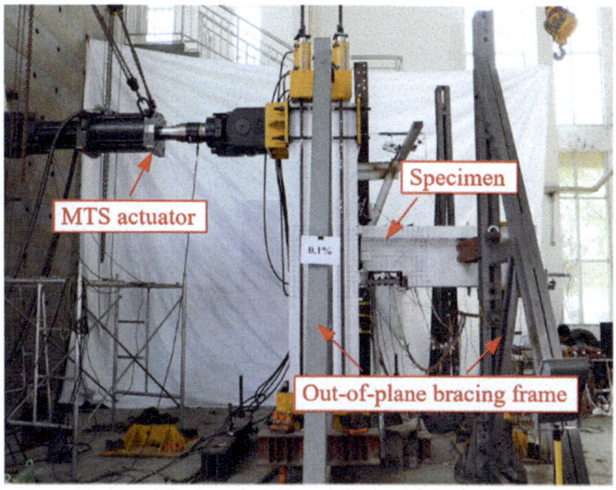

Fig. 2.17 Photograph of the test setup

Here, $V_{transfer,top}$ represents the shear capacity provided by the top shear transfer element.

The downward shear force of the SYPC connection is transmitted by the bottom shear transfer element. The design of the bottom shear transfer element can be designed based on Eq. (2.18), as follows:

$$V^{+} \leq V_{transfer,bottom} \tag{2.18}$$

In the above, $V_{transfer,bottom}$ represents the shear capacity provided by the bottom shear transfer element.

(4) Seismic performance

To investigate the hysteretic behavior of the SYPC connection under cyclic loading, a full-scale test specimen was fabricated, as shown in Fig. 2.17.

Owing to the single-yielding mechanism, most of the precast components (beam and column) other than the REDC basically remained elastic throughout the entire loading process. In view of this, four REDCs with three different thicknesses were replaced in a single beam-column specimen to conduct five tests, as shown in Table 2.3. The hysteretic responses from the five tests are shown in Fig. 2.18a–e. Except for Test 1 with an amplitude up to 2%, the other tests (Tests 2–5) were loaded until the bearing capacity dropped to 85% of the maximum strength, and then until the reverse loading reached the maximum drift. The lateral drift ratio was calculated by dividing the lateral displacement by the effective height of the column (H_c = 2995 mm). The force in the hysteretic curve was measured using the load cell of the MTS actuator. The hysteretic responses of all tests are characterized by stable behavior without signs of "pinching," and there is no evident strength degradation at a given drift level. Thus, the REDCs show excellent characteristics. In addition, there is an evident low-stiffness section (a small slip) near the zero force in the hysteretic

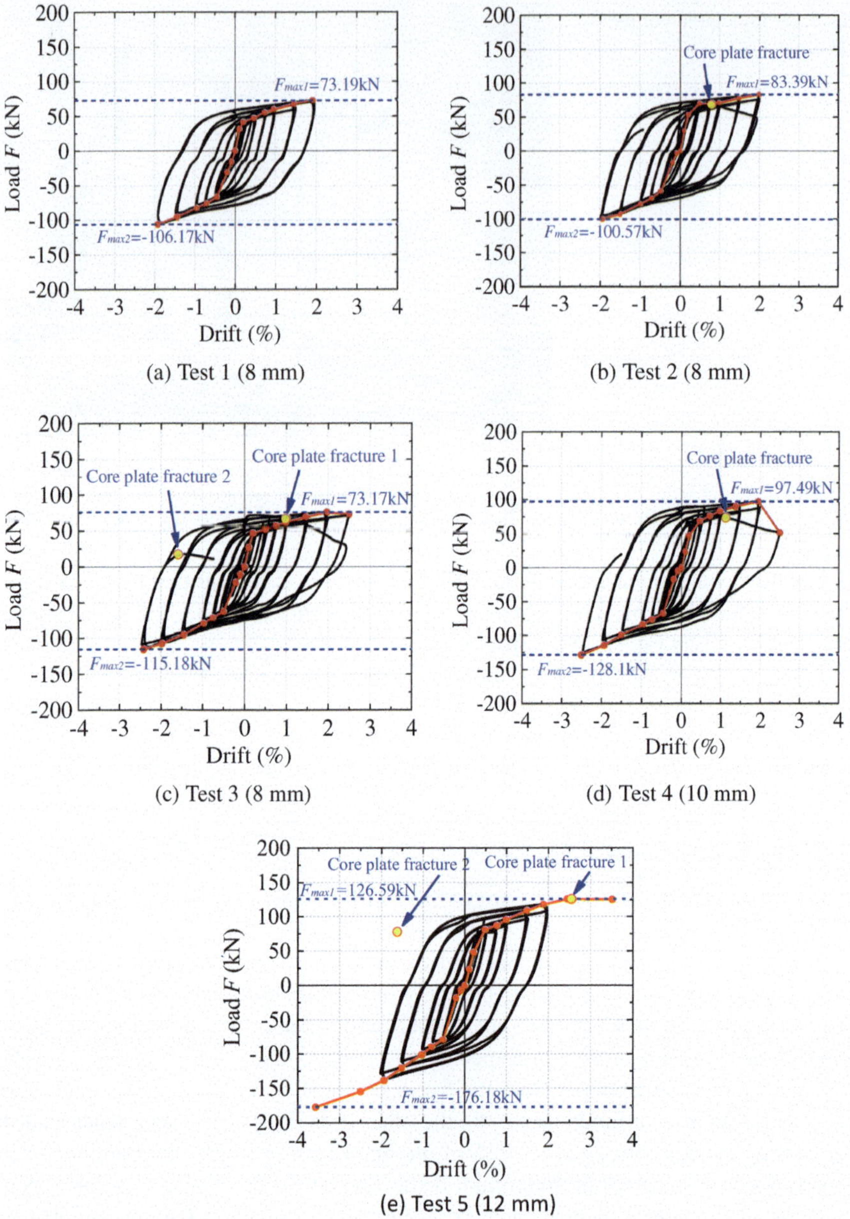

Fig. 2.18 Load versus drift response for the SYPC connection: **a** Test 1; **b** Test 2; **c** Test 3; **d** Test 4; **e** Test 5

Table 2.3 Test matrix

Test number	Thickness of the core plate (mm)	Whether to replace replaceable energy-dissipation connector (REDC) (Y/N)	Loading protocol
Test 1	8	/	2%
Test 2	8	N	Load to failure
Test 3	8	Y	Load to failure
Test 4	10	Y	Load to failure
Test 5	12	Y	Load to failure

loops; this is caused by the small gap between the pin shaft and lug plate of the doubled-hinged support.

The maximum positive drift ($\theta_{\max1}$), maximum negative drift ($\theta_{\max2}$), maximum positive force ($F_{\max1}$), maximum negative force ($F_{\max2}$), drift level in the loading protocol corresponding to the first fractured side of the REDC core plate ($\theta_{f1,level}$), drift level in the loading protocol corresponding to the second fractured side of the REDC core plate ($\theta_{f2,level}$), number of cycles at $\theta_{f1,level}$ (n_{f1}), and number of cycles at $\theta_{f2,level}$ (n_{f2}) are summarized in Table 2.4.

Compared with Tests 1 and 2, the maximum forces of Tests 1 and 2 are similar, indicating that the unrepaired SYPC connection can resist the severe earthquake (2% exceedance probability in 50 years) again. The maximum force of Test 1 does not decrease, indicating that the SYPC connection has quick repairability. As shown from the comparison of the bearing capacities of Tests 1, 4, and 5, the bearing capacity increases with the thickness of the REDC core plate.

The cracking patterns of the specimens in Tests 1–5 are shown in Fig. 2.19. Because there are no cracks in the column, only the cracks on the beam are shown. Detailed illustrations of the cracks at the top hinge are shown in Fig. 2.19a–e. The cracks at the top hinge are mainly flexural cracks originating near the edge of the vertical slots, and extend to the top of the slab. With the increase in the loading drift, diagonal cracks (cracks 13-5) appear at the top hinge under the negative bending

Table 2.4 The results of the three tests

Test number	$\theta_{\max1}$ (%)	$F_{\max1}$ (kN)	$\theta_{\max2}$ (%)	$F_{\max2}$ (kN)	$\theta_{f1,level}$ (n_{f1}) (%)	$\theta_{f2,level}$ (n_{f2}) (%)
Test 1	2.0	73.19	– 2.0	– 106.17		
Test 2	2.0	83.39	– 2.0	– 100.57	2.0 (3rd cycle)	
Test 3	2.5	73.17	– 2.5	– 115.18	2.5 (2nd cycle)	2.5 (3rd cycle)
Test 4	2.5	97.49	– 2.5	– 128.10	2.5 (1st cycle)	
Test 5	3.5	126.59	– 3.5	– 176.18	3.5 (2nd cycle)	3.5 (2nd cycle)

Notes $\theta f_{1,level}$ and $\theta f_{2,level}$ are the drift level in the loading protocol corresponding to the two side of the REDC core plate fractured; n_{f1} and n_{f2} are the number of the cycles at $\theta_{f1,level}$ and $\theta_{f2,level}$ respectively

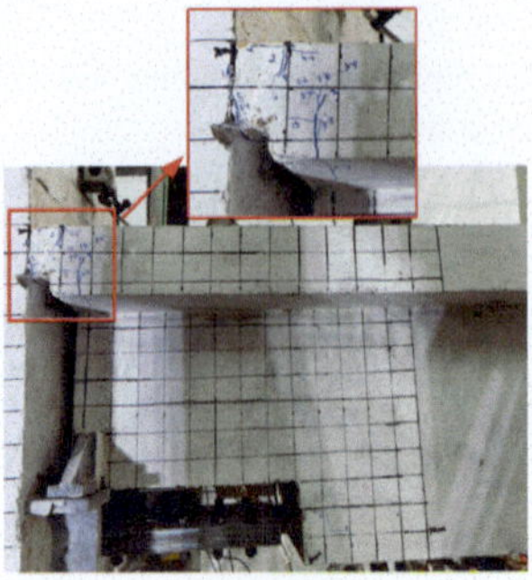

(a) Test 1 (8 mm)

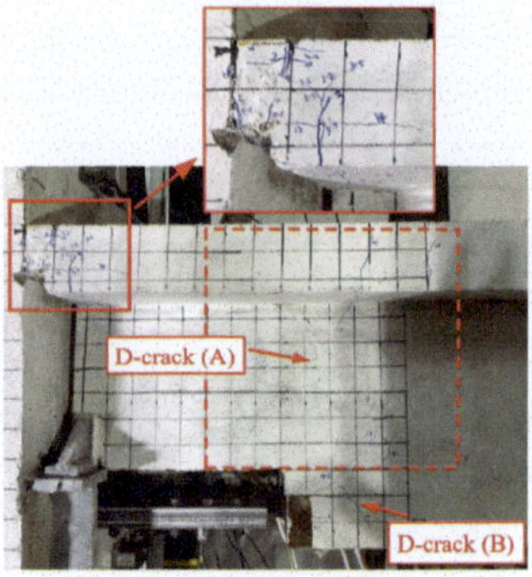

(b) Test 2 (8 mm)

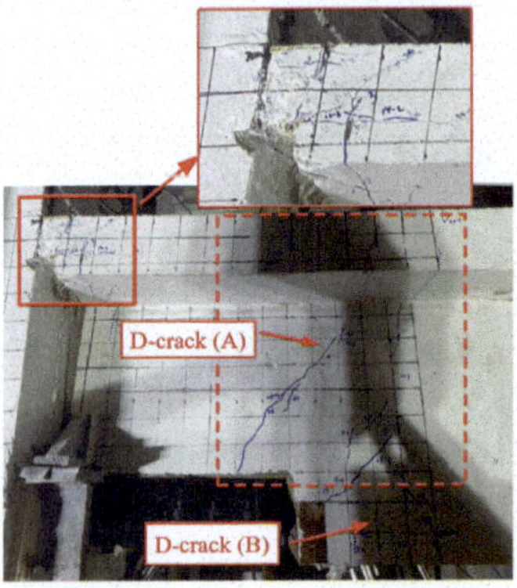

(c) Test 3 (8 mm)

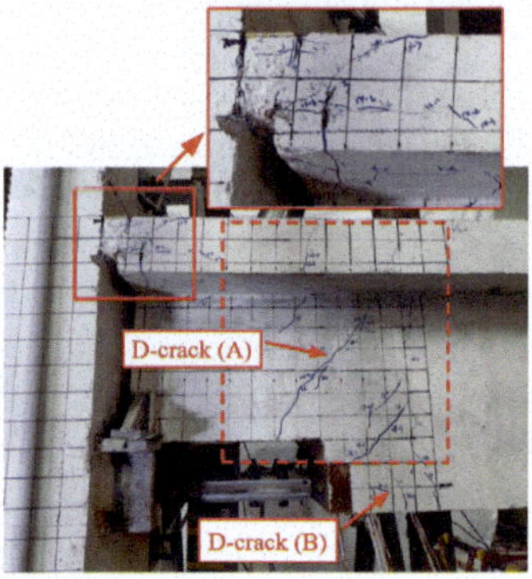

(d) Test 4 (10 mm)

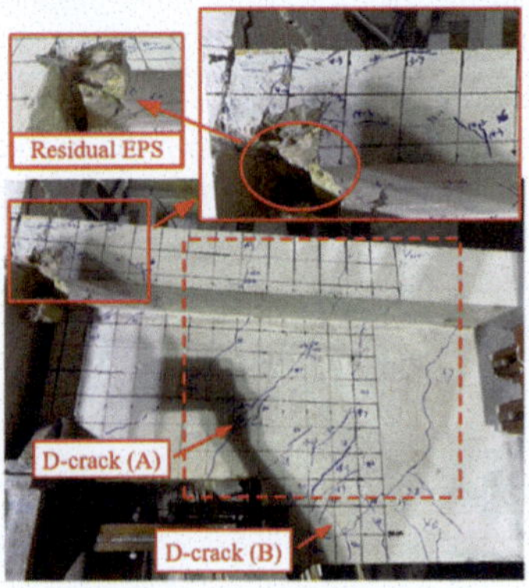

(e) Test 5 (12 mm)

Fig. 2.19 Cracking patterns at the ultimate condition connection: **a** Test 1; **b** Test 2; **c** Test 3; **d** Test 4; **e** Test 5

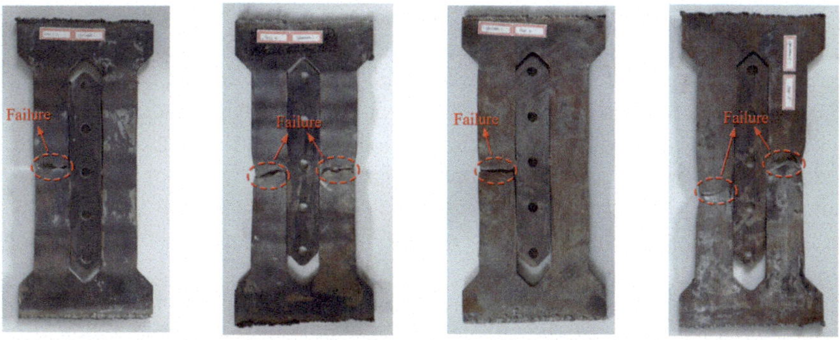

(a) Test 1 and 2 (8 mm) (b) Test 3 (8 mm) (c) Test 4 (10 mm) (d) Test 5 (12 mm)

Fig. 2.20 Failure pattern of the core plate: **a** Test 1 and 2; **b** Test 3; **c** Test 4; **d** Test 5

moment (negative drift), and extend from the edge of the vertical slot to the upper-right area of the beam. Notably, two horizontal cracks (cracks 13 and 14) and a vertical crack (crack 3) appear at the bottom of the beam hinge and gradually expand during the loading process, and concrete spalling occurs during the Test 5 loading process. This is attributed to a residual EPS block during the concrete vibration process, which makes the internal force unbalanced. However, on the back of the specimen, there is no similar concrete spalling and there are no cracks, indicating that the SYPC connection will not result in these types of cracks and concrete spalling at the top hinge. The diagonal cracks at the beam can be divided into two types: those under a negative drift (A), and those under a positive drift (B). The A-diagonal cracks are mainly concentrated in the red-dashed bordered rectangle. The B-diagonal cracks are mainly concentrated in the lower part of the beam. An intersection angle of approximately 90° between the two types of diagonal cracks appears in the lower part of the beam. In addition, no cracks appear in the joint area or column; this can be attributed to the controlled bending moment.

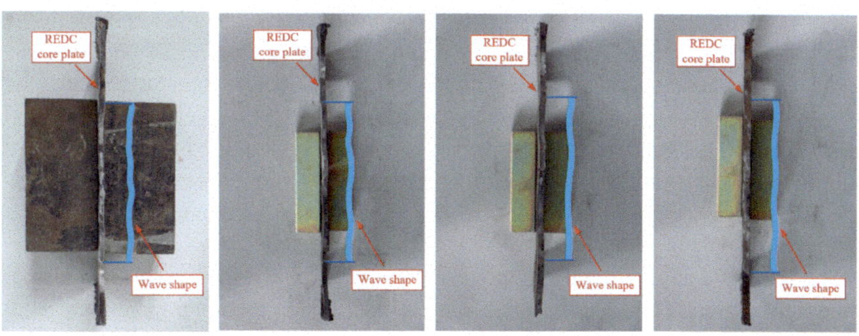

(e) Test 1 and 2 (8 mm) (f) Test 3 (8 mm) (g) Test 4 (10 mm) (h) Test 1 and 2 (12 mm)

Fig. 2.21 Wave shape of the core plate: **a** Test 1 and 2; **b** Test 3; **c** Test 4; **d** Test 5

Figure 2.20 shows the core plate failure patterns of Tests 1–5. Test 2 was tested on the basis of Test 1, i.e., without repair or replacement of the REDC. One or two strips of these four REDC core plates fracture completely, owing to low-cycle fatigue failures. The fracture position of the REDC core plate is in the yielding segment of the core plate. In addition, the ductile failure modes of the four REDC core plates can be determined, as evidenced by the apparent necking at their yielding segments. As shown in Fig. 2.21, all four REDC core plates present evident multi-wave buckling but the multi-wave amplitudes are very small, indicating that the buckling restraint effect is evident.

As shown in Fig. 2.20a, c, only one strip of the REDC core plate fractures in Tests 2 and 4. However, as shown in Fig. 2.20b, d, both strips of the REDC core plate fracture in Tests 3 and 5. This is attributed to the slight angle deviation in the welding process of the REDC, which causes the two strips of the REDC core plate to have slightly different strain amplitudes during the loading process. As a result, one strip of the REDC core plate fractures first owing to low-cycle fatigue, whereas the other strip of the REDC core plate fractures later. In Tests 2 and 4, under a positive bending moment, one strip fractures under tension. Then, when loading is continued again to the negative maximum drift and to positive loading, the other strip still does not fracture. Nevertheless, in Tests 2 and 4, when positive loading is repeated, the other strip is fractured in tension.

2.5 Dry-Connected Column Base

2.5.1 Construction and Mechanism Performance

As shown in Fig. 2.22, the precast concrete column base with REDCs (REDC-PCCB) is composed of the precast column, the concrete foundation and the REDCs. A reserved space could be observed on each side of the column base for fabricating the REDCs. All four sides and the bottom of the core concrete are wrapped with a square steel tube. The top plate of the square steel tube extends laterally to the outer edge of the upper concrete segment. The bottom of the precast column rests on the top of an embedded steel plate anchored in the concrete foundation; thus, the column base can rock in two orthogonal directions. The raised steel stoppers (RSSs) are attached to the four sides of the square steel tube and welded to the embedded steel plate on the top of the foundation on the outside. The inner sides of the RSSs have a certain slope to prevent them from restricting the horizontal rotation of the column base.

The transfer paths of axial force, shear force and bending moment are separate in the joint region. The concrete-filled square steel section mainly transfers the axial compressive force, while the embedded steel plate on the top of the foundation helps protect the concrete foundation in terms of local compression. The RSSs can effectively limit the horizontal sliding of the column base and contribute to the horizontal

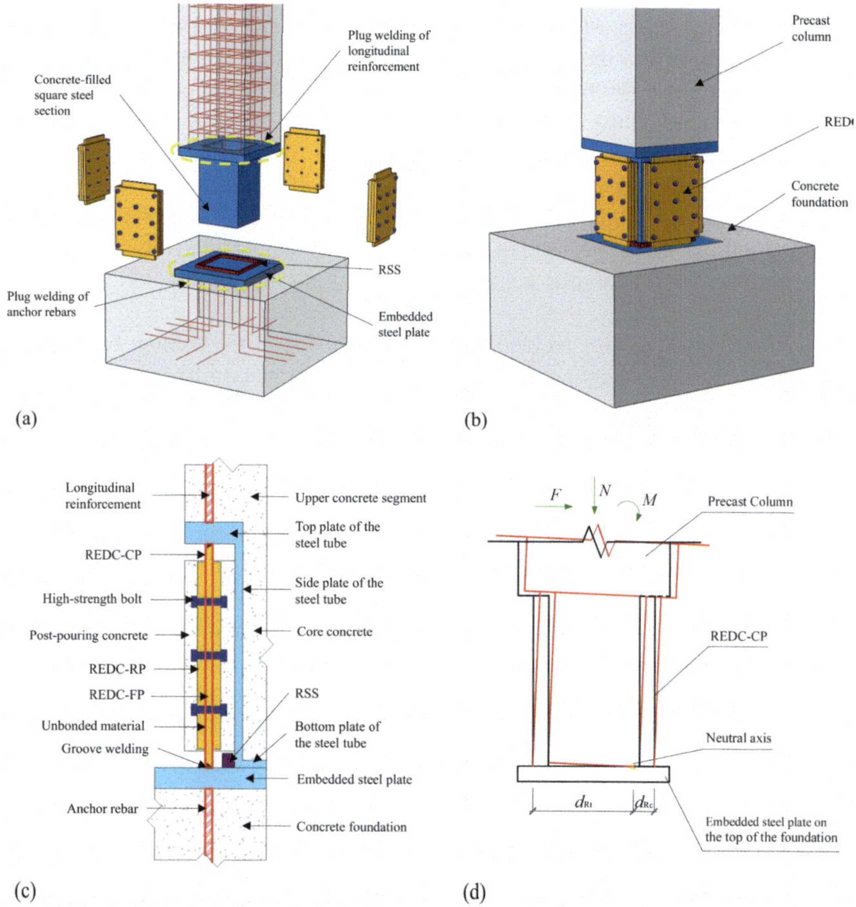

Fig. 2.22 Details of REDC-PCCB. **a** Explode view, **b** overall structure, **c** part construction profile and **d** rotation of REDC-PCCB

shear resistance. Furthermore, the REDCs act as short BRBs, which provide flexural bearing capacity for the column base and first yield under strong earthquakes to dissipate inputted seismic energy, effectively protecting the adjacent main structural components [13, 14].

2.5.2 Seismic Performance

A full-scale REDC-PCCB test specimen is designed for quasi-tests, which is separated from the part below the point of contraflexure in the ground story column of the prototype reinforced concrete frame. The total height of the test specimen is

2570 mm, while the precast column has a height of 1870 mm and a cross section of 500 mm × 500 mm. For the foundation, the total length is 2800 mm, while the cross section is 700 mm × 700 mm. The basic dimensions and reinforcement layouts of the test specimen are illustrated in Fig. 2.23.

Considering that the variation in the actual axial force applied to the column may be dramatic due to the axial deformation of the PT strands when the relative rotation of the column-foundation connection is large [15–18], a high-precision static servo-hydraulic console was used to automatically maintain a relatively constant oil pressure in the 500 t hydraulic center hole jack during the loading and unloading process.

A 20 mm high RSS was welded tightly on each side of the column base in the loading direction, the height of which exceeds the estimated lifting height of the column base under the target lateral drift ratio. The surface of the RSSs was polished in advance, with a slope of 15% reserved on the inside. Then, the REDCs need to be fabricated before loading. The REDC-CPs with grooves on the outer sides of both ends were connected to the top plates of the square steel tube and the embedded steel plates on the top of the foundation using carbon dioxide arc welding. A layer of butyl rubber was uniformly pasted on the surface of the yielding segments of the REDC-CPs as a type of unbonded material, reducing the friction coefficient of the steel contact surface. In this case, the uneven strain distribution of the REDC-CPs due to the difference between the transferred tension force and compression force during the cyclic loading process can be effectively relieved. Furthermore, butyl rubber also effectively reduces surface scratches and harsh noise due to wear [19]. The process

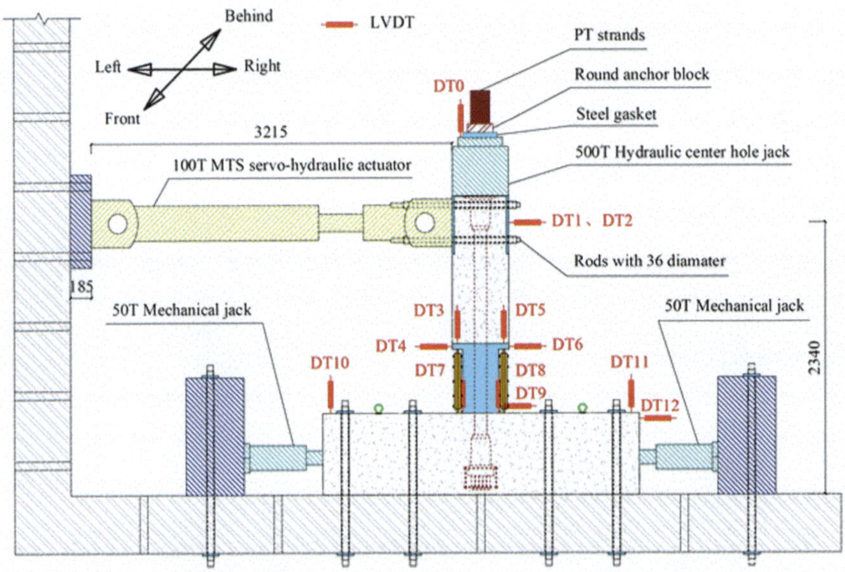

Fig. 2.23 Test setup of the column

Cut the weld by oxygen-acetylene flame | Polish the surface of the steel plate | Polished column base | Completing the installation of REDC

Fig. 2.24 Replacement of REDC after a test

of replacing the REDCs after a test is shown in Fig. 2.24. The remaining region of the column base was not filled with post-pouring concrete for the convenience of observing the test phenomena.

A total of 5 tests were carried out, which can be divided into three comparative groups as listed in Table 2.5. Among them, the first group consisted of two comparative tests before and after rehabilitation of the test specimen with a relatively low axial compression ratio (the test axial compression ratio equals to 0.19), which mainly verified the rationality of the load-carrying mechanism of the REDC-PCCB joint and the effect of post-earthquake rehabilitation. The second group included three comparative tests with different axial compression ratios to compare the force and deformation properties of the REDC-PCCB joint under three test axial compression ratios of 0.19, 0.32 and 0.50. The effects of this configuration, including damage concentration and easy rehabilitation even with a high axial compression ratio, were

Table 2.5 Test condition

Test name	Experimental axial compression ratio	Thickness of REDC-CPs (mm)	Test purpose
Test 0.19-16	0.19	16	The 1st test
Test 0.19-16R	0.19	16	Comparison of the seismic performance before and after rehabilitation
Test 0.32–16	0.32	16	Study the influence of the axial compression ratio
Test 0.50–16	0.50	16	Study the influence of the axial compression ratio
Test 0.50–12	0.50	12	Study the influence of the thickness of REDC-CP

Note 'R' represents the test after rehabilitation under the same conditions

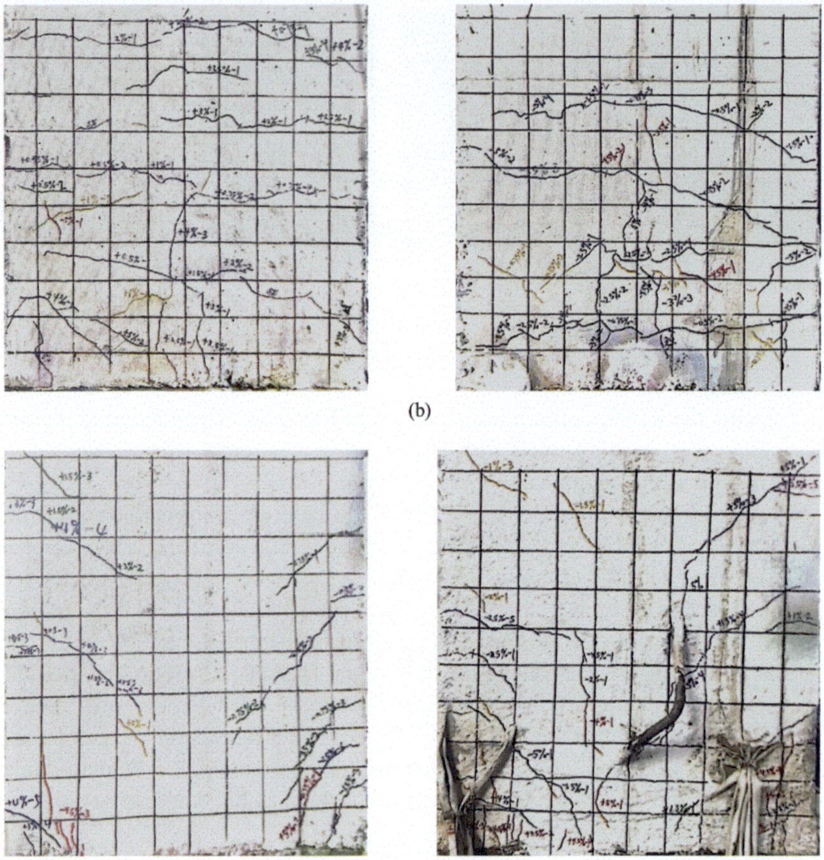

(a) (b) (c) (d)

Fig. 2.25 Crack patterns of the upper concrete segment. **a** Left side, **b** right side, **c** front side, and **d** back side

proven through these three tests. Finally, the third group contained two comparative tests by changing the thickness of the REDC-CPs to explore the section size of the energy-dissipating connectors on the load-carrying capacity, lateral stiffness and energy dissipation capacity of the test specimen.

As shown in Fig. 2.25, the cracks on the surface of the upper concrete segment were marked by lines in different colors for different tests. Four horizontal flexural through cracks formed on both the left and right sides after all five tests. The two ends extended toward the bottom of both the front and back sides. Several short vertical cracks developed at the interface between the upper concrete segment and the square steel tube. The cracks were related to the concentration of compressive stress caused by the rotation of the column base and the strengthening of the REDC-CPs. The number of vertical random cracks was limited. There was no obvious oblique shear crack on the surface of the upper concrete segment. On the whole, the distribution of the cracks was relatively uniform and symmetrical, the widths of which were basically less than 0.2 mm. All the cracks could close by themselves when the test specimen recentered its original position. The measurements of the strain gauges also confirmed that the longitudinal reinforcements and the stirrups of the column remained elastic all the time. The square steel tube merely partially yielded in compression with a high axial compression ratio, while the rest of it still remained elastic.

Since the out-of-plane slippage of the column base in Test 0.19-16 led to large relative movement between the upper and lower ends of the REDC-CPs, an additional horizontal shear force was generated and rapidly increased in the REDC-CPs, finally resulting in early tearing of the welding seams. However, the fractures of the REDC-CPs all appeared after more than 25 repeated cycles at a 5.0% lateral drift ratio in the following four tests, which was attributed to low-cycle fatigue. The fracture positions were either near the stopper in the middle of the yielding segment or at the junction of the yielding segment and the transition segment, as shown in Fig. 2.26. It can be observed from the extrusion traces of the pasted butyl rubber that the yielding segments exhibited obvious multi-wave buckling.

The load–displacement curves of the test specimen are listed in Fig. 2.27 according to the lateral load recorded by the MTS loading control system and the lateral

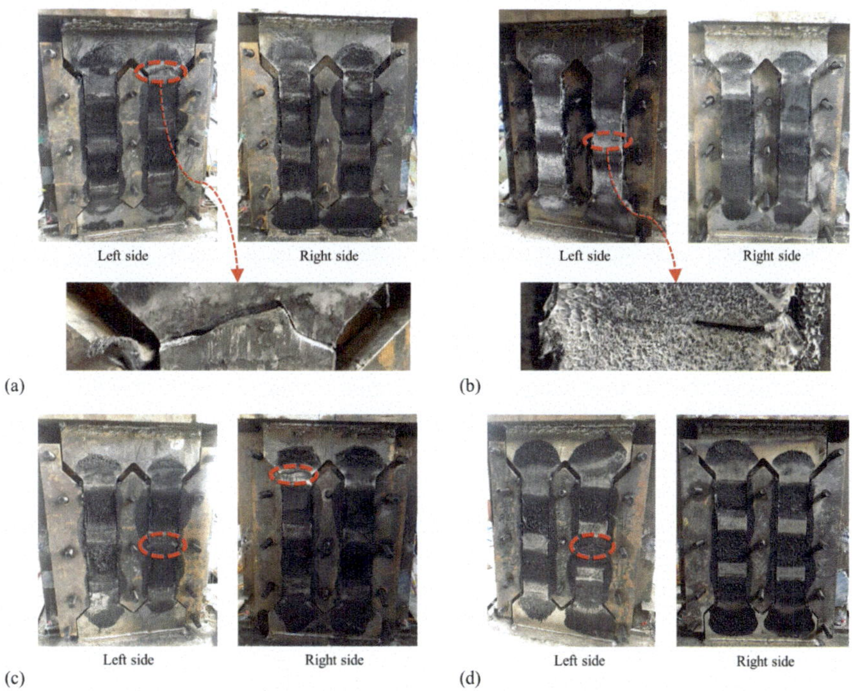

Fig. 2.26 Fracture mode of REDC-CP. **a** Test 0.19-16R, **b** Test 0.32-16, **c** Test 0.50-16, and **d** Test 0.50-12

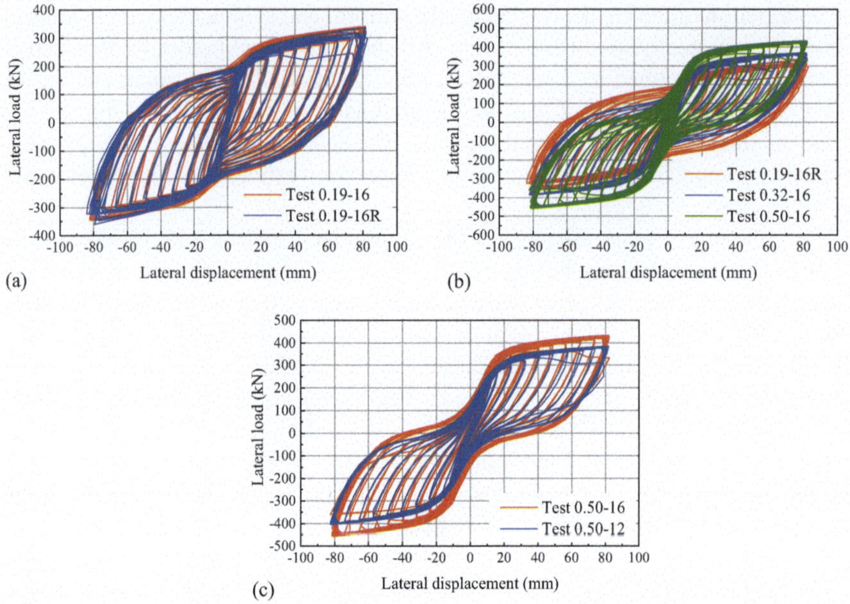

Fig. 2.27 Load–displacement curves. **a** Before and after rehabilitation with a low axial compression ratio, **b** difference in axial compression ratio, and **c** difference in the thickness of the REDC-CPs with a high axial compression ratio

displacement at the loading point measured by DT2. The three comparative groups included tests before and after rehabilitation of the test specimen with a low axial compression ratio, tests with different axial compression ratios and tests with differences in the thickness of the REDC-CPs with a high axial compression ratio.

The test hysteretic curves all have stable and plump loops. The bearing capacity and unloading stiffness of the test specimen still had no downward trend until the lateral drift ratio reached 5.0%, which ensures that the specimen could not be damaged even under extremely rare earthquakes. There was no obvious degradation of the bearing capacity in subsequent repeated cycles compared with the initial cycle at the same displacement level. Further analyses by comparison of the curves are illustrated below.

(1) The hysteretic curves of the test specimen before and after rehabilitation with a low axial compression ratio almost coincide, in which case the bearing capacity, post-yield stiffness and unloading stiffness are similar, indicating good repairability of the test specimen. The out-of-plane deformation of the column base after rehabilitation was effectively restrained in Test 0.19-16R compared with that in Test 0.19-16, so no additional horizontal shear force was generated at the ends of the REDC-CPs due to large relative displacement, improving the brittleness mechanical properties of the welding seams. Hence, the number of repeated loading cycles at a constant amplitude significantly increased, which is

more conducive to delivering the low cycle fatigue performance of the Q235B steel plates.

(2) With increasing the axial compression ratio of the column, the shape of the hysteretic loops changes from a plump shuttle shape to a bow shape with a certain flag-shaped feature. More serious pinching could be observed in the hysteretic loops with a larger axial compression ratio since the applied axial compressive force perpendicular to the top surface of the column could always enhance its anti-overturning stability. The larger the axial compression ratio of the column is, the larger the anti-overturning moment generated by the axial compressive force on the neutral axis of the column base.

(3) Reducing the thickness of the REDC-CPs when the axial compression ratio of the column remained constant resulted in a decrease in the load-carrying properties of the test specimen at the same deformation level because the decrease in the axial stiffness of the REDC-CPs led to a corresponding reduction in the rotational stiffness of the column base. However, the variation amplitude of the bearing capacity is significantly smaller than that of the thickness of the REDC-CPs due to the large contribution of axial compressive force to the bending resistance of the column base.

2.6 Design Method of Dry-Connected Precast Concrete Frame

2.6.1 Equal Displacement Rule

As pointed out by Chopra [20] the displacement demand of an elasto-plastic single-degree-of-freedom (EP-SDOF) system approximates to the displacement demand of an elastic single-degree-of-freedom (E-SDOF) system when the natural period of the EP-SDOF system determined from the response spectrum is greater than 0.7 s. Substantially, it is called the "equal displacement rule." As illustrated in Fig. 2.28, the force–displacement relation for the ideal EP-SDOF system and that for the E-SDOF system are both given. For an ideal EP-SDOF system, $\mu = \Delta_m/\Delta_y$ is defined as the ductility factor, where Δ_m and Δ_y represent the target displacement and the yield displacement of the EP-SDOF system, respectively. In addition, $R_y = V_e/V_y = \Delta_e/\Delta_y$ is defined as the strength reduction factor, where V_y represents the yield base shear of the EP-SDOF system, and V_e and Δ_e represent the elastic base shear and the elastic displacement of the corresponding E-SDOF system, respectively. Moreover, R_y can be calculated by the relationship of R_y-μ-T established by Newmark and Hall [21]:

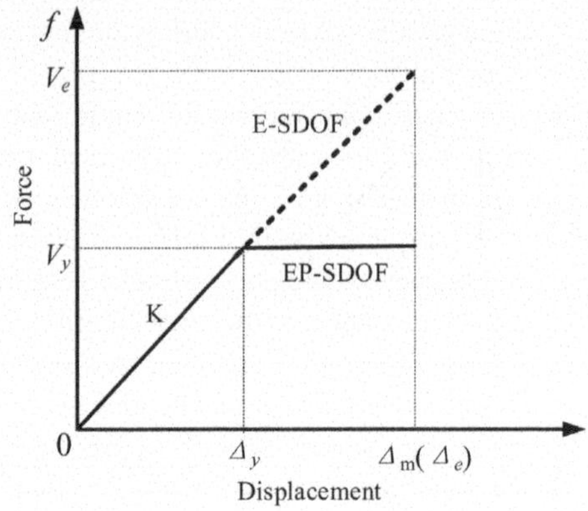

Fig. 2.28 Force–displacement curves of EP-SDOF system and E-SDOF system

$$R_y = \begin{cases} 1 & T < T_a \\ (2\mu - 1)^{\beta/2} & T_a \le T < T_b \\ \sqrt{2\mu - 1} & T_b \le T < T_c' \\ \mu - \frac{T}{T_c} & T_c' \le T < T_c \\ \mu & T \ge T_c \end{cases} \tag{2.19}$$

where T is the structural natural period, T_a and T_b equal to 0.0303 s and 0.125 s, respectively, T_c equals 0.666 s when the damping ratio is 5%, $T_c' = T_c(2\mu - 1)^{0.5}/\mu$, and $\beta = \ln(T/T_a)/\ln(T_b/T_a)$. In Eq. (2.19), R_y will equal to μ when $T > T_c$. Subsequently, $V_y = V_e/\mu$ and $\Delta_m = \Delta_e$ can be obtained, which implies that the displacement demand of the EP-SDOF system can be estimated by an E-SDOF system with the same natural period (see Fig. 2.29). Furthermore, the curve corresponding to $\mu = 1$ plotted in Fig. 2.29, represent the elastic displacement design spectrum in *Chinese Code for Seismic Design of Buildings* (GB 50,011–2010), and the curves corresponding to $\mu = 2, 4, 6, 8$ is the elastic–plastic displacement design spectrum calculated using Eq. (2.19). As shown in Fig. 2.29 ($T_d = 4.12$ s), when $T > T_c$, the peak deformation demand of EP-SDOF is equal to that of E-SDOF.

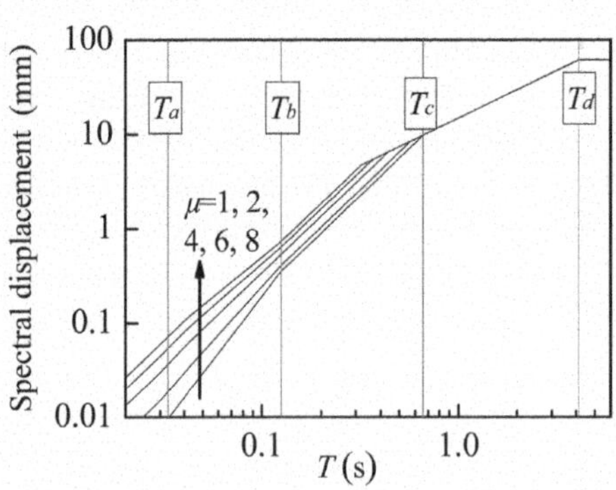

Fig. 2.29 Displacement design spectrums with different ductility factors

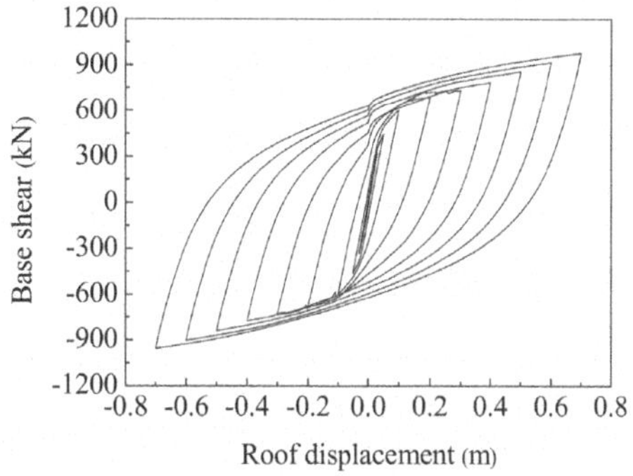

Fig. 2.30 Load–displacement hysteresis curve for REEDC-PCF system

Gupta et al. [22] analyzed 9 steel frames from 3 to 20 stories (which natural periods in the range of 1.01–3.15 s), finding that the ratios of the inelastic displacement responses to the elastic displacement responses for frames with longer periods were in the range of 0.7–0.8. The study by Feng et al. [23] showed that the plastic-elastic displacement ratio of an ideal EP-SDOF system is in the range of 0.8–1.2 when its natural period is in the range of 0.5–4 s. The plastic-elastic displacement ratio of an EP-SDOF with the second stiffness ratio in the range of 0.1–0.8 is found to be between 0.8 and 1.1. It implies that the equal displacement rule has been widely applied in the case of medium-period structures. Moreover, Priestley et al. [24] proposed that the equal displacement rule was suitable for the medium-period structures with full hysteretic characteristics but not for short-period and long-period structures.

Figure 2.30 illustrates the full hysteretic behavior of an example 5-story concrete frame with REDCs (The details of this example structure will be described in Sect. 2.6.3), having a natural period of 1.1 s, indicating that REDC-PCF system is applicable to the equal displacement rule. Therefore, it is an appropriate practice to estimate the elasto-plastic displacement demand of the REDC-PCF system by its elastic displacement demand. Accordingly, it is also suitable for establishing the direct displacement-based design method based on the equal displacement rule.

2.6.2 Design Procedure

In this section, an elastic displacement spectrum-based design method for REDC-PCF systems has been developed according to the equal displacement rule, where the influence of the beam-to-column connection rigidity on the structural lateral stiffness is taken into account. In particular, the design process can be divided into three stages, as shown in Fig. 2.31. The first stage (Step 1 to Step 4) is to determine the target period directly based on the elastic displacement design spectrum without considering the inelastic properties of the structure. Moreover, the second stage (Step 5 to Step 7) is to calculate the fixity factor p based on the target period and determine the sectional area and the length of the REDC-CP yield segment. The last stage (Step

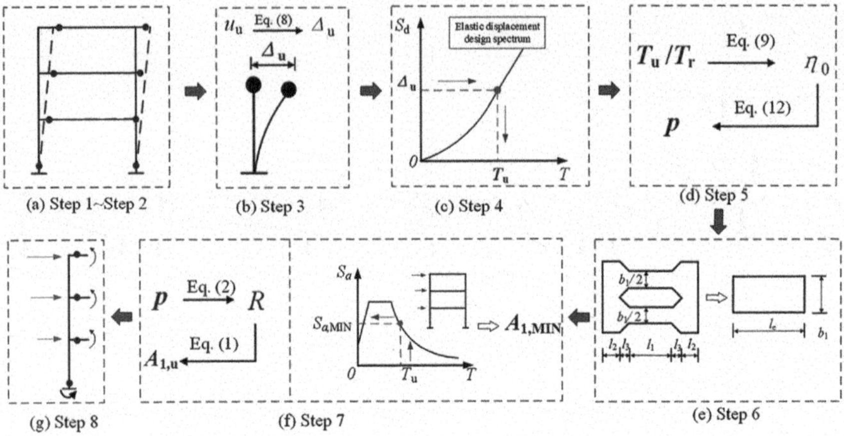

Fig. 2.31 Flow chart of the design of the REDC-PCF

8) is to design the concrete beams and columns. Furthermore, the design procedure can be illustrated as follows:

Step 1: Determine the target inter-story drift, φ_u, and the yield mechanism

According to code requirements, determine the corresponding target inter-story drift, φ_u, based on the selected performance objective. It is assumed that plastic deformations of the REDC-PCF system subjected to extreme loads occur on predetermined members (REDC-CPs placed at the beam-ends and the column bases) while other structural members remain elastic (refer to Fig. 2.31a).

Step 2: Determine the configuration of the structure

In this step, the structural parameters need to be determined according to the code for seismic design of building, including the number of stories, the building height, the span, and the sectional size of concrete beams and concrete columns.

Step 3: Calculate the target displacement of the equivalent single-degree-of-freedom (SDOF) system, Δ_u

A regular multistory structure with less significantly higher mode effects can be equivalent to SDOF system according to its first mode. The target displacement of the equivalent SDOF system, Δ_u, can be expressed as:

$$\Delta_u = \frac{\{\zeta\}^T[M]\{\zeta\}}{\{\zeta\}^T[M]\{1\}} u_u \tag{2.20}$$

where u_u denotes the target roof displacement of the REDC-PCF, $[M]$ is the mass matrix of multi-degree-of-freedom system, and $\{\zeta\}$ denotes the first mode of the REDC-PCF system. The value of *uu* can be obtained based on the first mode $\{\zeta\}$ and the target inter-story drift φ_u. The value of $\{\zeta\}$, which might not be accurately calculated prior to the determination of the fixity factor p, is taken as the first mode of the structure with rigid beam-to-column connections. This treatment is based on the fact that the variation of p has little impact on the first mode of the REDC-PCF system when p exceeds 0.3, as illustrated in Sect. 2.6.3.

Step 4: Determine the target period of REDC-PCF, T_u

Based on the equal displacement rule, the target period T_u can be directly determined using the elastic displacement design spectrum with a damping ratio of 5% as shown in Fig. 2.31c. In addition, it is noteworthy that the inelastic properties of the structure do not have to be considered when determining T_u. If $T_r > T_u$, where T_r represents the natural period of the frame with rigid beam-to-column connections, the displacement demand of the REDC-PCF system will exceed the target displacement. Consequently, we should go back to Step 2, and appropriately adjust the structural parameters until T_r becomes less than T_u.

Step 5: Determine the fixity factor, p

In Fig. 2.32, Frame A and Frame B represent the rigid frame and semi-rigid frame with natural period of T_r and T_u, respectively. The stiffness reduction factor, η_0, is defined as the ratio of the lateral stiffness of Frame B, K_u, to that of Frame A, K_r, to consider the reduction in beam-to-column connection rigidity on structural lateral stiffness. Thus, η_0 can be solved by the ratio between T_r and T_u:

$$\frac{T_r}{T_u} = \sqrt{\frac{K_u}{K_r}} = \sqrt{\frac{\eta_0 K_r}{K_r}} = \sqrt{\eta_0} \tag{2.21}$$

Then, establish the relationship between η_0 and p to solve the value of p. D-value method [25, 26] provides a simple method to calculate the story lateral stiffness in which the effect of flexible beams are taken into account. The modified lateral stiffness of a column is:

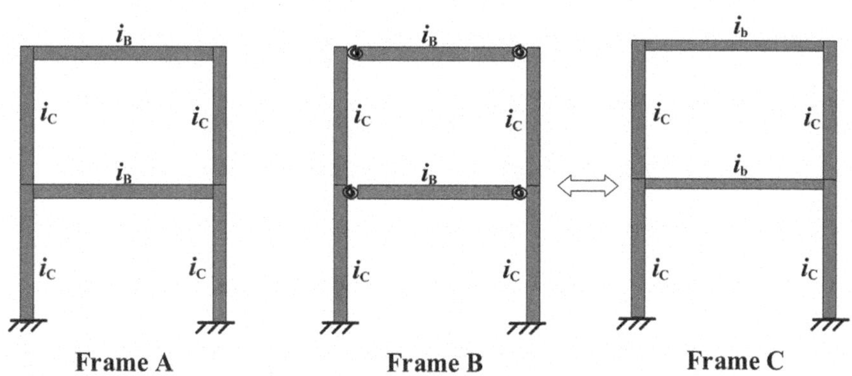

Fig. 2.32 Equivalent frame

$$D = \alpha_c \frac{12 i_C}{L_C^2} \tag{2.22}$$

where α_c represents the modified factor for inflection-point method. Then the story lateral stiffness is the sum of the lateral stiffness of all columns of the story. It should be noted that Eq. (2.22) represents the case that frame with rigid beam-to-column connections. By using Eq. (2.23), the beam linear stiffness i_B in Frame B can be reduced by the stiffness reduction factor β to obtain the equivalent rigid frame with lower beam linear stiffness i_b named Frame C [27].

$$i_b = \beta i_B = \frac{p}{2 - p} i_B \tag{2.23}$$

Therefore, according to the D-value method, the story lateral stiffness of Frame A and Frame C can be expressed as $\sum D_A$ and $\sum D_C$, respectively, representing the sum of all the column lateral stiffness of the story. Moreover, for a vertical regular frame, $\sum D_A$ and $\sum D_C$ are proportional to K_u and K_r, respectively. Thus, the stiffness reduction factor η_0 can be related to fixity factor p as:

$$\eta_0 = \frac{K_u}{K_r} = \frac{\sum D_C}{\sum D_A} = \frac{\frac{\beta i_{BC}}{1+\beta i_{BC}} + \frac{2\beta i_{BC}}{2+\beta i_{BC}}}{\frac{i_{BC}}{1+i_{BC}} + \frac{2 i_{BC}}{2+i_{BC}}} \tag{2.24}$$

Step 6: Pre-select REDC-CP parameters

There are four parameters including l_1, l_2, l_3, and b_2/b_1 which need be predetermined for the REDC-CP in this step, as is seen in Fig. 2.33. l_2 should be determined to provide a sufficient operational space for installing the REDC-CP. l_3 is taken as $l_2/2$ to avoid stress concentration between the yield segment and the connection segment. The value of b_2/b_1 should be determined such that the elasticity of the connection segment under the target inter-story drift can be ensured. In order to avoid fatigue failure, the length of the REDC-CP yield segment, l_1, should be determined by limiting the strain of the REDC-CP within 3% under the target inter-story drift. As shown in Fig. 2.33, ensuring the axial stiffness unchanged, the REDC-CP can be equivalent to a rectangular member with an equivalent length of l_e and an equivalent sectional area A_e (A_e is taken as $A_1 = b_1 t$). Only the sectional area of the yield segment of the REDC-CP, A_1, can be regarded as an unknown parameter.

Step 7: Calculate the sectional area for the REDC-CP yield segment, A_1

To achieve the target displacement, the required fixity factor p has been given in Step 5, and the corresponding sectional area demand of the REDC-CP yield segment, $A_{1,u}$. For the REDC-PCF with regular vertical distribution of stiffness, strength, and mass, the REDCs can be placed along with the story with the same strength, which makes it easier to produce and install. In addition, according to the method proposed by Mazza [28] that, for a vertical irregular frame, the structural vertical arrangement should be determined to obtain a structure globally regular in terms of stiffness and strength.

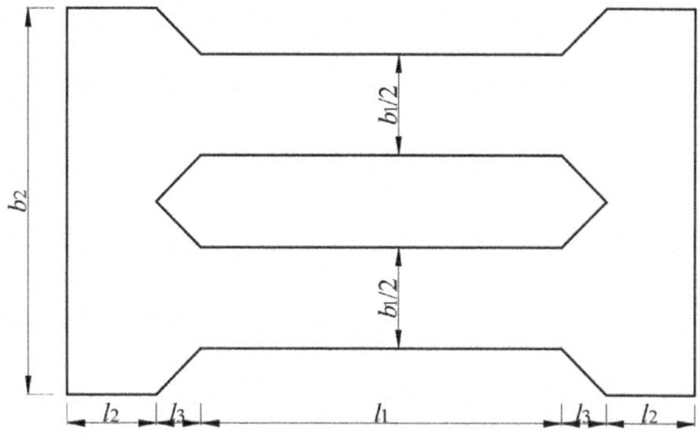

Fig. 2.33 Parameters of REDC for analysis

The strength vertical distribution pattern should be designed that their yielding occur at every story simultaneously. And the stiffness distribution of REDC along the story is calculated ensuring a constant value of the inter-story drift at each story.

Based on the structural strength capacity required for minor earthquakes (with the exceedance probability of 63% in 50 years), the sectional area demand of the REDC-CP yield segment, $A_{1,\mathrm{MIN}}$, should also be calculated to ensure the elasticity of the REDC-PCF system. Therefore, A_1 is determined as the maximum value of $A_{1,\mathrm{min}}$ and $A_{1,\mathrm{u}}$:

$$A_1 = \max\left(A_{1,u}, \mathrm{A}_{1,\mathrm{min}}\right) \tag{2.25}$$

If A_1 equals to $A_{1,\mathrm{min}}$, the length of the yield segment of the REDC-CP, l_1, should be increased, which is to keep the fixity factor p unchanged (the rotational stiffness at the beam ends remains constant).

Step 8: Design of concrete beams and columns

The design target of the concrete beams is to keep the longitudinal reinforcements in the beams from being plastic during the process of achieving the target inter-story drift. Thus, the required moment capacities of the concrete beams can be obtained by:

$$M_{\mathrm{B}} = \lambda A_1 d f_{\mathrm{y}} \tag{2.26}$$

where λ denotes the over-strength factor, which considers the impact of the strain hardening and the compression over-strength of the REDC-CP, and f_{y} denotes the yield strength of the REDC-CP. In this study, λ is 1.52 for the axial force–deformation curve of the REDC, which corresponds to 3% strain of the REDC-CP.

In addition, the concrete columns should be designed to ensure that their plastic hinges only developed at the column bases under the strong earthquakes. Thus, the sectional longitudinal reinforcements of column bases can be determined according to the strength capacity required for minor earthquakes (with the exceedance probability of 63% in 50 years). While other sections of concrete columns should be

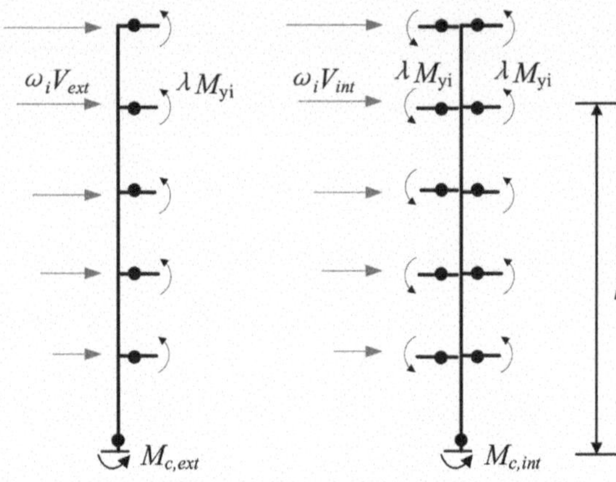

Fig. 2.34 Equilibrium analysis for external column and the internal column

designed according to the equilibrium between the external force and the internal force of the structure under the target inter-story drift, as shown in Fig. 2.34, to ensure their elasticity under the effect of post-yield over strength of REDC on the column. Then, V_{ext} and V_{int} representing the total force required to maintain the equilibrium of the external column and the internal column can be calculated as:

$$V_{ext} = \frac{\sum_{i=1}^{n} \lambda M_{yi} + M_{c,ext}}{\sum_{i=1}^{n} \omega_i h_i}$$
$$V_{int} = \frac{2\sum_{i=1}^{n} \lambda M_{yi} + M_{c,int}}{\sum_{i=1}^{n} \omega_i h_i}$$
$$\omega_i = \frac{G_i h_i}{\sum_{j=1}^{n} G_j h_j} \tag{2.27}$$

where M_{yi} is the yield moment provided by REDC at i_{th} story, $M_{c,ext}$ and $M_{c,int}$ are the required moment capacity under the minor earthquakes according to in Chinese Code for Seismic Design of Buildings (GB 50011–2010), h_i is the height from the ground to the i_{th} story, ω_i is the lateral force distribution factor for the column given in Chinese Code for Seismic Design of Buildings (GB 50011–2010), and G_i is the vertical load acted at i_{th} story.

2.6.3 *Case Study*

(1) The Example Structure

The rationality of the proposed design method is validated through a case study of a 5-story REDC-PCF. Table 2.6 shows the key parameters for designing the REDC-PCF

Table 2.6 Key design parameters for REDC-PCF

Item	REDC-FCF	Step
Target inter-story drift, φ_u/%	1.8	Step 1
First mode of the frame, $\{\zeta\}$	{0.161, 0.435, 0.692, 0.885,1.000}	Step 3
Natural period of rigid frame, T_r/s	0.84	
Target roof displacement of REDC-PCF, u_u/m	0.256	
Target displacement of the equivalent SDOF system, Δ_u/m	0.196	
Target period, T_u/s	1.1	Step 4
Fixity factor, p	0.68	Step 5
REDC-CP yield segment area, A_1/mm^2	2370	Step 7

system. Moreover, the geometric dimensions of the REDC-CPs and the sectional longitudinal reinforcement areas for beams and columns on each story are listed in Tables 2.7 and 2.8, respectively. The details of the design process are presented as follows:

Step 1: Select the target inter-story drift, φ_u, and the yield mechanism

According to the Chinese Code for Seismic Design of Buildings (GB 50,011–2010), the REDC-PCF system being designed is located at the area of Seismic Intensity 9, Seismic Group 1, and Soil type III, with a peak ground acceleration (PGA) of

Table 2.7 Geometric parameters for REDC-CPs at each story

Story	A_1/mm^2	l_1/mm	l_2/mm	l_3/mm	b_1/mm	b_2/mm	t/mm
1	2370	240	50	25	170	320	14
2	2370	240	50	25	170	320	14
3	2370	240	50	25	170	320	14
4	2370	240	50	25	170	320	14
5	2370	240	50	25	170	320	14

Table 2.8 Sectional longitudinal reinforcement areas for columns and beams of REDC-PCF

Story	Concrete column			Concrete beam		
	Section/mm	Reinforcement/mm^2		Section/mm	Reinforcement/mm^2	
		External	Internal		Top	Bottom
5	550 × 550	2000	3200	550 × 300	2400	2400
4	550 × 550	2800	4000	550 × 300	2400	2400
3	550 × 550	2800	4000	550 × 300	2400	2400
2	550 × 550	2300	3000	550 × 300	2400	2400
1	550 × 550	2000	2000	550 × 300	2400	2400

0.62 g for the maximum considered earthquake (MCE) (2% exceedance probability in 50 years). Furthermore, the performance targets for the REDC-PCF system are no serious damages and no more than 0.9 times the plastic structural deformation limit under the MCE. Therefore, the target inter-story drift φ_u is calculated to be 1.8%, and the expected yield mechanism can be obtained.

Step 2: Determine the configuration of the structure

A 5-story REDC-PCF of three spans is designed. In particular, each story has the height of 3900 mm and each span is 6000 mm as shown in Fig. 2.35. The dead load applied on each story is 45.62 kN/m, and the live loads applied on the roof story and other stories are 1.2 kN/m and 12 kN/m, respectively. According to the Chinese Code for Seismic Design of Buildings (GB 50011–2010), the cross-sections of the columns are determined as 550 × 550 mm to ensure $N/N_y < 0.65$, where N and N_y represent the axial force applied to the column and the axial compression capacity of the columns, respectively. And, the height of the beam section should be controlled in the range of 1/12 and 1/8 of the span. Then, cross-section demission of 550 × 300 mm is assigned to the concrete beams. HRB400 steel with a yield strength of 360 MPa as the yield strength of HRB400 is 360 MPa in GB50010-2010 and Q235 steel with a yield strength of 235 MPa are utilized for the reinforcements and the REDCs, respectively. A C40-grade concrete with the compressive strength of 40 MPa is used for both beams and columns.

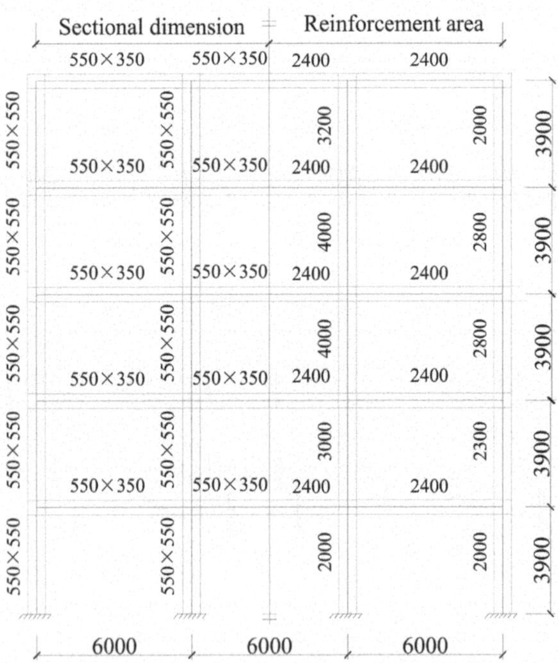

Fig. 2.35 Example structure

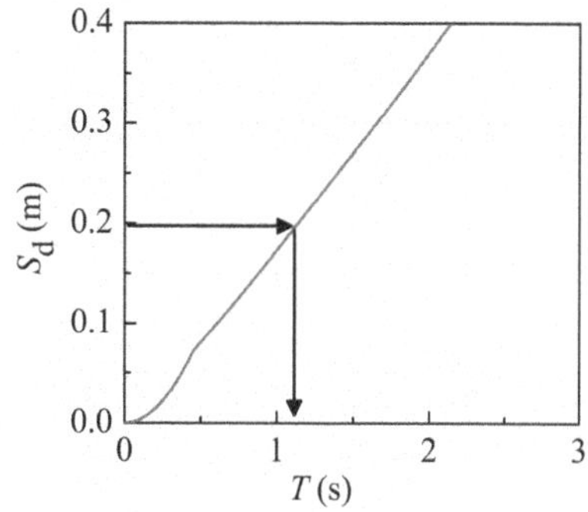

Fig. 2.36 Elastic displacement design spectrum in GB 50011–2010

Step 3: Calculate the target displacement of the equivalent SDOF system, Δ_u

The concrete frame with the structural parameters given in Step 2 is assumed with rigid beam-to-column connections. It has a first mode eigenvector, $\{\varsigma\}$, of $\{0.161, 0.435, 0.692, 0.885, 1.000\}$ and a natural period, T_r, of 0.84 s. This indicates that the maximum inter-story drift controlled by the first mode occurs at the second story. The target roof displacement of the structure, u_u, is calculated to be 0.256 m according to the predetermined target inter-story drift of 1.8%. Subsequently, the target displacement for the equivalent SDOF system, Δ_u, is 0.196 m.

Step 4: Determine the target period of REDC-PCF, T_u

Given the target displacement Δ_u from Step 3, the target period, T_u, is determined to be 1.1 s based on the elastic displacement design spectrum presented in the Chinese Code for the Seismic Design of Buildings (GB 50011–2010) (Fig. 2.36). Since T_u is longer than T_r, there will be no need to go back to Step 2 for readjusting the configuration of the structure.

Step 5: Determine the fixity factor, p

Based on the ratio of T_u to T_r, η_0 can be calculated to be 0.58 by Eq. (2.21). Next, p is obtained as 0.68 by Eq. (2.23). Moreover, Fig. 2.37a, b presents the first modes of the structure and the roof displacements of the equivalent SDOF system at different values of p, respectively. It can be concluded that the values of p variation from 0.3 to 1.0 have a minor impact on the first mode and the roof displacement of the equivalent SDOF system.

Step 6: Pre-select several REDC-CP parameters

Some parameters of the REDC-CP are pre-selected, which include $l_2 = 50$ mm, $l_3 =$ 25 mm, and $b_2/b_1 = 2$. Next, the distance between the upper and the lower REDC-CPs at the beam-end, d, is determined to be 450 mm with respect to the height of the beam section. Finally, $l_1 = 180$ mm is chosen to meet the fatigue performance requirement. Based on the determined parameters of the REDC-CP, the equivalent length of REDC-CP, l_e, is calculated to be 284 mm. At this step, only A_1 is unknown.

Step 7: Determine the sectional area for the REDC-CP yield segment, A_1

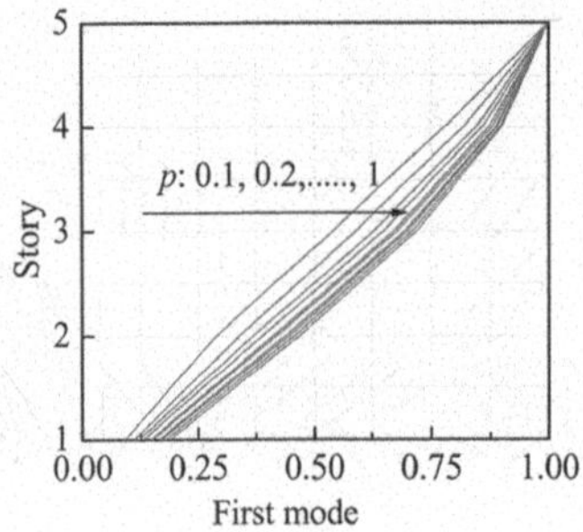

(a) First mode for structure with different fixity factor

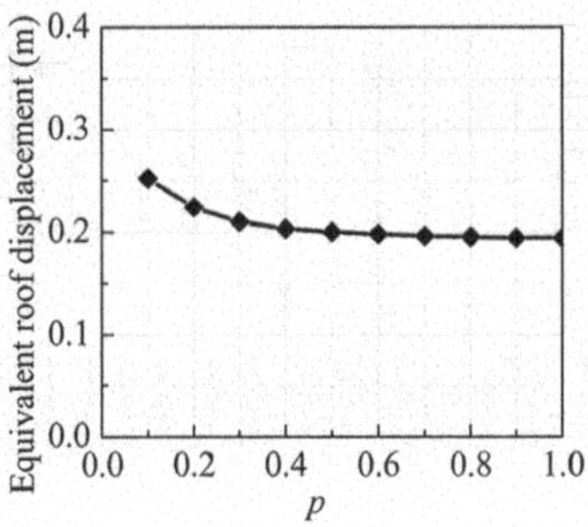

(b) Roof displacement of equivalent SDOF system with different fixity factor

Fig. 2.37 Structure with different fixity factors

The rotational stiffness R can be calculated to be 1.43E+11 N/mm; subsequently, $A_{1,u}$ is calculated to be 1918 mm^2 with the given value of p. To ensure strength capacity of REDC-PCF under the minor earthquakes, $A_{1,\,MIN}$ is determined to be 2370 mm^2 based on the Chinese Code for Seismic Design of Buildings (GB 50011–2010). Therefore, the required A_1 is 2370 mm^2, which is the maximum value of $A_{1,\,u}$ and $A_{1,\,MIN}$. On the other hand, the equivalent length of the REDC-CP, l_e, should be recalculated to keep the fixity factor p unchanged. The corresponding length of the REDC-CP yield segment, l_1, is 240 mm. For a vertical regular frame, the REDC can be distributed along with the story with the same strength and stiffness. Table 2.7 shows the determined parameters of the REDC-CPs placed on each story.

Step 8: Design of concrete beams and columns

Table 2.8 and Fig. 2.35 show the designed sectional longitudinal reinforcement details (including the total areas at each side, the number, and the diameter of the reinforcement) for the concrete beams and the concrete columns, which is to ensure the predetermined yield mechanism under target inter-story drift.

(2) The structural modeling approach

A 2-D finite element model of the REDC-PCF system, as shown in Fig. 2.38, is developed using Opensees [29] for structural seismic analyses. In particular, the simulation of concrete beams and concrete columns is carried out using nonlinear beam-column elements with fiber sections. Meanwhile, material Concrete02 and material Steel02 are employed to simulate concrete and reinforcement, respectively. Truss elements with material Steel02 are utilized to represent the REDCs at the top and the bottom of the beam ends. Therefore, the transfer of the bending moments between concrete beams and concrete columns can be realized. The REDCs are connected to the beams and the columns via rigid link elements, which is to ensure the plane section assumption. In addition, the pin-connected members at the beam-ends are simplified into hinge nodes that can transfer the shear forces. It should be noted that the section of the REDC-CP varies along the length, so the equivalent length, l_e, is assigned to the length of the truss elements in the model, ensuring the consistency between the rotational stiffness of the connections in the numerical model and those of the actual structure. Furthermore, the P-Δ effect is taken into account for the numerical model, and a 5% Rayleigh proportional damping is assumed for the first and second modes. Figure 2.39 shows the comparison results between the cyclic force–deformation curve of the REDC for the physical model test and that for the numerical simulation. It can be concluded that the numerical model for REDC used in this paper can ultimately reflect its hysteretic characteristics.

Moreover, a finite element model is also established for the RCF, which is to compare the seismic performance of the REDC-PCF with that of the RCF. The structural configuration for the RCF is identical to that for the REDC-PCF, and the moment-resistant capacities of the beams and the columns are the same for the two systems.

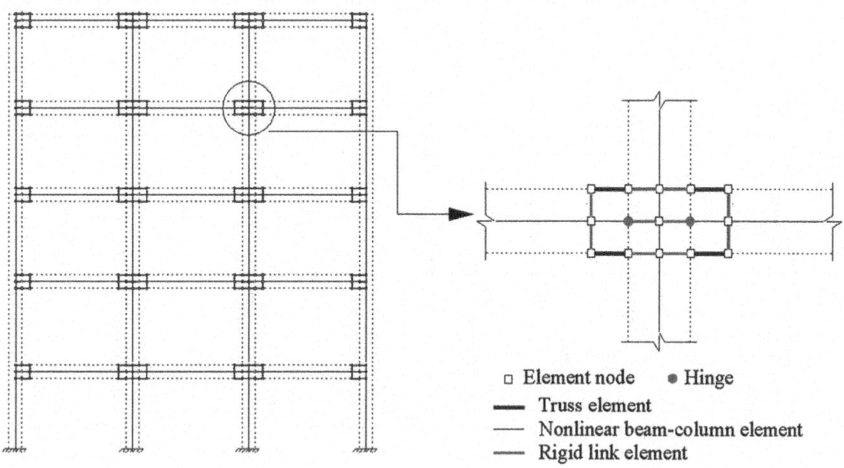

Fig. 2.38 Numerical model for REDC-PCF

Fig. 2.39 Confirmation of numerical model of REDC-CP

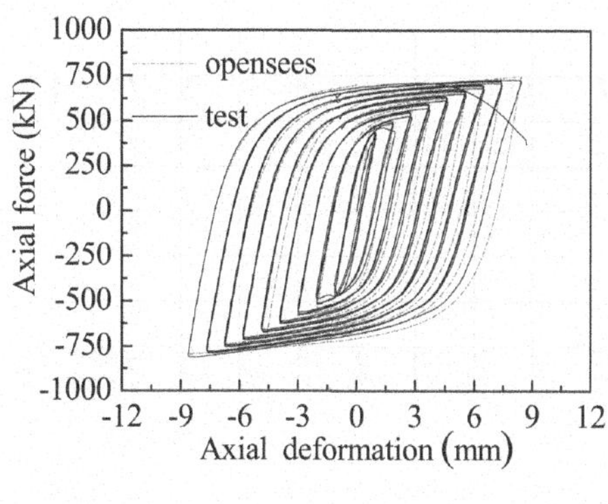

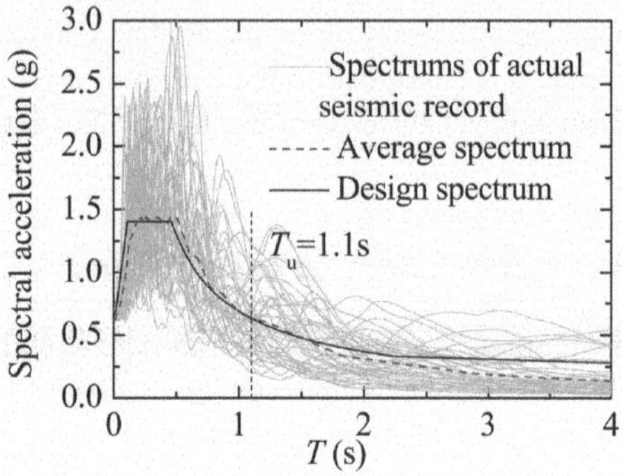

Fig. 2.40 Design spectrum and spectrum of the selected seismic records

(2) Selection of Ground Motions

Forty-four far-field ground motions were used for nonlinear time-history analyses for both the REDC-PCF system and the RCF system as recommended by FEMA-695. All records are scaled to a *PGA* of 0.62 g, representing an MCE. The scaled spectrums of the actual seismic records and the average spectrum is shown in Fig. 2.40. At the target period T_u of 1.1 s, an error of 0.12% is observed between the average spectrum and the design spectrum presented in Chinese Code for the Seismic Design of Buildings (GB50011-2010).

(3) Analytical Validation

To validate the rationality of the proposed design procedure, nonlinear time-history analyses are conducted for the dynamic responses of the REDC-PCF. In Fig. 2.41a, the average maximum inter-story drift under the selected seismic records is 1.72%, which is close to but not exceeds the target inter-story drift φ_u of 1.8% (with an error of 4.4%). Thus, it can be concluded that the design objective has been achieved, and the equal displacement rule can be validated. It should be noted that the inelastic displacement of a REDC-PCF system can be assumed to be equal to its elastic displacement, which means that the target period can be directly determined by the elastic displacement design spectrum. However, this process does not take the inelastic performance of the structure into consideration, which would simplify the seismic design method. Furthermore, the average dynamic responses of the REDC-PCF, including the inter-story drift, the roof displacement, the drift concentration factor (DCF), the base overturning moment (OTM), the base shear and the floor acceleration, are compared with those of the RCF as shown in Table 2.9. The DCF can be obtained by [30]:

$$DCF = \frac{\varphi_{\max}}{\left(u_{roof}/h_n\right)} \tag{2.28}$$

where $\varphi_{\max}$, u_{roof} and h_n represent the maximum inter-story drift, the maximum roof displacement and the total height of the structure, respectively. Compared with the

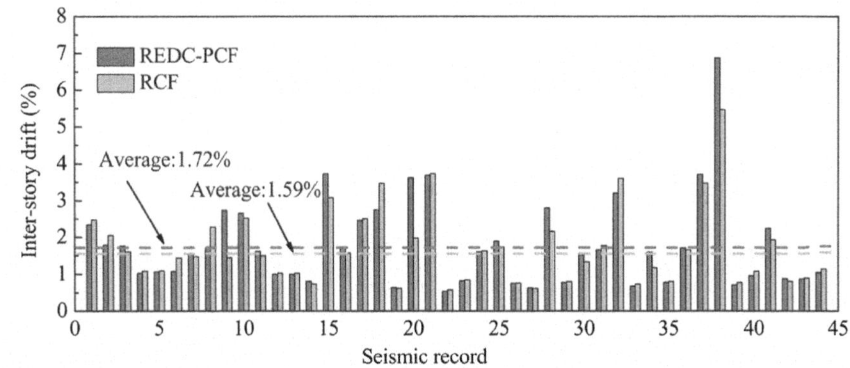

(a) Inter-story drift

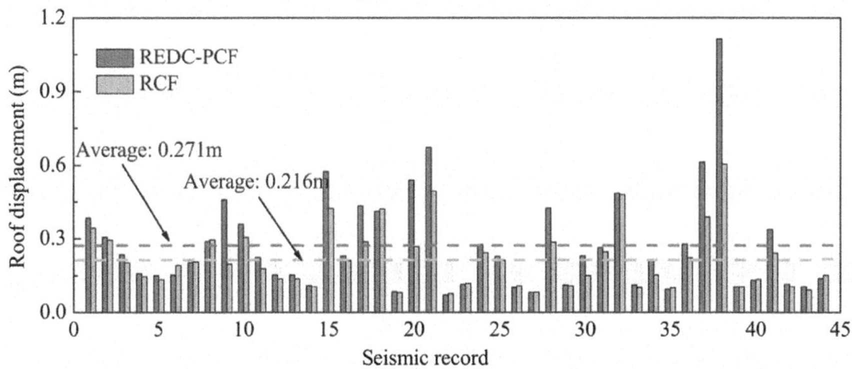

(b) Roof displacement

Fig. 2.41 Maximum deformation under the selected seismic records for REDC-PCF and RCF

Table 2.9 The average response of REDC-PCF and RCF under the selected seismic records

Items	Maximum inter-story drift/%	Roof displacement/m	DCF	Base OTM/kN.m	Base shear/kN	Maximum floor acceleration/g
REDC-PCF	1.72	0.271	1.24	11,708	810	0.83
RCF	1.55	0.213	1.42	14,384	1001	1.16
REDC-PCF/RCF	1.11	1.26	0.87	0.81	0.8	0.72

RCF system, the deformation of each story is enlarged for the REDC-PCF system as the result of the decrease in the lateral stiffness as shown in Figs. 2.41 and 2.42. More specifically, the maximum inter-story drift and the roof displacement of the REDC-PCF are on average 1.11 and 1.26 times those of the RCF, respectively (Table 2.9). However, the average inter-story drift for the REDC-PCF is still controlled

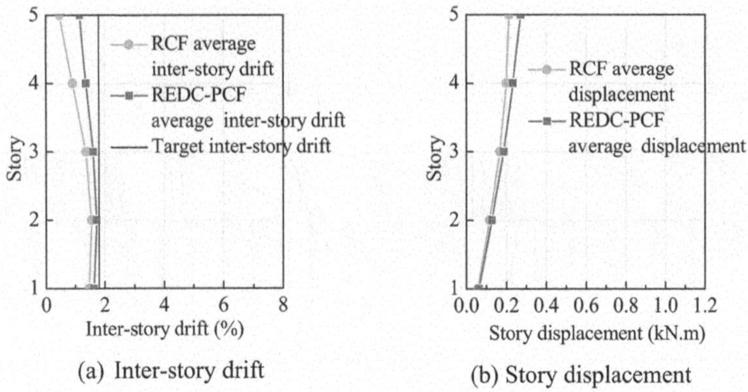

Fig. 2.42 Average value of maximum story deformation under the selected seismic records for REDC-PCF and RCF

within the target range. It indicates that it is indispensable to consider the impact of the connection stiffness on the structural lateral stiffness during the design process to avoid the target inter-story drift being exceeded and achieve accurate deformation control. In addition, the average DCF of the REDCF-CF is less than that of the RCF as shown in Table 2.9, indicating that a uniform structural deformation is achieved for the REDC-PCF, and the soft stories are avoided in the system.

Figure 2.43 shows the maximum base shear and the maximum base OTM under the selected seismic records for both REDC-PCF and RCF. Meanwhile, Fig. 2.44 shows the average maximum story shear and the average maximum story OTM for both systems. It can be observed that the internal force demands of the REDC-PCF system under the selected seismic records are reduced to some extent compared with the RCF system. As shown in Table 2.9, the ratios of REDC-PCF to RCF are on average 0.81 and 0.8 for the maximum base shear and the maximum base OTM, respectively. It implies that the precast concrete frame with semi-rigid connections can effectively control the structural internal forces under seismic actions by reducing the lateral stiffness of the system. With the consideration of the impact of the connection rigidity on the structural seismic responses, the proposed design procedure can limit the structural displacement demands under the seismic actions within the target range and reduce the seismic internal force demands of the structure. Moreover, the reduction in the lateral stiffness for the REDC-PCF system leads to the decrease in the floor acceleration, and the ratio of REDC-PCF to RCF for the maximum floor acceleration is 0.72 on average as shown in Table 2.9. Therefore, the damages of nonstructural members sensitive to the acceleration will be reduced for the REDC-PCF system.

Figure 2.45 shows the maximum reinforcement stresses encountered by the exterior columns and the interior columns in REDC-PCF system under the selected seismic records, respectively. It can be seen from the figures that the maximum stresses in the reinforcements in columns exceed the yield strength of 360 MPa

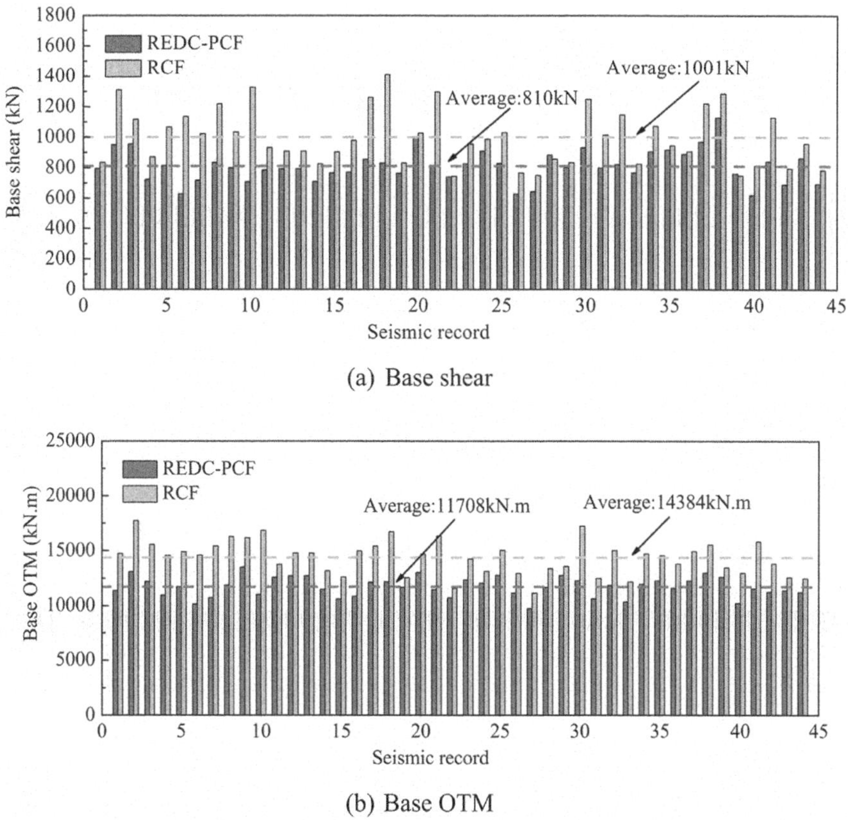

Fig. 2.43 Maximum internal force under the selected seismic records for REDC-PCF and RCF

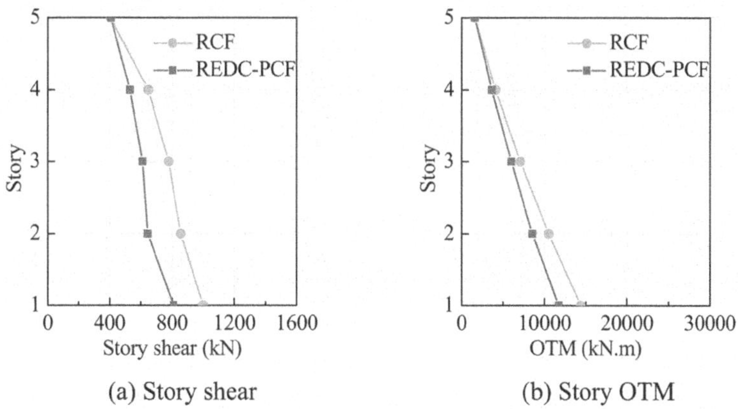

Fig. 2.44 Average value of maximum story internal force under the selected seismic records for REDC-PCF and RCF

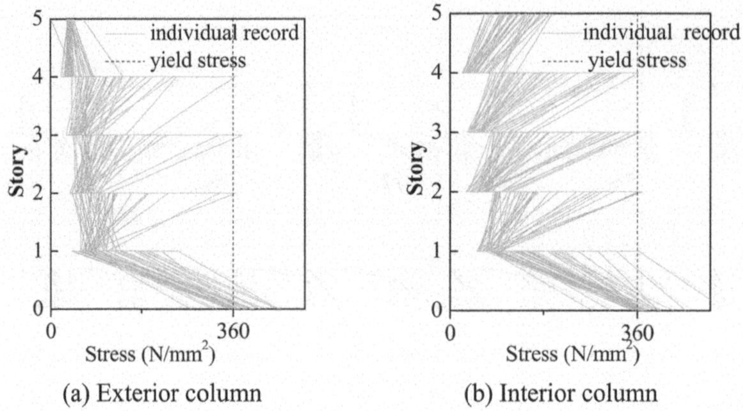

Fig. 2.45 Maximum stress of reinforcements in the columns of REDC-PCF under the selected seismic records

under two seismic records at the second, the third, and the fourth stories. Meanwhile, only the maximum stresses in the reinforcements at the column bases exceed the yield stress under the other seismic records. Therefore, it can be concluded that the predetermined yield mechanism is achieved for the design method developed in this paper because the unfavorable failure of the column hinge can be avoided with a 95.46% probability for the present method.

The average maximum strain of the REDC-CPs under the selected seismic records is 2.57%, which is within the designing stain limit of 3% as shown in Fig. 2.46. In addition, only the maximum strains for 5 out of the 44 selected seismic records exceed 3%, with the probability of 11.3%. Moreover, a cumulative plastic deformation (CPD) limit value of 200 for BRB is recommended in ANSI/AISC 341–10 (AISC 2010). The average maximum CPD calculated for the REDC-CP under the selected seismic records is 133, which is less than the limit value recommended in ANSI/AISC 341–10. As shown in Fig. 2.47, only CPDs for 6 out of the 44 seismic records exceed 200, indicating a probability of 86.4% to avoid a low cycle fatigue fracture. This implies that the present design method can avoid fatigue failure under the target displacement by controlling the deformation of the REDC-CP.

2.7 Conclusions

In this Chapter, a novel type of replaceable energy-dissipating connectors (REDCs) and its dry-connected precast concrete frames are proposed. The mechanical property of REDCs are experimentally studied, and the beam-column joints and column base with the REDCs are also tested to analyze the seismic performance. According to the equal displacement rule, a simplified seismic design method is proposed based on the elastic displacement design spectrum, and an example REDC-PCF is designed to

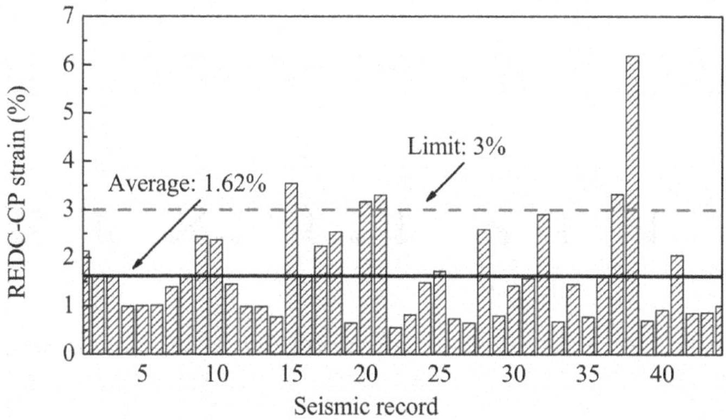

Fig. 2.46 Maximum strain for REDC-CP under the selected seismic records

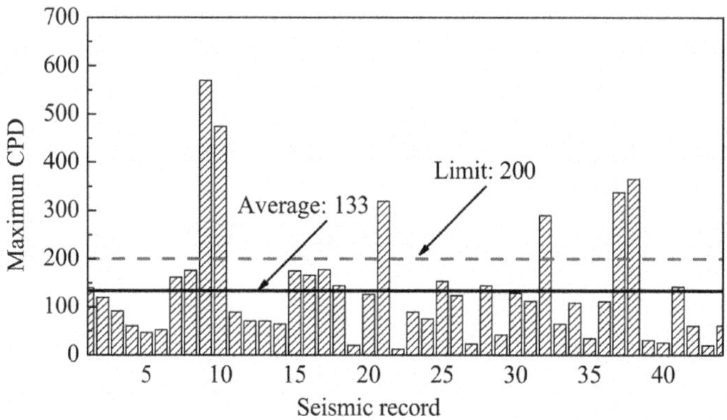

Fig. 2.47 Maximum CPD for REDC-CP under the selected seismic records

validate the proposed design procedure, and then its seismic performance is compared with an RCF. The following conclusions can be drawn:

(1) The REDCs are fabricated in the top and bottom of the beam end, which are located under the concrete cover and connected to the reinforced bars through transition blocks by welding. The REDCs can be regarded as a part of the reinforced bars in the beam and can transmit the bending moment of the beam end under serviceability-limit states. Once the frame suffers a strong earthquake, the REDCs yield first and dissipate the inputted energy, while the main structural members remain elastic, such as the beam and column. The REDC-precast concrete frame (REDC-PCF) could be rehabilitated by replacing the damaged REDCs after strong earthquakes, which is very convenient.
(2) An experimental study of the mechanical behavior of the REDC under cyclic reverse loading was conducted, in which the axial stiffness, hysteretic behavior and energy dissipation capacity were analyzed. The results reveal that REDCs

manufactured from Q235 steel could produce stable hysteretic properties under cyclic reverse loading conditions, and the corresponding cumulative plastic deformation capacity is sufficient for the energy dissipation demand under strong earthquakes, which could meet the requirement of AISC341-10.

(3) The test results of REDC-PCF joints reveal that the plastic deformation under large-amplitude rotation is concentrated in the REDC specimens, while the beams and columns remain elastic. The rehabilitation test also confirms the convenient seismic rehabilitation process. The hysteretic curves of the tests are plump, without any pinching. The fracture type is low-cycle fatigue fracture in the REDC fuse member, while the fracture position is concentrated in the yielding segment.

(4) The proposed REDC-PCF system with full hysteretic characteristics conforms to the equal displacement rule. For the elastic displacement spectrum-based seismic design method, the elastic displacement design spectrum can be used to directly obtain the target period without additional consideration of other plastic parameters. Ultimately, the sectional area of the REDC-CP yield segment can be easily calculated.

(5) Based on the time-history seismic analyses of a 5-story REDC-PCF, the average maximum inter-story drift of REDC-PCF is close to (but not exceeding) the target inter-story drift and the relative error is 4.4%. This indicates that the elastic displacement can be used to predict the elasto-plastic displacement demand of a system with a medium period. Regarding the displacement-based design method, it is not necessary to consider additional ductility factor or the equivalent-damping ratio during the design process, and the displacement responses under strong earthquakes can be estimated very accurately only using the elastic response spectrum.

(6) The seismic performances of REDC-PCF calculated by time history analyses are compared with those of RCF. Compared with the RCF system, the REDC-PCF has a lower lateral stiffness due to the semi-rigid beam-column connections, which helps to reduce the internal force demands; however, the maximum inter-story drift of the REDC-PCF is enlarged. Therefore, it is necessary to consider the impact of reduction in connection rigidity between the beam and the column in the design process in order to reduce the structural internal force demand and to accurately control the structural deformation at the same time. In addition, the maximum floor accelerations of REDC-PCF are reduced under seismic actions, which is advantageous for reducing the damage in acceleration-sensitive members. The average maximum strain and the average maximum CPD of the REDCs derived from the nonlinear time history analyses are both controlled within the acceptable ranges specified in Codes of Practice, which indicates that good fatigue behavior of REDCs can be achieved by controlling the length of REDC-CPs in the design stage.

References

1. Ye LP, Ma QL, Miao ZW (2010) Study on weak beam-strong column design method of RC frame structures. Gongcheng Lixue/Eng Mech 27:102–113
2. ACI Committee 374 (2013) Guide for testing reinforced concrete structural elements under slowly applied simulated seismic loads (374.2 R-13). ACI, Farmington Hills (MI)
3. Huang L, Zhou Z, Huang X, Wang Y (2019) Variable friction damped self-centering precast concrete beam–column connections with hidden corbels: experimental investigation and theoretical analysis. Eng Struct:110150
4. Englekirk RE (2002) Design-construction of the paramount a 39-story precast prestressed concrete apartment building. PCI J 47(4):56–71
5. Tianyou W, Xiaozu S, Jiangsheng F (2007) Application of STMs for calculation of reinforced concrete frame joints. China Civ Eng J 40(11):36–40
6. Xie LQ, Wu J, Huang Q et al (2019) Analysis of the seismic demand of high-performance buckling-restrained braces under a strong earthquake and its aftershocks. Adv Civ Eng:1482736. https://doi.org/10.1155/2019/1482736
7. Wang CL, Liu Y, Zhou L (2018) Experimental and numerical studies on hysteretic behavior of all-steel bamboo-shaped energy dissipaters. Eng Struct 165:38–49
8. Tateishi K, Hanji T, Kitoh K et al (2005) A prediction model for extremely low cycle fatigue strength of welded materials. J Struct Eng JSCE:1275–1281. (in Japanese)
9. Nakamura H, Takeuachi T, Maeda Y et al (2000) Fatigue properties of practical-scale unboned braces. Nippon Steel Tech Rep 7(82):51–57
10. Xie LQ, Wu J, Huang Q (2018) Experimental study on low-cycle fatigue performance of weld-free buckling-restrained braces. J Earthquake Eng 22(8):1392–1414
11. ANSI/AISC 360-10 (2010) Seismic provisions of structural steel buildings. American Institute of Steel Construction, Chicago, Illinois
12. Englekirk RE (1996) An innovative design solution for precast prestressed concrete buildings in high seismic zones. PCI J 41(4):44–53
13. Park R (1989) Evaluation of ductility of structures and structural assemblages from laboratory testing. Bull New Zealand Natl Soc Earthq Eng 22(3):155–166
14. Li C et al (2020) Elastic displacement spectrum-based design approach for precast concrete frame with replaceable energy-dissipating connectors. J Earthq Eng:1762805
15. Lu L, Liu X, Chen J et al (2017) Seismic performance of a controlled rocking reinforced concrete frame. Adv Struct Eng 20(1):4–17
16. Cai X, Pan Z, Zhu Y et al (2021) Experimental and numerical investigations of self-centering post-tensioned precast beam-to-column connections with steel top and seat angles. Eng Struct:1–17. https://doi.org/10.1016/j.engstruct.2020.111397
17. Jia J, Zhang K, Wu S et al (2020) Seismic performance of self-centering precast segmental bridge columns under different lateral loading directions. Eng Struct:1–19. https://doi.org/10.1016/j.engstruct.2020.111037
18. Liu Y, Guo Z, Liu X et al (2019) An innovative resilient rocking column with replaceable steel slit dampers: experimental program on seismic performance. Eng Struct 183(Mar 15):830–840
19. Xie L (2020) Research on the mechanism of replaceable energy dissipation connector and its seismic performance in precast concrete frame joint. D. SC. thesis. Nanjing, China, Southeast University
20. Chopra AK (1995) Dynamics of structures: theory and applications to earthquake engineering, Pearson Education, Prentice Hall, Englewood Cliffs, New Jersey
21. Newmark NM, Hall WJ (1982) Earthquake spectra and design. Earthquake Engineering Research Institute, Oakland, CA
22. Gupta A, Krawinkler H (2000) Estimation of seismic drift demands for frame structures. Earthq Eng Struct Dynam 29:1287–1305. https://doi.org/10.1002/1096-9845(200009)29:9%3c1287::AID-EQE971%3e3.0.CO;2-B

23. Feng YL, Wu J, Meng SP (2016) Elastic displacement spectrum-based design approach for buckling-restrained braced frames. J Earthquake Eng 20:841–860. https://doi.org/10.1080/13632469.2015.1104762
24. Priestley MJN, Calvi GM, Kowalsky MJ (2007) Displacement-based seismic design of structures. IUSS Press, Pavia, Italy
25. Lu Y, Gu X, Wei J (2009) Prediction of seismic drifts in multi-storey frames with a new storey capacity factor. Eng Struct 31(2):345–357. https://doi.org/10.1016/j.engstruct.2008.09.005
26. Muto K (1974) A seismic design analysis of buildings. Maruzen, Tokyo
27. Liu QP, Li GQ, Wang JF (2008) A simplified method for side sway analyze of semi-rigid composite beam frame under serviceability limit states. Chin Q Mech 29:448–454 In Chinese
28. Mazza F (2019) A simplified retrofitting method based on seismic damage of a SDOF system equivalent to a damped braced building. Eng Struct 200. Article number 109712. https://doi.org/10.1016/j.engstruct.2019.109712
29. McKenna F, Fenves GL, Scott MH (2007) Open system for earthquake engineering simulation (Opensees). University of California, Berkeley, Pacific Earthquake Engineering Research Center
30. Song LL, Guo T, Cao ZL (2015) Seismic response of self-centering prestressed concrete moment resisting frames with web friction devices. Soil Dyn Earthq Eng 71:151–162. https://doi.org/10.1016/j.soildyn.2015.01.018

Chapter 3
Prestressed Precast Concrete Frame with External Dissipaters

Abstract This chapter deals with the subject of Prestressed Precast Concrete Frame with External Dissipators, such a precast concrete frame consists of non-emulative connections where precast frame members are connected with post-tensioned tendons and externally mounted dissipators. The external dissipators provide additional reinforcing and damper, improving the energy dissipation of the proposed self-centering prefabricated system. This chapter also presents the development of all-steel fuse-type dissipators and evaluates the hysteresis behavior of precast post-tensioned frame connections. Additionally, previous research is introduced in terms of the configuration of fuse-type dissipators, characterizing experiments, theoretical studies, and component tests, providing theoretical and technical support for the promotion and application of prestressed precast concrete frames with external dissipators.

Keywords Prestressed precast concrete frame · External dissipator · Self-centering · Fuse-type dissipator · Component tests · Hysteresis behavior

3.1 Introduction

Prestressed precast concrete frame consists of non-emulative connections where precast frame members are connected with post-tensioned tendons and externally mounted dissipaters. As a substitute for a cast-in-place concrete structure, a precast concrete structure is beneficial because it protects the environment, produces less carbon dioxide, dust and construction waste and reduces labor costs, which comply with the modern requirements for structural systems, including cost efficiency, fast construction, resilience and mass production. The external dissipaters provide additional reinforcing and damper, improving the energy dissipation of the proposed self-centering prefabricated system. This chapter presents the development of all-steel fuse-type dissipaters and evaluates the hysteresis behavior of precast post-tensioned frame connections. Additionally, previous researches are introduced in terms of the configuration of fuse-type dissipaters, characterizing experiments, theoretical studies and component tests, providing theoretical and technical supports for the promotion and application of prestressed precast concrete frames with external dissipaters [1–4].

G. Wu et al., *Novel Precast Concrete Structure Systems*, Springer Tracts in Civil Engineering, https://doi.org/10.1007/978-981-19-6821-1_3

3.2 External Replaceable Fuse-Type Dissipaters

3.2.1 Concept and Configuration of Bamboo-Shaped Dissipaters

As shown in Fig. 3.1, the bamboo-shaped energy dissipater (BED) is composed of an inner core and a partially restraining tube. Typically, the core consists of yielding segments, elastic slubs, two clamped ends and one stopper, as depicted in Fig. 3.1a. All components mentioned above in the core are machined from a round bar via computer numerical control lathe processing. The yielding segments, designed to dissipate energy through plastic deformation, are milled from the round bar along the longitudinal direction. Openings 3.0 mm in diameter are gouged in the middle of the stopper, as well as in the middle of the partially restraining tube. The relative movement between the core and the partially restraining tube along the longitudinal direction is prevented by a nail with a radius of 1.4 mm and a length of 40 mm going through the openings in the stopper and the tube. The surface of the yielding segment is sprayed with red paint to show the contact conditions between the yielding segments and the restraining tube from the scratches of the paint.

As demonstrated in Fig. 3.2, the working mechanism of the proposed bamboo-shaped energy dissipater can be idealized as a laterally restrained bar loaded under compression. By comparing Fig. 3.2a, b, the critical buckling load of the bar increases with the addition of the lateral supports. Besides, the lateral deformation of the bar can be significantly decreased and effectively controlled. The proposal of the bamboo-shaped energy dissipater just initiated from the fact, where lateral supports externalized as slubs. The components between slubs, called segments, can be ideally viewed as short bar, which is favorable to prevent buckling of the segment by adjusting the length of the segment. After the prevention of the buckling of the dissipater under compression, the energy dissipater can be effectively applied in structures both in tension and compression.

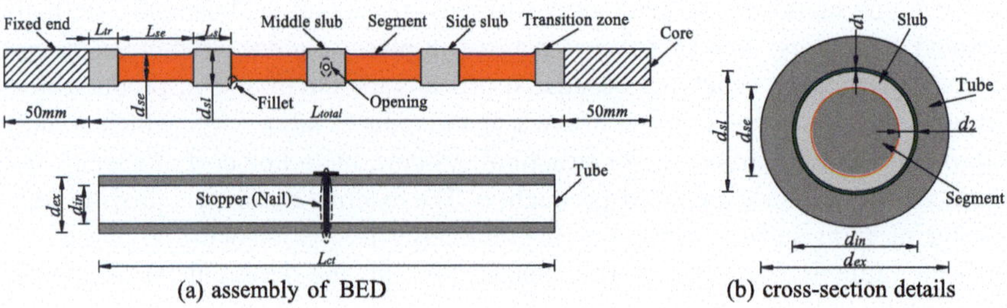

Fig. 3.1 Details of a representative BED

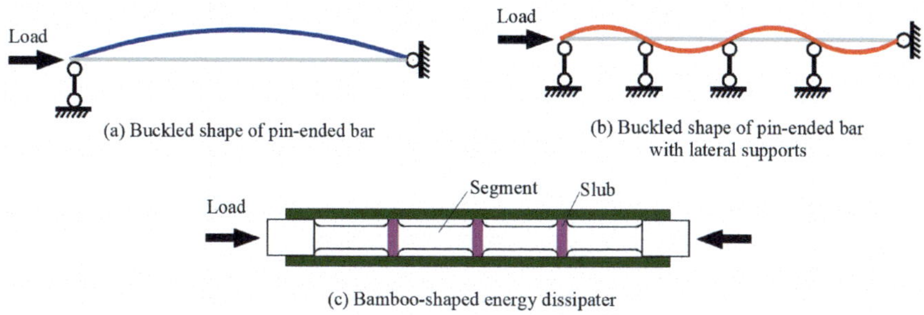

Fig. 3.2 The original concept of the proposed bamboo-shaped dissipater

3.2.2 All-Steel Bamboo-Shaped Dissipaters

(1) Dimensions of BEDs

The Q235b steel bars were adopted in the fabrication of the bamboo-shaped cores via computer numerical control lathe processing. The average material property obtained from coupon tests is listed in Table 3.1. Measured geometric dimensions of BED specimens are given in Table 3.2.

To control the slub to remain elastic during loading histories, the ratio of the sectional area of the slub, A_{sl}, to the sectional area of the segment, A_{se}, should be larger than the ratio of σ_u to σ_y:

$$A_{sl}/A_{se} \geq \sigma_u/\sigma_y \tag{3.1}$$

When Poisson's ratio, υ, is taken into consideration, the incremental diameter of the slub after loaded, Δ_d, should be limited less than d_1. Then, Eq. (3.2) as follows should be fulfilled:

$$\Delta_d = 0.5 \cdot d_{sl} \cdot \varepsilon_x = 0.5 \cdot d_{sl} \cdot (-\upsilon \cdot \varepsilon_y) < d_1 \tag{3.2}$$

where ε_x is the strain in the diameter direction of the slub; ε_y is the strain in the length direction of the slub. In this experimental protocol, the design of cores satisfied the requirement of Eq. (3.2). Finally, two 50-mm fixed ends in each specimen are prepared for actuators to achieve enough holding force. All tubes have 20.0-mm inner diameter (d_{in}) and 30.0-mm external diameter (d_{ex}). In this experimental program, the design of tubes satisfies the requirements of the overall buckling proposed by Usami et al. [5].

Table 3.1 Material properties of Q235b

Grade	Young's modulus E (GPa)	Yield stress σ_y (MPa)	Yield strain ε_y (%)	Ultimate stress σ_u (MPa)	Breaking elongation ε_u (%)
Q235b	208.4	284	0.136	416	39.43

Table 3.2 Measured dimensions of BEDs

No.	Specimens	d_{sl} (mm)	$L_{sl,1}/L_{sl,2}$ (mm)	d_{se} (mm)	L_{se} (mm)	L_{tr} (mm)	L_{total} (mm)	L_{ct} (mm)	d_1 (mm)
S1	L40S20-C1	18.9	20.1/20.1	14	40.2	30.2	281.5	260.0	0.55
S2	L40S20-C2	19	19.8/19.8	14	40.3	29.8	280.2	260.0	0.50
S3	L40S20-C3	18.9	19.6/19.6	13.8	40.4	30.1	280.6	260.0	0.55
S4	L40S20-C4	19.1	19.9/19.9	13.9	40.1	29.9	279.9	260.0	0.45
S5	L60S20-C2	19	20.0/20.0	13.9	60	29.9	359.8	340.0	0.55
S6	L60S20-C3	19	19.9/19.9	14	60	29.8	359.3	340.0	0.55
S7	L40S5-V1	19	10.0/5.0	14	40.5	25.1	232.2	210.0	0.50
S8	L60S5-V1	19	10.0/5.0	14.5	60.4	25.2	312	290.0	0.50
S9	L80S5-V1	19	10.0/5.0	14	80.1	24.9	390.2	370.0	0.50
S10	L40S20-V1	18.9	19.8/19.8	14.1	40.5	29.8	281	260.0	0.55
S11	L60S20-V2	18.6	20.2/20.2	14.1	60.4	29.9	362	340.0	0.70
S12	L70S20-V2	19	20.6/20.2	13.8	69.4	29.8	399	380.0	0.50

Note d_{sl} is the diameter of the slub; d_{se} is the diameter of the segment; L_{sl} is the length of the slub; L_{se} is the length of the segment; L_{tr} is the length of the transition zone; L_{total} is the total length of the core (excluding two fixed ends); L_{ct} is the length of the tube; d_1 is the width of the air gap; The symbols listed in the table are illustrated in Fig. 3.1

(2) Loading protocols

The BED specimen was installed in the hydraulic servo universal testing machine MTS 810, which is capable of producing up to 250 kN loading and maximum displacement of ±300 mm. During a typical test, the axial displacement and force of the actuator were automatically collected by a digital data acquisition system. Three different low-cycle loading patterns illustrated in Fig. 3.3 were applied to BED specimens. All loading patterns were controlled by the nominal axial strain of the core which is derived from the axial displacement of the actuator divided by total length of segments, $L_{se,t}$. As shown in Fig. 3.3, the strain amplitude, $\Delta\varepsilon$, is the absolute value of the peak or valley strain in each cycle.

(3) Low-cycle fatigue performance

The experimental stress–strain hysteresis curves after calibration are presented in Fig. 3.4. The tensile state of the BED specimen is displayed in the positive direction. The abscissa is the nominal axial strain of the bamboo-shaped core defined as the calibrated displacement of the BED specimen divided by the total length of segments, while the ordinate is the nominal axial stress derived from the axial force of the actuator divided by A_{se}. All tested BED specimens demonstrate stable and repeated hysteretic capacity, without any local and overall buckling during loading histories even at the maximum strain amplitude. Test results of all BED specimens are summarized in Table 3.3.

From the comparison between specimens S1–S4, including L40S20-C1 with N_f of 114, L40S20-C2 with N_f of 47, L40S20-C3 with N_f of 29, L40S20-C4 with N_f of

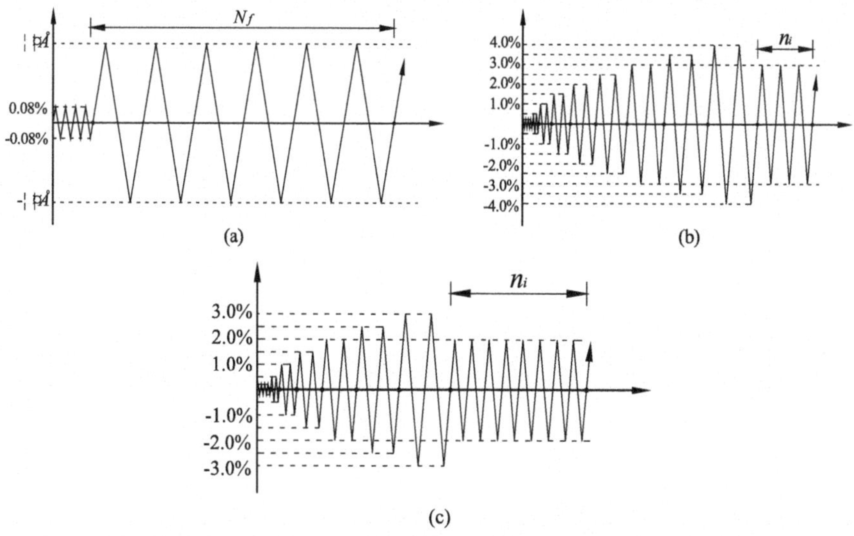

Fig. 3.3 Loading protocols: **a** constant strain amplitude (CSA), **b** variable strain amplitude (VSA1) and **c** variable strain amplitude (VSA2)

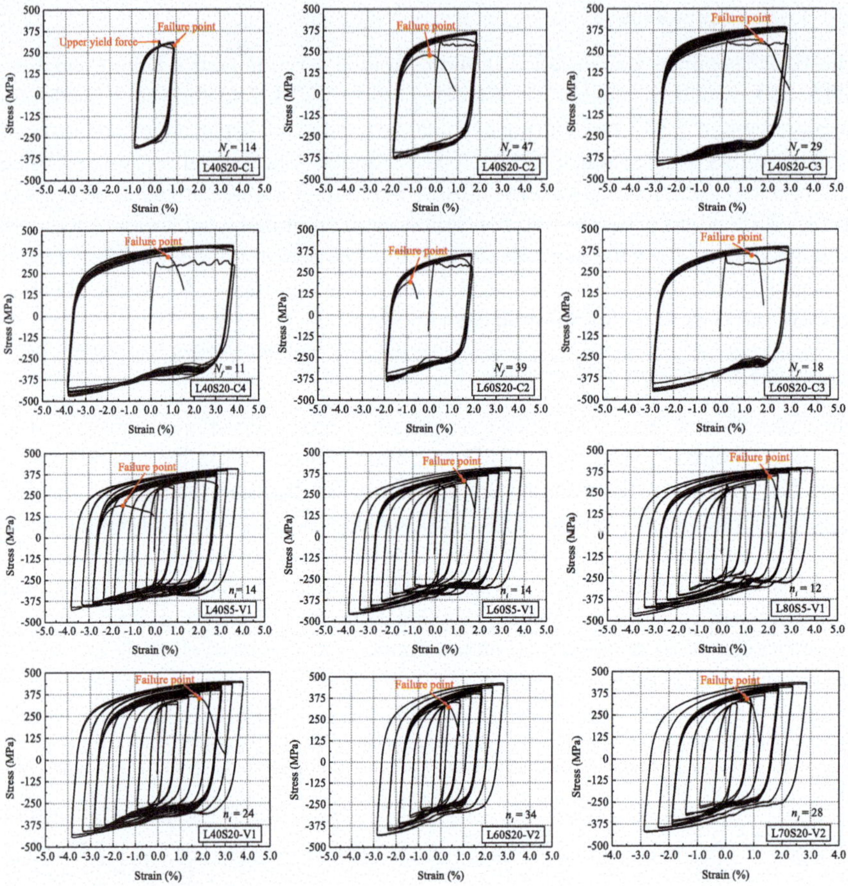

Fig. 3.4 Stress–strain hysteresis curves of BED specimens

11, the low-cycle fatigue life of BED specimens in terms of the number of the failure cycles, N_f, were found to be related with the amplitudes of the constant strain. The increasing constant strain amplitude decreased the low-cycle fatigue cycles of BED specimens. Two additional tests on L60S20-C2 with N_f of 39 and L60S20-C3 with N_f of 18 also addressed the same conclusion.

To evaluate the effect of the different lengths of segments on the low-cycle fatigue life of BED specimens with 5-mm slubs, L40S5-V1 with four 40-mm segments, L60S5-V1 with four 60-mm segments and L80S5-V1 with four 80-mm segments were tested and compared. The numbers of cycles at additional constant strain amplitude, n_i, were 14, 14 and 12 for the three BED specimens, respectively. Similar low-cycle fatigue life was observed in specimens L40S5-V1 and L60S5-V1, which showed that the low-cycle fatigue life of BED specimens had little relationship with the lengths of the segments for BED specimens with 5-mm slubs and the stress concentration around the fillet accounted for the failure mechanism. However, a

Table 3.3 Test results of BED specimens

Series	Specimen	$\Delta\varepsilon$ (%)	N_f	n_i	CPD (%)	Contact conditions
S1	L40S20-C1	1	114	–	2612.2	No
S2	L40S20-C2	2	47	–	2444.2	No
S3	L40S20-C3	3	29	–	2418.1	No
S4	L40S20-C4	4	11	–	1218.7	No
S5	L60S20-C2	2	39	–	2081.8	Yes
S6	L60S20-C3	3	18	–	1476.7	Yes
S7	L40S5-V1	–	–	14	2031.7	Yes
S8	L60S5-V1	–	–	14	2080.4	Yes
S9	L80S5-V1	–	–	12	1942.2	Yes
S10	L40S20-V1	–	–	24	2839.8	Yes
S11	L60S20-V2	–	–	34	2226.4	Yes
S12	L70S20-V2	–	–	28	1969.1	Yes

Note N_f is number of failure cycles; n_i is number of cycles at additional constant strain amplitude; CPD is cumulative inelastic deformation [6] $= \sum_i |\Delta_{pi}|/\Delta_y$, where $|\Delta_{pi}|$ is the plastic deformation in ith cycle and Δ_y is the yielding deformation

decrease of low-cycle fatigue life for specimen L80S5-V1 was observed due to torsion, compared with L40S5-V1 and L60S5-V1.

Furthermore, the effect of the different lengths of segments on the low-cycle fatigue life of BED specimens with 20-mm slubs was discussed by the comparison between specimens L60S20-V2 with four 60-mm segments and L70S20-V2 with four 70-mm segments. The number of cycles at additional constant strain amplitude, n_i, was 34 for specimen L60S20-V2 and 28 for specimen L70S20-V2. The low-cycle fatigue cycles of BED specimens with 20-mm slubs decreased with the increase of the length of the segment.

(4) Deformation and failure modes of BEDs

The failure pictures of BED specimens shown in Fig. 3.5 demonstrated failure positions, contact conditions and deformation patterns of BED specimens. As shown in Fig. 3.5a–d, the lateral deformation of BED specimens increased with the constant strain amplitude. The shape of the specimen L40S20-C1 nearly kept in a straight line after being loaded while the specimen L40S20-C4 exhibited a relatively severer deformation. No contact between segments and the restraining tube was observed in BED specimens with 40-mm segments and 20-mm slubs under constant strain amplitudes of 1, 2, 3 and 4%. When comparing Fig. 3.5e, f, the lateral deformation of specimen L40S20-V1 with larger slubs was smaller than that of specimen L40S5-V1 with smaller slubs. It proved that the weaker rotation capacity of slubs decreased the lateral deformation of the segments. As shown in Fig. 3.5e–h, the contact in L40S5-V1 was viewed as the point contact and the contact conditions in L60S5-V1 and L80S5-V1 were the line contact, while the contact areas of the three BED

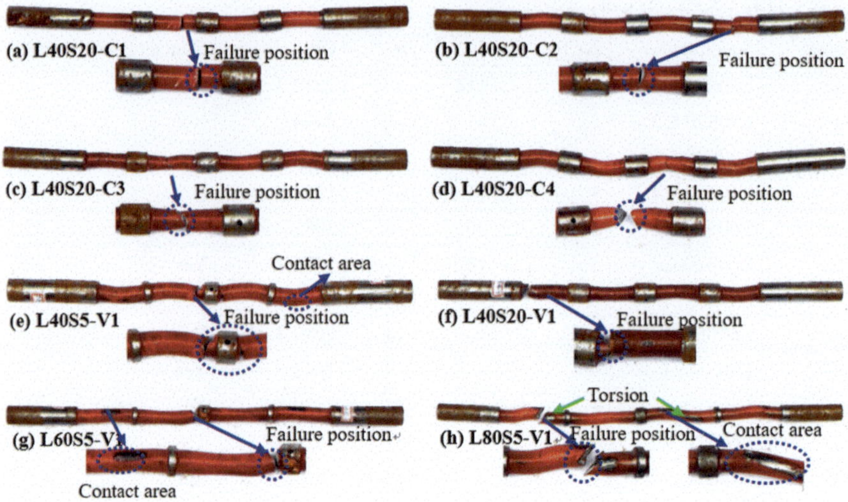

Fig. 3.5 Failure modes of BEDs

specimens gradually increased with the length of segments. Especially, a torsional deformation was observed in specimen L80S5-V1 via the scratches of the red paint on the surface of segments.

3.2.2.1 Aluminum Alloy Energy Dissipater

To favor the application of the BEDs under a corrosive environment, an study was also conducted on the hysteresis performance of Aluminum alloy Bamboo-shaped Energy Dissipaters (ABEDs) experimentally. A series of tests including 12 aluminum alloy bamboo-shaped energy dissipater specimens was performed to compare the key design parameters and address the low-cycle fatigue performance. The test results are as follows.

(1) Specimens and test setup

Measured geometric dimensions of cores and tubes are given in Table 3.4. Obviously, A_{sl}/A_{se} listed in Table 3.4 satisfies Eq. (3.1). In this chapter, two batches of A6061-T6 aluminum alloy labelled as A6061-S1 and A6061-S2 were adopted in the fabrication of cores in the present test. The material properties achieved from coupon tests are listed in Table 3.5.

The adopted two loading protocols are shown in Fig. 3.6.

(2) Low-cycle fatigue life of ABEDs

The experimental stress–strain hysteresis curves are presented in Figs. 3.7 and 3.8, classified according to different loading patterns. The tensile state of BED

Table 3.4 Measued dimensions of ABEDs

Series	Specimens	d_{sl} (mm)	d_{se} (mm)	$L_{sl,1}/L_{sl,2}$ (mm)	L_{se} (mm)	L_{tr} (mm)	L_{total} (mm)	L_{ct} (mm)
S1	S1-L40S20G1-C1	19.0	14.1	19.3/19.3	40.7	30.1	280.9	260.0
	S1-L40S20G1-C2	19.0	14.0	19.9/19.9	40.4	30.1	281.5	260.0
	S1-L40S20G1-C3	19.0	14.1	19.4/19.4	41.0	30.0	282.2	260.0
	S1-L40S20G2-C1	19.5	14.0	19.8/19.8	40.8	29.9	282.4	260.0
	S1-L40S20G2-C2	19.4	14.0	19.8/19.8	40.3	29.8	280.2	260.0
	S1-L40S20G2-C3	19.5	14.0	19.4/19.4	41.6	30.1	284.8	260.0
S2	S2-L40S5G1-V	18.9	13.8	10.0/5.0	40.2	24.9	230.6	210.0
	S2-L40S5G1-V(6)	19.0	13.9	10.0/5.0	40.1	37.5	345.6	325.0
	S2-L60S5G1-V	19.0	13.8	9.8/5.0	60.2	42.5	345.6	325.0
	S2-L80S20G1-V(2)	19.1	13.8	20.0/20.0	80.1	25.1	230.4	210.0
	S2-L40S20G1-V	18.9	14.0	20.2/20.2	39.6	10.2	239.4	230.0
	S2-L60S20G1-V	18.9	13.9	20.1/20.1	59.7	22.5	344.1	325.0

Note d_{sl} is the diameter of the slub; d_{se} is the diameter of the segment; L_{sl} is the length of the slub; L_{se} is the length of the segment; L_{tr} is the length of the transition zone; L_{total} is the total length of the core (excluding two fixed ends); L_{ct} is the length of the tub

Table 3.5 Material properties of A6061-T6

Grade	E (GPa)	$\sigma_{0.2}$ (MPa)	σ_0 (MPa)	$\varepsilon_{0.2}$ (%)	ε_0 (%)	σ_u (MPa)
A6061-S1	67.32	335.08	301.57	0.67	0.43	359.47
A6061-S2	70.92	294.43	264.99	0.44	0.37	335.23

Note E is initial Young's modulus; $\sigma_{0.2}$ is 0.2% nominal yield stress; σ_0 is $0.9\sigma_{0.2}$ [5]; $\varepsilon_{0.2}$ is 0.2% nominal yield strain; ε_0 is strain corresponding to the stress σ_0; σ_u is ultimate tensile strength

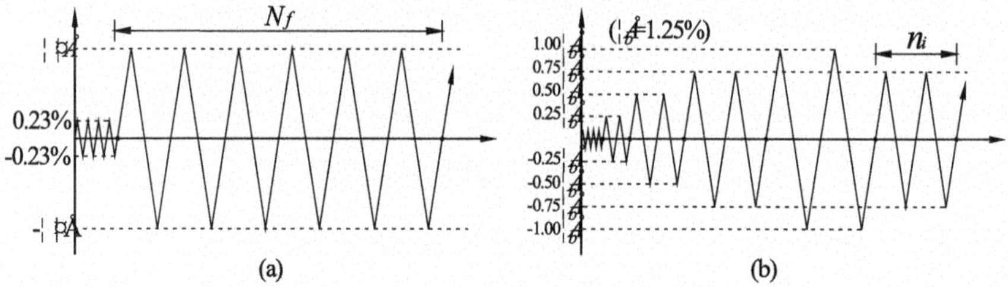

Fig. 3.6 Loading protocols: **a** constant strain amplitude (CSA) and **b** variable strain amplitude (VSA)

specimens is displayed in the positive direction. The abscissa is the average axial strain of the core, while the ordinate is the average axial stress derived from the axial force of the core divided by A_{se}. All testing specimens demonstrate s table and repeated hysteretic capacity, without any local and overall buckling during loading histories even at the maximum strain amplitude. Test results of all specimens are summarized in Table 3.6.

The test results of some specimens listed in Table 3.6, such as S1-L40S20G1-C1 and S1-L40S20G2-C1, reveals that under comparatively small strain amplitude, the number of failure cycles, N_f, increases 49% with the decrease of the gap between slubs and the tube, d_1. However, the N_f of S1-L40S20G1-C2 and S1-L40S20G2-C2 is similar even if different widths of gap exist in the two specimens and the same rule can be discovered in S1-L40S20G1-C3 and S1-L40S20G2-C3. Obviously, the narrower gaps provide better low-cycle fatigue behavior for BED specimens. This phenomenon is perfectly elaborated under the relatively small strain amplitude but is kind of hidden when the strain amplitude becomes comparatively large.

As listed in Table 3.6, the number of cycles at additional constant strain amplitude, n_i, of S2-L40S5G1-V with four segments is only 2 cycles more than that of S2-L40S5G1-V(6) with six segments, which demonstrates that the increase of the number of segments only has limited influence on the hysteretic behaviors of BED specimens. Specimen S2-L80S20G1-V(2) can be viewed as moving side slubs of S2-L40S5G1-V to the middle of the core, it is found that the n_i of the former is 2 cycles less than the latter one. Even if only slight advantages of low-cycle behaviors existed in S2-L40S5G1-V compared with S2-L80S20G1-V(2), the necessity of distribution of slubs cannot be ignored especially when the number of segments becomes more.

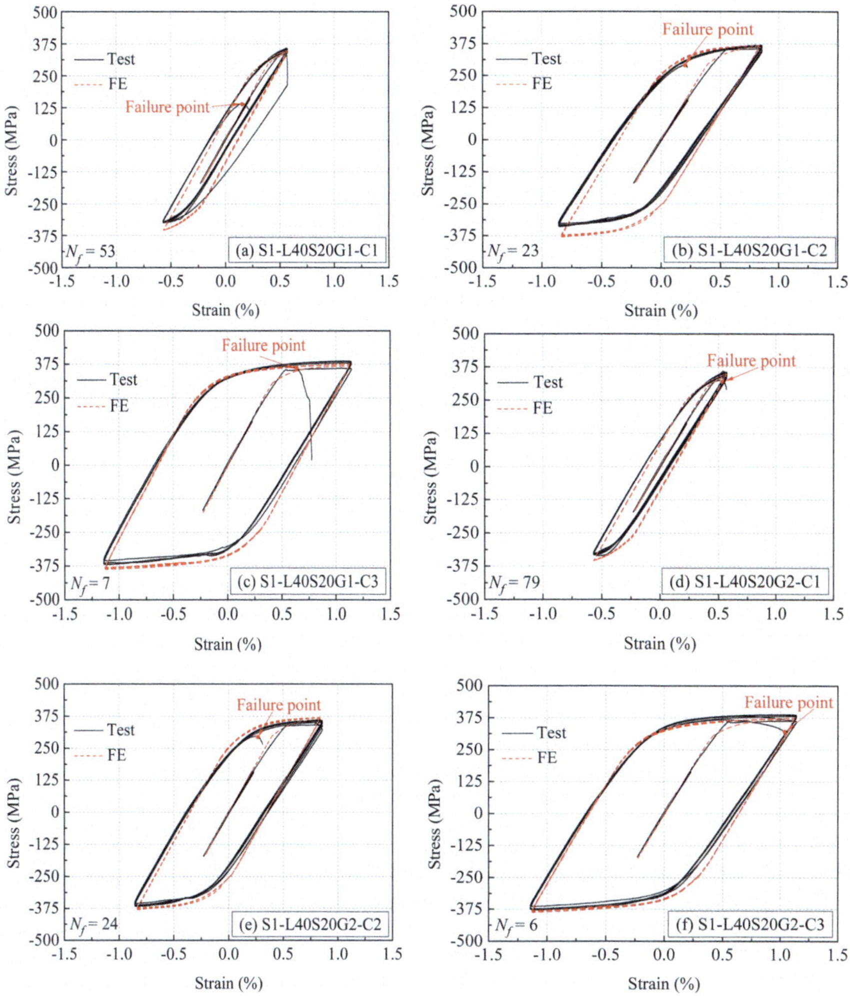

Fig. 3.7 Stress–strain hysteresis curves of ABED specimens under CSA

(3) Failure modes of ABEDs

Failure modes of all specimens are presented in Fig. 3.9. The rupture of specimens with 20-mm slubs concentrated on the end of the segment except S1-L40S20G2-C3 which apparently necked in the middle of the segment. The rupture of specimens with 5-mm side slubs always initiated from the end of the segment or the middle of the segment depending on the length of the segment. For specimens with 5-mm side slubs and 40-mm segments, no contact occurred between the segment and the tube and the failure positions located at the end of the segment. However, the contact of the specimen with 5-mm side slubs and 60-mm segments was observed between the segment and the tube. The rupture position accordingly shifted to the middle of the

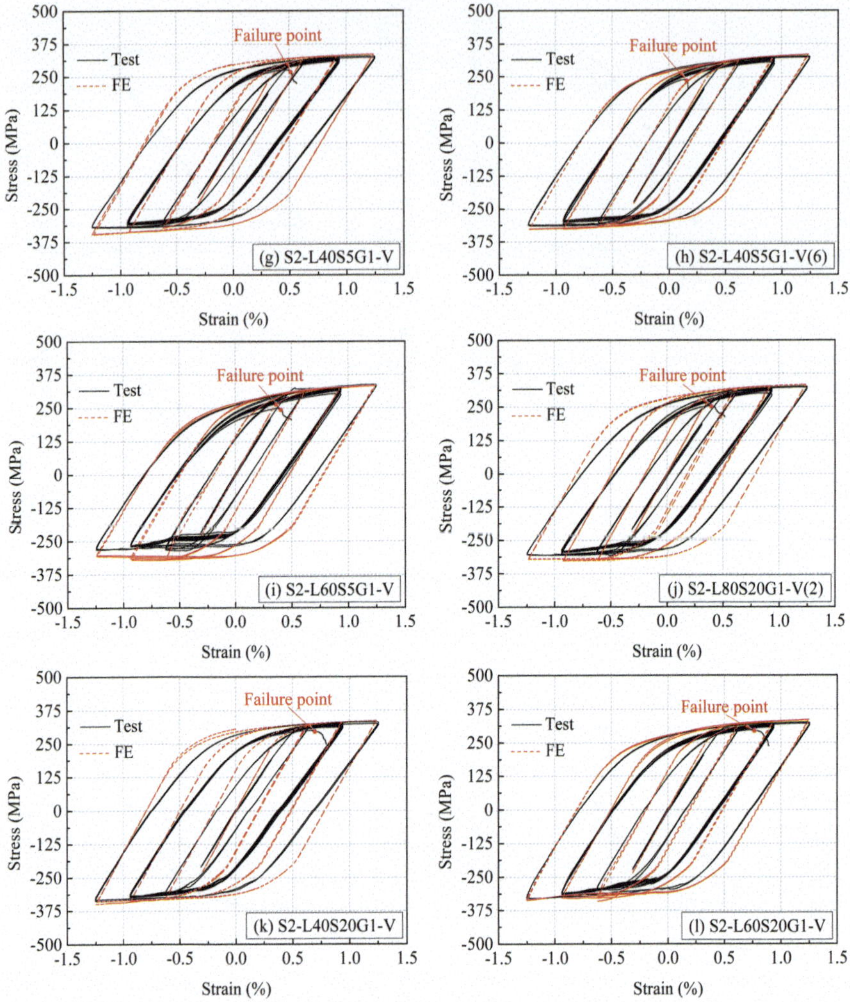

Fig. 3.8 Stress–strain hysteresis curves of ABED specimens under VSA

segment which was just the failure position. Furthermore, no significant deformation and failure occurred around the opening, so the stopper was proved to be effective and reliable having no effect on the failure of the ABED. All fracture surfaces of testing specimens were extraordinarily rough depicted in Fig. 3.9.

From the final loops of some ABED specimens shown in Fig. 3.10, a sudden loss of strength was observed so that the failure of BED specimens was regarded as brittle fracture. The same brittle failure was observed in extruded aluminum alloy BRB specimens discussed by Wang et al. [9].

(4) Compressive strength adjustment factor

Table 3.6 Test results of ABEDs

Series	Specimen	$\Delta\varepsilon$ (%)	N_f	n_i	CID (%)	CED (N·m)	Contact conditions
S1	S1-L40S20G1-C1	0.57	53	–	69.1	2657.5	No
	S1-L40S20G1-C2	0.86	23	–	76.0	4404.9	No
	S1-L40S20G1-C3	1.14	7	–	37.1	2539.4	No
	S1-L40S20G2-C1	0.57	79	–	103.0	4181.4	No
	S1-L40S20G2-C2	0.86	24	–	79.3	4713.0	No
	S1-L40S20G2-C3	1.14	6	–	31.8	2219.8	No
S2	S2-L40S5G1-V	–	–	14	39.8	3385.4	No
	S2-L40S5G1-V(6)	–	–	12	35.8	4966.6	No
	S2-L60S5G1-V	–	–	20	51.7	6615.6	Yes
	S2-L80S20G1-V(2)	–	–	12	35.8	2951.5	No
	S2-L40S20G1-V	–	–	20	51.7	4671.3	No
	S2-L60S20G1-V	–	–	14	39.8	5391.1	Yes

Note N_f is number of failure cycles; n_i is number of cycles at additional constant strain amplitude; CID is cumulative inelastic deformation [7]; CED is cumulative energy dissipation [8]

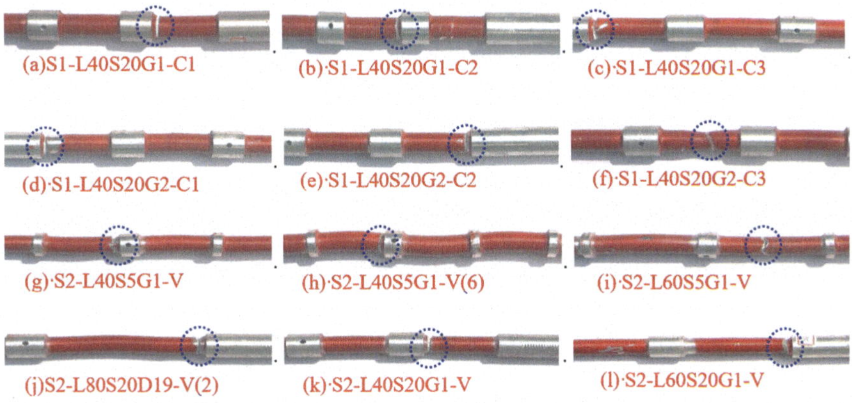

Fig. 3.9 Failure modes of ABEDs

The compression-strength adjustment factor, β, denoted by the ratio of the maximum compression strength to the maximum tension strength of the test specimen is specified by the AISC 341-10 [10] for the BRB specimen. The β value should be less than 1.3 according to the AISC 341-10 [10] to avoid unbalanced force resulting in the fracture of beam. The β value of some ABED specimens is calculated and the results are depicted in Fig. 3.11. The overall trend of β values is that the β values are relatively high at first several cycles because of the cyclic hardening effect [11] and then the β values slowly increase with the progress of loading step. In the present study, all ABED specimens satisfy the limitation of β value required in AISC. The

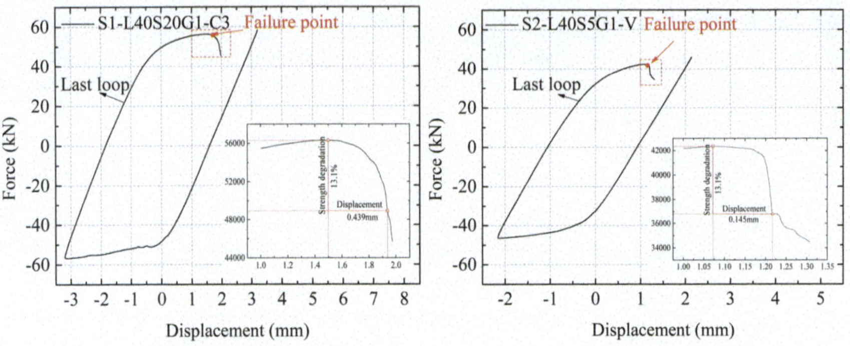

(a) S1-L40S20G1-C3 **(b) S2-L40S5G1-V**

Fig. 3.10 Last loops of certain BED specimens

β values of most specimens are around 1.0 except S2-L60S5G1-V. This property means that the proposed ABED specimens have symmetric compression and tension strength, which is attractive to the application of ABED specimens in the practical engineering.

Firstly, specimens with 20-mm slubs are compared. As shown in Fig. 3.11, the β value of S2-L60S20G1-V is slightly less than that of S2-L40S20G1-V and the β value of S2-L80S20G1-V(2) is least among S2-L40S20G1-V, S2-L60S20G1-V and S2-L80S20G1-V(2). According to the comparison above, it is concluded that the β value decreases with the increase of L_{sl} when slubs are 20 mm. Then, comparisons are made among specimens with 5-mm side slubs. As discussed in the previous section, one side slub along with its adjacent segments can be viewed as a whole part. Therefore, the concept of the equivalent length of the segment is adopted herein. The equivalent lengths of the segment of S2-L40S5G1-V, S2-L40S5G1-V(6) and S2-L60S5G1-V are further calculated as 80, 120 and 120 mm, respectively. It is

Fig. 3.11 Compression-strength adjustment factor of some BED specimens

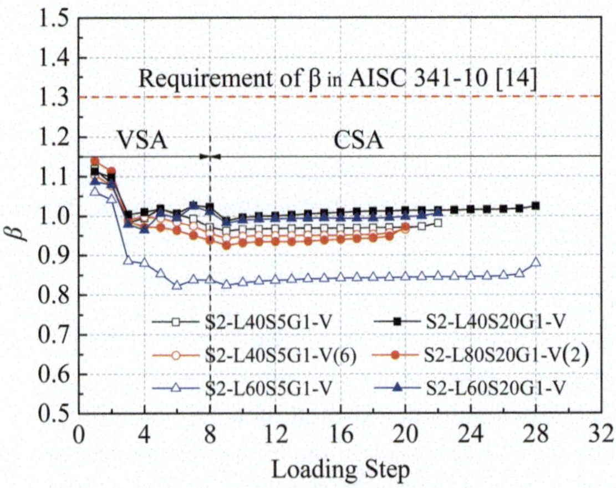

found that the β value of S2-L40S5G1-V is distinctly more than the β value of S2-L60S5G1-V, which demonstrates that the larger equivalent length of the segment leads to less β value. Furthermore, the comparison between S2-L40S5G1-V(6) and S2-L60S5G1-V with same equivalent length of the segment reveals that the β value of S2-L40S5G1-V(6) is distinctly more than that of S2-L60S5G1-V. This difference just illustrates the following conclusion of interest: 5-mm side slubs reduce the extent of the bending deformation of the segment and further the β value is increased due to the existence of side slubs. Finally, specimens with different lengths of slubs are also taken into consideration. Comparing the β value of S2-L40S20G1-V and the β value of S2-L40S5G1-V, it is found that the β value of the specimen with larger slub is more than the β value of the specimen with smaller slub. Same phenomenon can be observed in the comparison between S2-L60S20G1-V and S2-L60S5G1-V.

Based on the comparison of β values above, in order to control β values and design ABED specimens with relatively symmetric compression and tension strength, it is recommended that ABED should be designed with relatively larger length of slub, shorter length of segment and shorter equivalent length of segment. Besides, when slubs are small in an ABED specimen, the number of side slubs should be increased.

3.2.3 Partially Restrained Energy Dissipaters

For the recently developed energy dissipaters without grouting and welding, such as the bamboo-shaped energy dissipater, the ratio of the elastic portion without energy dissipation capacity to the plastic portion was relatively high, which reduced the utilization of the material. To quantitatively evaluate the material utilization in the energy dissipater, a new partially restrained energy dissipater (PED) is further proposed by Wang et al. [12].

(1) Conceptual proposal

A factor called the material utilization factor, U_m, is introduced, as defined in Eq. (3.3).

$$U_m = V_p/(V_p + V_e) \tag{3.3}$$

where V_p is the volume of the plastic portion and V_e is the volume of the elastic portion. The material utilization factor was approximately 0.42 for a BED with four 40-mm segments and three 20-mm slubs. Less than half of the material was utilized in a typical BED in terms of the material utilization factor, which is uneconomical.

An upgraded conceptual proposal, shown in Fig. 3.12, was thus introduced herein to solve the disadvantage observed for the BED. Compared with the typical bamboo-shaped energy dissipater, the yielding segments in Fig. 3.12b were designed without slubs in the new proposal. By discarding the elastic slubs, the material utilization factor, U_m, can be largely increased. However, a problem with the deformation control occurred due to the absence of slubs, which were employed to control the lateral deformation of the bamboo-shaped core in BEDs. Inspired by the new restraining system (see Fig. 3.12a), a similar mechanism was employed to provide effective restraining of yielding segments in the new proposal. The edges of the yielding segments were partially restrained by the partially restraining tube, so the possible lateral deformations of the yielding segments can be limited within the air gap (see Fig. 3.12b) between the corner of the yielding segment and the inner surface of the partially restraining tube. The yielding segments are designed to dissipate energy by entering the plastic regime while the other parts remain in the elastic regime during loading.

(2) Configuration of the PED

As shown in Fig. 3.13, the partially restrained energy dissipater (PED) is composed of an inner core and a partially restraining tube. Typically, the core consists of two yielding segments, two elastic transitional segments, two clamped ends and one stopper, as depicted in Fig. 3.13a. All components mentioned above in the core are machined from a round bar via computer numerical control lathe processing. The yielding segments, designed to dissipate energy through plastic deformation, are milled from the round bar along the longitudinal direction. The surface of the yielding segment is sprayed with red paint to show the contact conditions between the yielding segments and the partially restraining tube from the scratches of the paint. After processing, the profile of the PED is as explicitly demonstrated in Fig. 3.13b. An air gap is thus provided between the corner of the yielding segment and the inner surface of the partially restraining tube.

All PED specimens have two 50-mm clamped ends to achieve sufficient holding force during tests. Part of the transitional segment is set inside the partially restraining tube, and the rest is outside the tube. Such a configuration ensures that the transitional segment will not be pulled out of the partially restraining tube under tension and will maintain sufficient space for an actuator's clampers moving back and forth. Fillets are added to alleviate the stress concentration via smoothing of the sudden variation

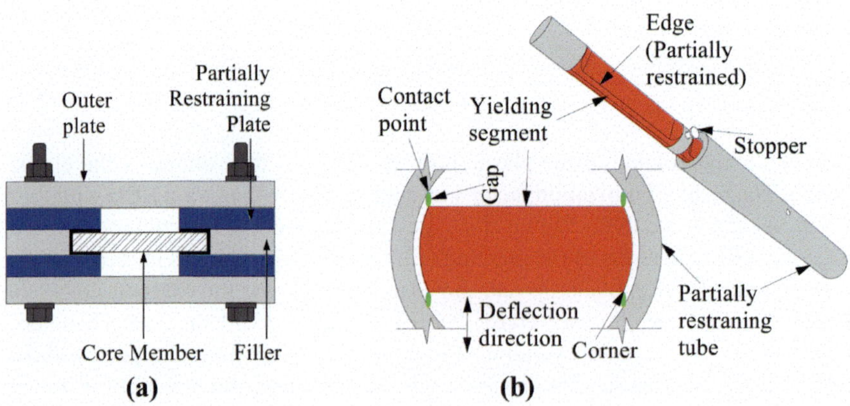

Fig. 3.12 Sketches for **a** a partially buckling-restrained brace and **b** the PED

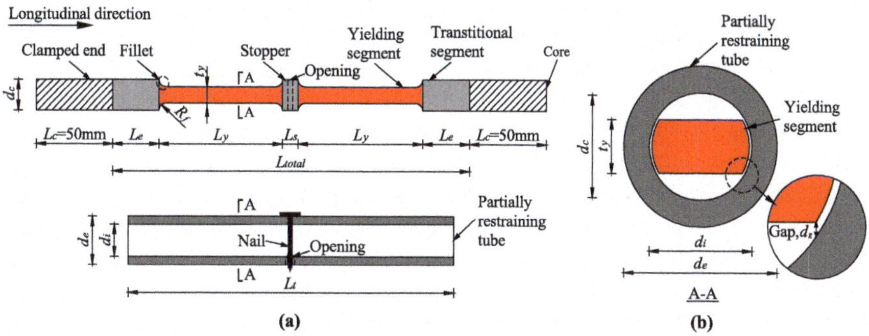

Fig. 3.13 Details of a representative PED

of the cross-section between the yielding segment and the elastic transitional segment or the stopper. Openings 3.0 mm in diameter are gouged in the middle of the stopper, as well as in the middle of the partially restraining tube. The relative movement between the core and the partially restraining tube along the longitudinal direction is prevented by a nail with a radius of 1.4 mm and a length of 40 mm going through the openings in the stopper and the tube.

The cores and the partially restraining tubes of tested PED specimens were fabricated out of Q235b steel [13]. The average material properties of Q235b steel determined from coupon tests are listed in Table 3.7.

(3) Design guidelines

The yielding segment is designed to bend about the weak axis, while torsional buckling with respect to the longitudinal axis should be intentionally avoided. Because torsional-flexural buckling is generally not expected in a specimen with a double-symmetric section, analysis of the prevention of torsional buckling for a PED is simplified as a decoupling analysis of torsional buckling and flexural buckling. The computation graph for the prevention of torsional buckling is shown in Fig. 3.14. The torque and moment equilibrium differential equations are expressed in Eqs. (3.4) and (3.5) [14], respectively.

$$EI_{\omega}\varphi''' + (Pi_0^2 - GI_t)\varphi' = 0 \tag{3.4}$$

$$EI_x y^{IV} + Py'' = 0 \tag{3.5}$$

Table 3.7 Material properties of Q235b steel

Grade	E (GPa)	σ_y (MPa)	ε_y (%)	σ_u (MPa)
Q235b	206.5	260.2	0.126	404.2

Note E is the initial Young's modulus; σ_y is the yield stress; ε_y is the yield strain; σ_u is the ultimate tensile stress where I_ω is the warping moment of inertia; φ is the torsion angle; i_0 is the polar radius of gyration about the shear center of the yielding segment's cross section, defined as the square root of the ratio of I_ω to A_y; G equal to $E/2(1 + \upsilon_e)$ is the shear modulus, with υ_e being the elastic Poisson ratio, identified as 0.3 [15]; I_t is the torsion constant, defined as the sum of the moment of inertia about the weak axis (X axis) of the yielding segment, I_x, and the moment of inertia about the strong axis (Y axis), I_y; P is the applied axial force; and y is the lateral deformation in the Y direction. The boundary conditions are $\varphi = 0$ and $\varphi' = 0$ for Eq. (3.4) and $y = 0$ and $y' = 0$ for Eq. (3.5). The torsional buckling load, P_ω, and flexural buckling load, P_{cr}, for the yielding segment are further calculated as:

$$P_\omega = \frac{1}{i_0^2}\left[\frac{\pi^2 E I_\omega}{(\mu_0 L_y)^2} + G I_t\right] \tag{3.6}$$

$$P_{cr} = \frac{\pi^2 E I_x}{(\mu_0 L_y)^2} \tag{3.7}$$

where μ_0 equal to 0.5 is the effective length factor and L_y is the length of the yielding segment shown in Fig. 3.14. I_ω is nearly zero for the solid cross section, similar to the yielding segment's section [16]. The derivation of I_x and I_y is further given as follows:

$$I_x = \frac{d_c^4}{32}[\theta_1 - 0.25\sin(4\theta_1)] \tag{3.8}$$

$$I_y = \frac{t_y}{12}(d_c^2 - t_y^2)^{\frac{3}{2}} + I_x \tag{3.9}$$

where θ_1 represents arcs in (t_y/d_c); d_c is the diameter of the transitional segment or the stopper depicted in Fig. 3.14; and t_y is the thickness of the yielding segment shown in Fig. 3.13. Therefore, without considering friction forces caused by the contact between the yielding segments and the partially restraining tube, the torsional buckling occurring before the flexural buckling can be prevented by making P_ω greater than P_{cr}. The required length of the yielding segment should thus satisfy the relationship proposed in Eq. (3.10).

$$L_y \geq 10\sqrt{\frac{i_0^2 I_x}{I_t}} \tag{3.10}$$

(4) Nominal and measured geometric dimensions

The main research objectives of the present study were to investigate the hysteretic behaviors and failure mechanisms of PEDs with various geometric properties. Geometric properties, including the length of the yielding segment L_y, the thickness of the yielding segment t_y and the width of the air gap d_g, were taken into

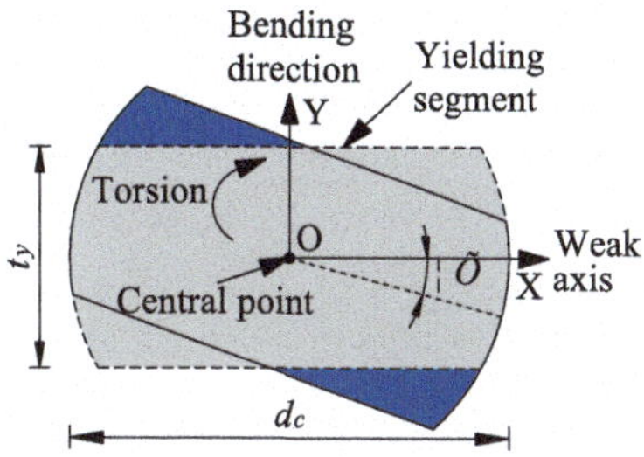

Fig. 3.14 Computation graphs for torsion prevention

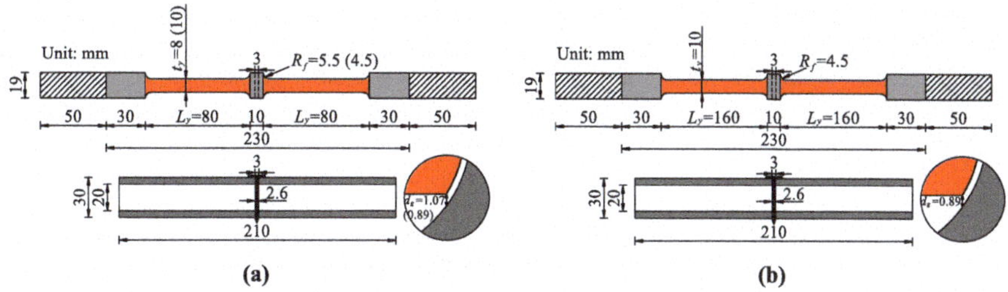

Fig. 3.15 Nominal geometric dimensions of PEDs (all dimensions marked)

consideration. The nominal geometric dimensions of all tested PED specimens are shown in Fig. 3.15, and a summary of the measured geometric dimensions of all tested PED specimens is reported in Table 3.8. As shown in Fig. 3.15a, 80-mm long and 8-mm or 10-mm thick yielding segments were employed in the PED specimens. As demonstrated in Fig. 3.15b, 160-mm long and 10-mm thick yielding segments were adopted.

(5) Test setup and loading protocols

The PEDs are tested is the same test setup with BEDs. The two types of testing protocols demonstrated in Fig. 3.16 were involved. As shown in Fig. 3.16a, three constant strain amplitude (CSA) loadings with strain amplitudes, $\Delta\varepsilon$, of 1.0%, 2.0% and 3.0% were employed, which were labelled C1, C2 and C3, respectively. The counting of loading cycles, C_i, for the CSA loading began subsequently after the completion of four testing cycles. Another loading protocol, labelled V, combined variable strain amplitude (VSA) loading and constant strain amplitude (CSA) loading, as depicted in Fig. 3.16b. The VSA loading stage consisted of stepwise strain amplitudes, $\Delta\varepsilon$, of 0.5, 1.0, 1.5, 2.0, 2.5 and 3.0%, where each strain amplitude was applied to the tested PED specimen for two cycles. The CSA loading stage continued after the VSA loading, and the strain amplitude, $\Delta\varepsilon$, was returned to 2.0%. Notably, all testing protocols were imposed cyclically to the tested PED specimens until failure.

(6) Low-cycle fatigue performance

The calibrated stress–strain hysteretic curves of the PED specimens are presented in Fig. 3.17. The tensile state of the PED specimen is displayed in the positive direction.

Table 3.8 Measured geometric dimensions of PEDs

Specimen	L_y	L_e	L_s	t_y	d_c	L_t	L_{total}	d_i	d_e
L80T8-C1	80.00	29.95	10.04	8.00	18.97	210.00	229.94	20.00	30.00
L80T8-C2	80.03	30.02	9.96	7.98	18.96	210.00	230.06	20.00	30.00
L80T8-C3	80.05	30.01	10.01	7.99	18.98	210.00	230.13	20.00	30.00
L80T8-V	80.10	30.04	9.98	8.00	18.96	210.00	230.26	20.00	30.00
L80T10-C1	80.11	30.02	9.99	9.96	18.98	210.00	230.25	20.00	30.00
L80T10-C2	80.00	30.01	10.04	9.99	18.94	210.00	230.06	20.00	30.00
L80T10-C3	79.95	29.97	10.03	9.97	18.96	210.00	229.87	20.00	30.00
L160T10-C1	160.03	30.03	9.98	9.95	18.92	370.00	390.10	20.00	30.00
L160T10-C2	160.01	29.98	10.01	9.95	18.93	370.00	389.99	20.00	30.00
L160T10-C3	160.10	29.98	10.05	9.96	18.96	370.00	390.21	20.00	30.00
L160T10-V	160.12	29.95	10.03	9.99	18.93	370.00	390.17	20.00	30.00

Note L_y is the length of the yielding segment; L_e is the length of the transitional segment; L_s is the length of the stopper; t_y is the thickness of the yielding segment; d_c is the diameter of the transitional segment or the stopper; L_t is the length of the partially restraining tube; L_{total} is the length of the core excluding the two clamped ends; d_i is the inner diameter of the partially restraining tube; d_e is the external diameter of the partially restraining tube

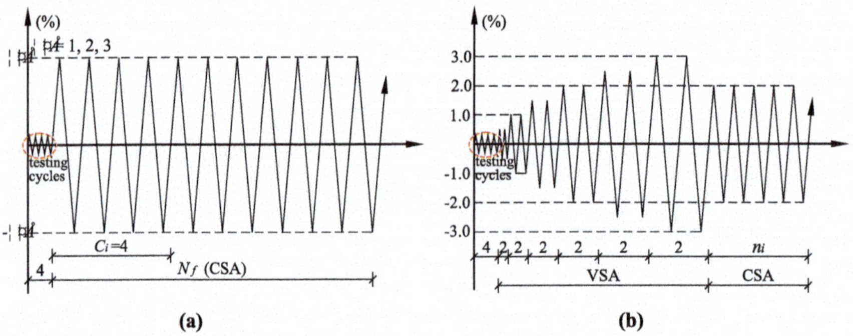

Fig. 3.16 Loading protocols: **a** CSA and **b** V

The abscissa of the hysteretic curve is the nominal axial strain of the core, defined as the calibrated displacement of the PED specimen, u_c, divided by the total length of the yielding segments, $2L_y$, while the ordinate is the nominal axial stress calculated as the measured axial force of the PED specimen, P_p, divided by the cross-sectional area of the yielding segment, A_y. All tested PED specimens demonstrated stable and repeated hysteretic performance. The test results are summarized in Table 3.9.

From the comparison between PED specimens L80T8-C1 with an N_f of 394, L80T8-C2 with an N_f of 75 and L80T8-C3 with an N_f of 13, the failure cycle numbers of the PED specimens were related to the strain amplitude applied in the CSA loading protocol. The failure cycle number decreased sharply with the increase

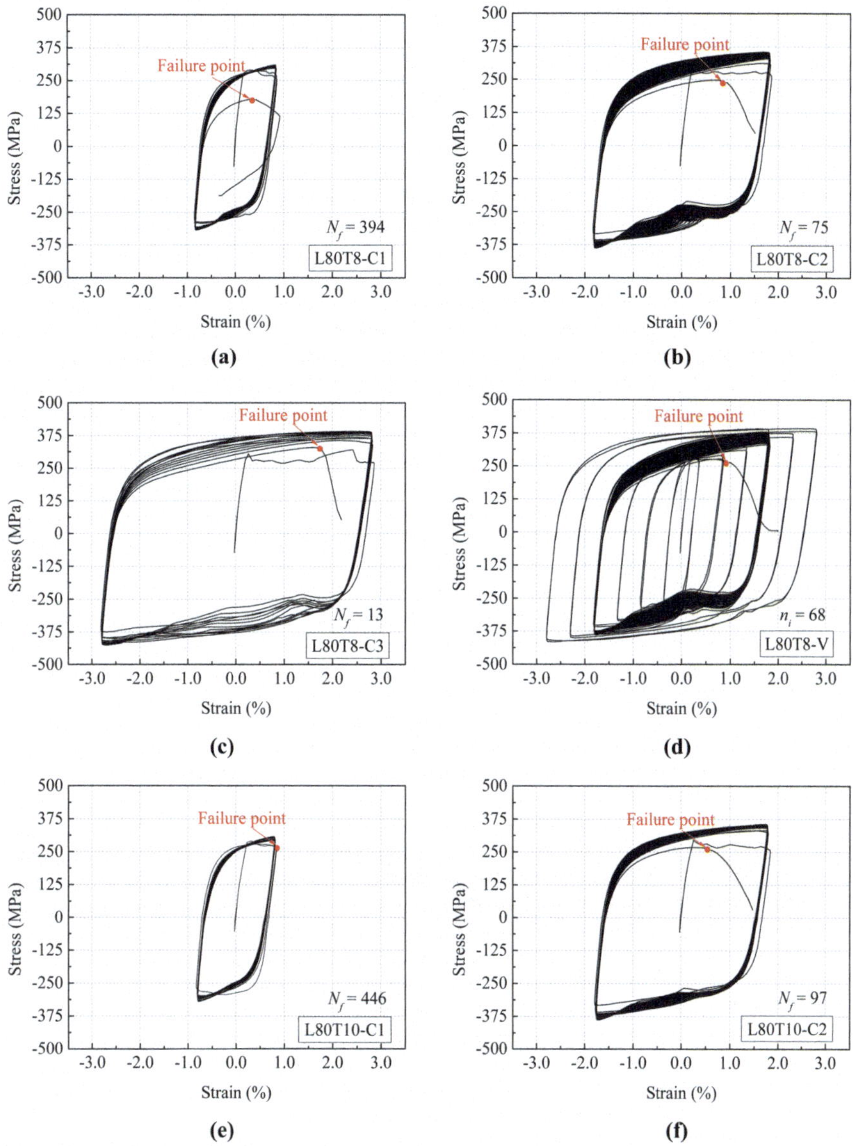

Fig. 3.17 Calibrated stress–strain hysteretic curves

of the constant strain amplitude for the PED specimens. A similar phenomenon was also observed in the PED specimens of L80T10-C1 with an N_f of 446, L80T10-C2 with an N_f of 97 and L80T10-C3 with an N_f of 20. Additionally, the PED specimens of L160T10-C1 with an N_f of 270, L160T10-C2 with an N_f of 48 and L160T10-C3 with an N_f of 10 showed the same variation of failure cycle number.

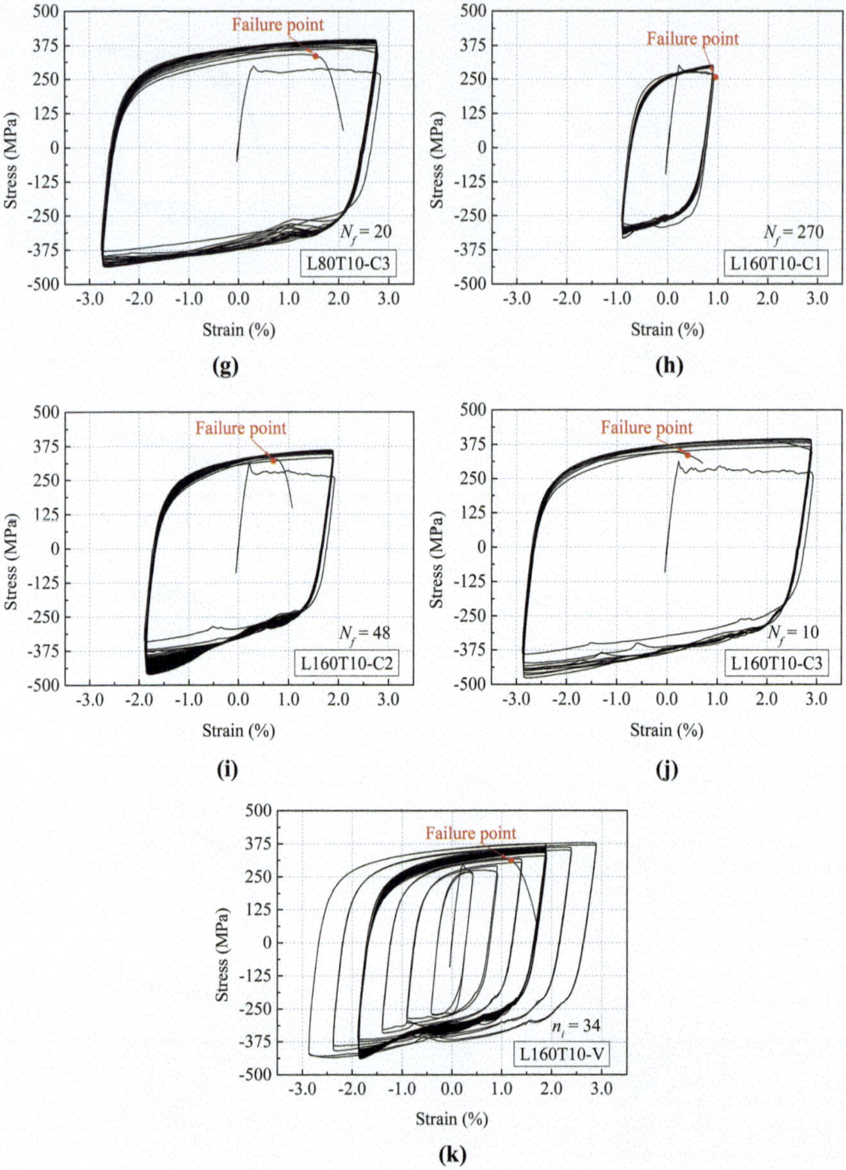

Fig. 3.17 (continued)

Table 3.9 Test results of all PED specimens

Specimen	$\Delta\varepsilon$ (%)	N_f	n_i	CPD	β_{max}	Loading protocol
L80T8-C1	1	394	–	10,547.1	1.04	CSA
L80T8-C2	2	75	–	4315.4	1.24	CSA
L80T8-C3	3	13	–	1148.0	1.19	CSA
L80T8-V	–	–	68	4510.8	1.18	V
L80T10-C1	1	446	–	11,939.1	1.07	CSA
L80T10-C2	2	97	–	5581.2	1.17	CSA
L80T10-C3	3	20	–	1766.2	1.18	CSA
L160T10-C1	1	270	–	7227.7	1.10	CSA
L160T10-C2	2	48	–	2761.8	1.30	CSA
L160T10-C3	3	10	–	883.1	1.21	CSA
L160T10-V	–	–	34	2554.5	1.29	V

Note N_f is the failure cycle number; n_i is the failure cycle number at the CSA stage in testing protocol V; CPD is the cumulative plastic deformation [6] $= \sum|\Delta_{pi}|/\Delta_y$, where $|\Delta_{pi}|$ is the plastic deformation in the ith cycle and Δ_y is the yielding deformation; β_{max} is the maximum compression-strength adjustment factor, β, of the PED specimen

To estimate the effect of the width of the air gap, d_g, provided between the corner of the yielding segment and the inner surface of the partially restraining tube, a comparison was conducted between PED specimens with 1.07-mm and 0.89-mm wide gaps. The failure cycle number of specimen L80T8-C1 with a 1.07-mm wide gap was 13.2% less than that of specimen L80T10-C1 with a 0.89-mm wide gap, both of which were loaded under a constant strain amplitude of 1%. Furthermore, the failure cycle number increased 29.3% at a constant strain amplitude of 2% and 53.8% at a constant strain amplitude of 3% when comparing specimens L80T8-C2 and L80T10-C2 and comparing specimens L80T8-C3 and L80T10-C3, respectively. It is concluded that the failure cycle numbers of the PED specimens were significantly affected by the width of the air gap, and a wider gap resulted in a lower failure cycle number. This is explained by the reduced lateral deformation of the yielding segment that could be observed in the narrower gap.

The length of the yielding segment was also considered, the effect of which was evaluated by comparing PED specimens with different lengths of the yielding segment. The failure cycle number of specimen L80T10-C1 with 80-mm yielding segments was nearly 1.65 times that of specimen L160T10-C1 with 160-mm yielding segments, both of which were loaded under a constant strain amplitude of 1%. The difference in failure cycle numbers was 2.02 times and 2 times at constant strain amplitudes of 2% and 3%, respectively, when comparing specimens L80T10-C2 and L160T10-C2 and comparing specimens L80T10-C3 and L160T10-C3. The discussion above demonstrated that a shorter yielding segment increased the failure cycle number of the PED specimens.

(7) Failure modes

The deformation and failure modes of the PED specimens are shown in Fig. 3.18, in which contact conditions, deformation patterns and failure features are demonstrated. The scratches of the red paint in the yielding segments showed that the proposed partially restraining mechanism worked effectively, and the yielding segments were partially restrained at the edges. No local failure of the transitional segment outside the partially restraining tube and torsion of the yielding segment occurred during the loading histories. In-plane contact at position $t_y/2$ between yielding segments and the partially restraining tube was also prevented.

As depicted in Fig. 3.18a–c, the deformation of the yielding segments increased with the constant strain amplitude. The same conclusion can be drawn from the comparisons between Fig. 3.18d–f. Based on a comparison of Fig. 3.18a, d, as well as of Fig. 3.18b, e and Fig. 3.18c, f, it is concluded that the lateral deformation of the yielding segments increased with the width of the air gap. When the width of the gap was controlled, the length of the yielding segment had a significant effect on the deformation pattern. For example, in specimens L80T10-C1 and L160T10-C1, more waves formed in the longer yielding segment.

(8) Mechanical properties

β values of tested PED specimens are shown in Fig. 3.19, and all β values were less than 1.3, satisfying the requirement in AISC 341-16 [6]. Similar to that of all-steel BEDs, the general trend of the β values' variation of PEDs was divided into the following stages:

1. Initial adjustment stage. In the first few cycles, the β values were not stable due to the cyclic hardening effect [7] and a relatively high upper yield force.

Fig. 3.18 Deformation and failure modes of PED specimens

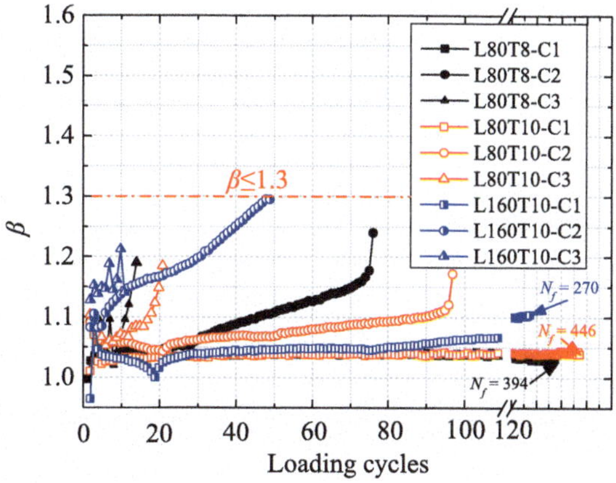

Fig. 3.19 β values for PED specimens

2. Stably increasing stage. When the strain amplitude was relatively low (e.g., 1%), the β values remained almost unchanged in specimens L80T8-C1 and L80T10-C1. The gentle contact of the two specimens shown in Fig. 3.18a, d accounted for this phenomenon. The β values increased quickly with the loading cycles when the strain amplitude grew up to 2 and 3%. A wider air gap and a longer yielding segment increased the β values.
3. Sudden change stage. A sharp increase of the β values in some specimens, such as L80T8-C2 and L80T10-C2, at the last loading step occurred because of a distinct decrease of the maximum tension force before failure.

The adjusted PED compression strength is expressed in Eq. (3.11) [6]:

$$F_{ca} = \beta\omega F_y \tag{3.11}$$

where ω is the strain hardening adjustment factor and F_y equal to $\sigma_y A_y$ is the axial yield strength of the PED specimen. ω is estimated by the ratio of σ_u to σ_y from coupon tests and is 1.55 for the Q235b material adopted in the present study.

(9) Buckling responses

By removing the partially restraining tube, the buckling responses could be inspected by distinguishing the polishing positions shown in Fig. 3.20. However, all PED specimens failed at the tension state, which indicated that the buckling wavelength was not reliable [17]. Numerical analyses were thus conducted to evaluate the yielding segment's buckling responses, including wave numbers, locations and wavelengths. Numerical results of specimens L80T10-C3 and L160T10-C2 were taken at the maximum compressive strain of the sixth cycle. As shown in Fig. 3.20, the positions of the test and numerical buckling wave peaks and valleys were compared for out-of-plane deformations. It was found that the established numerical model can effectively predict the wave numbers and the locations of wave peaks and valleys.

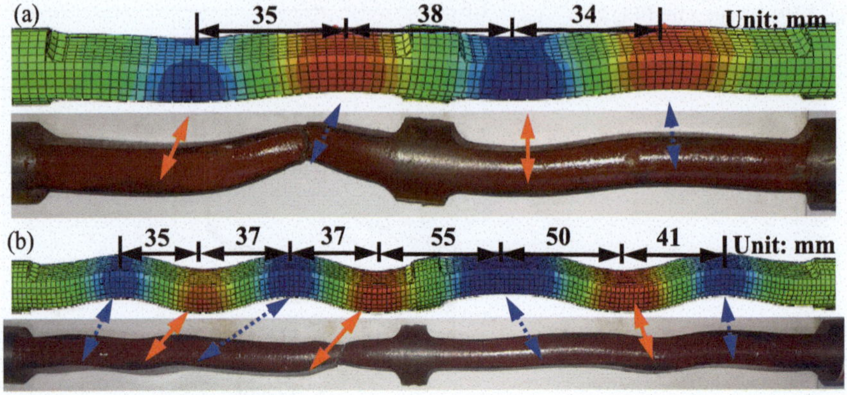

Fig. 3.20 Buckling responses at the maximum compressive strain of the 6th cycle for PED specimens: **a** L80T10-C3 and **b** L160T10-C2

The wavelengths of specimens L80T10-C3 and L160T10-C2 were 56.2 mm and 57.5 mm, respectively. The values of P for specimens L80T10-C3 and L160T10-C2 were obtained from the maximum compression force in the sixth cycle of the test results. The measured buckling half-wavelength from the numerical analyses is shown in Fig. 3.20. Unevenly distributed half-wavelengths could be observed due to the larger compressive force in the end [11], where the half-wavelengths at the two ends were less than that in the middle. The average wavelengths for specimens L80T10-C3 and L160T10-C2 were 71 mm and 85 mm, respectively, from numerical analyses. Compared with computed wavelengths, the wavelengths from numerical analyses were larger. This may be caused by the existence of the stopper. The rotation of the stopper was blocked by the partially restraining tube, which impeded the wave development and made the wavelength larger.

3.3 Post-tensioned Precast Concrete Connections with External All-Steel Dissipaters

3.3.1 Test Specimens

In this chapter, a total of 5 Precast Post-tensioned concrete connections with all-steel bamboo-shaped Energy Dissipaters, abbreviated as PPED, are tested under quasi-static cyclic loading. The effects of the initial prestressed force, the geometric parameters of the SBEDs and the number and configuration of the SBEDs on the seismic performance and self-centering capacity of the precast concrete connection were systematically evaluated. In addition, any servere damages on the frame column and beams are avoided by proper designs. The reliability of the proposed anchorage measures for the SBEDs was checked, and the end bending of the SBED resulting from the rotation of the connection was analyzed.

The reinforcement details and dimensions of the test specimens are shown in Fig. 3.21. The pinned connected support of the column corresponds to the frame member's reversal point. The cross section of the column was 400 mm × 400 mm, with a height of 1800 mm. The clear span of the beam was 1800 mm, while the width was 250 mm and depth was 400 mm. The average axial compressive strength of the concrete used for the precast connection was 31 MPa. As shown in Fig. 3.21b, c, twelve longitudinal rebars with different diameters were employed in the beam as well as twelve longitudinal rebars with a diameter of 16 mm in the column, as shown in Fig. 3.21d, e. Eight mm transverse rebars with varied spacing were arranged over the full length of both the beam and column, as shown in Fig. 3.21(a). All rebars, including longitudinal and transverse rebars, were embedded in the precast connection and were recognized as HRB 400 with a nominal yield stress of 400 MPa per GB50010-2010 [18]. No extra shear keys were employed in the construction of the proposed precast connection, satisfying ACI T1.2-03 [19]. The vertical shear at the interface between the beam and column for gravity and loadings can only be transferred by the friction induced by the post-tensioning force. To accomplish this, four prestressed tendons with a diameter of 15.2 mm were concentrically post-tensioned in the 55 mm pre-laid ducts. The ultimate tensile strength of a prestressed tendon was 1860 MPa.

As shown in Fig. 3.21a, four 10 mm thick steel plates were located at the top and bottom ends of the column and at the top and bottom surfaces of the right end of the beam to sustain the loading during tests. In addition, two 10 mm thick steel angles were placed at the upper and lower left corners of the beam to protect the beam corner from damage. Compared to the steel jackets adopted by Song et al. [20] at the corner of the beam, the use of the steel angle had advantages, including the need for less material and faster and cheaper construction. A 15 mm grout layer was filled between the beam-column interface to guarantee good contact between the beam and column.

As shown in Fig. 3.22a, four SBEDs were installed in the precast connection as a source of energy dissipation. One end of the SBED was bolted through a connection plate embedded in the beam using four M19 screw nuts, while the other end was screwed into the embedded sleeve in the column, as shown in Fig. 3.22a. The SBED was installed in the horizontal direction with the centerline 40 mm away from the beam surface. The locations of the embedded sleeves and connection plates are demonstrated in Fig. 3.22b, c. As shown in Fig. 3.22d, two SBEDs with different geometric parameters were employed in this paper, labelled as BED-1 and BED-2. BED-1 had six 60 mm segments and five 20 mm slubs, while BED-2 had four 60 mm segments and three 20 mm slubs.

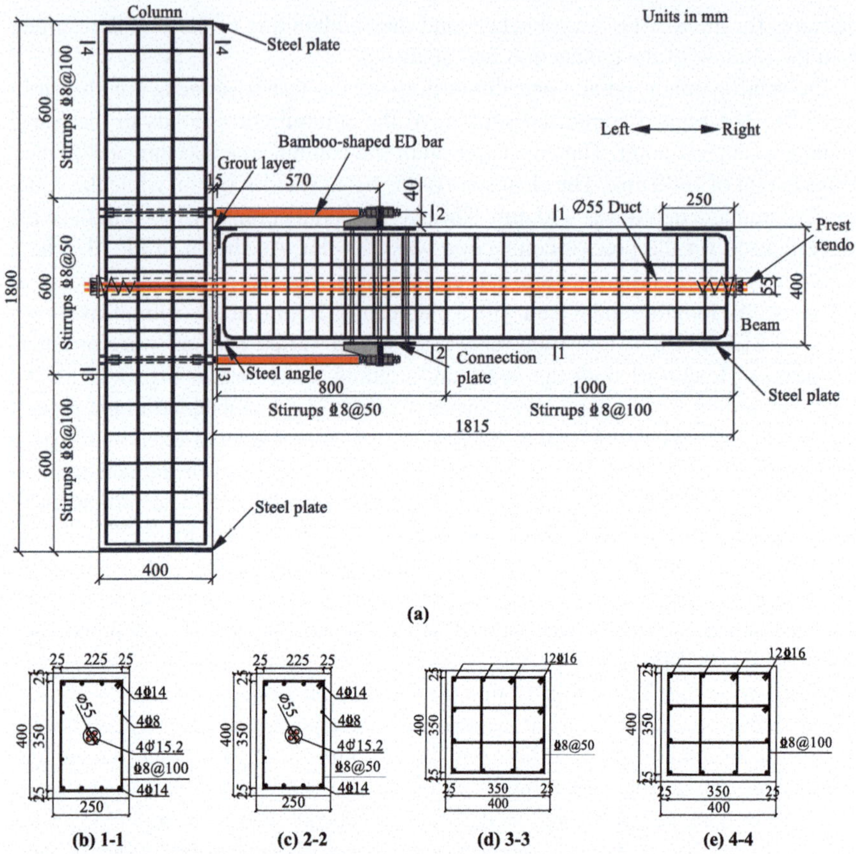

Fig. 3.21 Reinforcement details and dimensions (units in mm)

3.3.2 *Test Setup and Loading Protocols*

A series of tests on the precast post-tensioned beam-column connections with a SBED was conducted in the test setup as illustrated in Fig. 3.23. Two different test procedures were adopted in the current experimental protocol, named standard loading and repetition loading. Standard loading consisted of three symmetric load cycles of a load amplitude of 20 kN at the pregap opening to check the workability of the loading system and the proper assembly of the specimen. Displacement amplitudes representing drift ratios of 0.25, 0.35, 0.5, 0.7, 1.0, 1.5, 2.0, 2.5 and 3.5% were then applied to the precast connection for three cycles of each design drift ratio. Repetition loading was conducted based on standard loading, such that the starting point of repetition loading was the residual drift ratio produced by standard loading. The displacement amplitudes adopted in repetition loading were the same as those used in standard loading (Fig. 3.24).

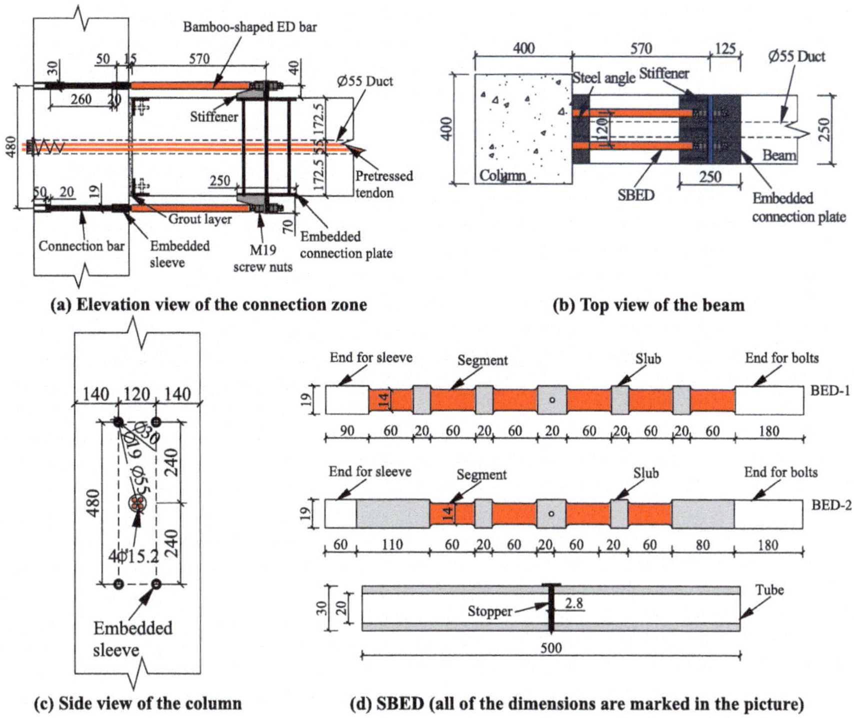

Fig. 3.22 Configuration of the SBED (units in mm)

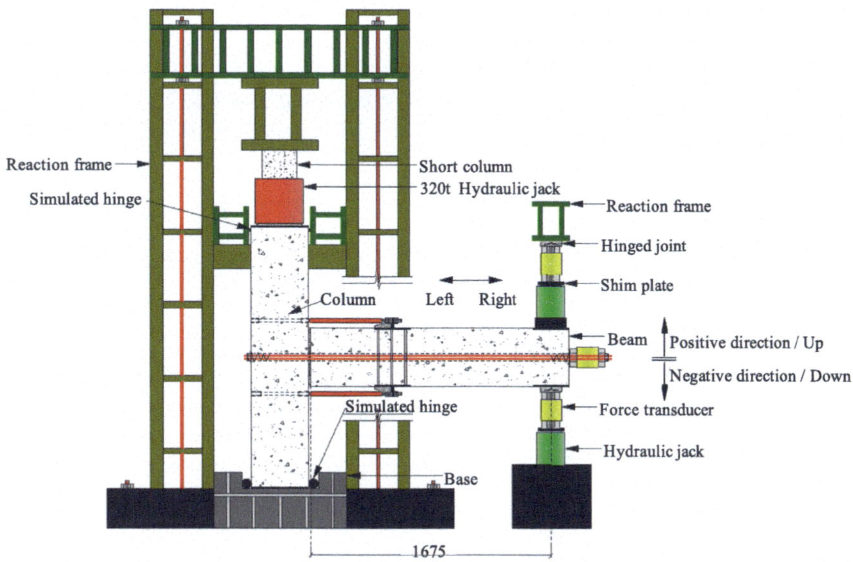

Fig. 3.23 Test setup

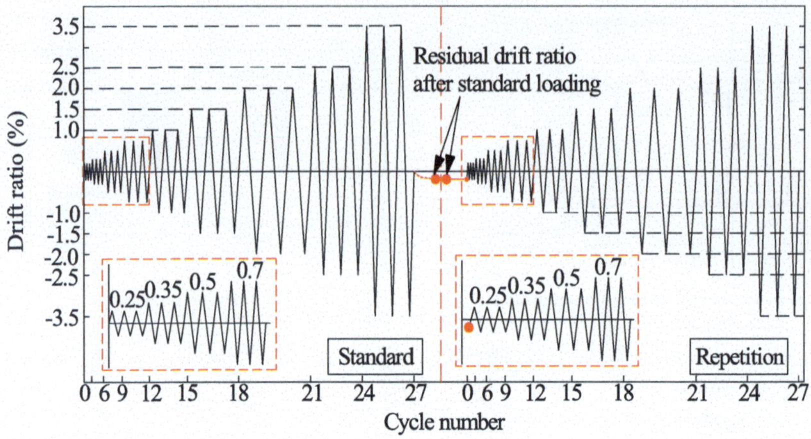

Fig. 3.24 Loading protocol

3.3.3 Test Program

The test matrix is identified in Table 3.10. Five tests were conducted on the proposed precast connection to evaluate the influence of different experimental parameters on the seismic behavior of the connection. Various parameters discussed herein include: the different loading protocols, initial prestressed force, total length of the segments in a SBED and configuration of the SBEDs. The specimen PPED was set as the base case and tested under the standard loading protocol. The total length of the segments in a SBED, L_{se}, was 360 mm, and a design level of the prestress, R_0^D, of 0.35 was applied, which was defined as the design's total initial prestressed force, T_0^D, normalized by the total ultimate force of the prestressed tendons, T_u. The variable (i.e., the only difference against the base case) is indicated by the bold in Table 3.10.

3.3.4 Experimental Observations and Results

(1) Observed deformation and damage

The cracking patterns for specimen PPED during the loading histories at various drift ratios are shown in Fig. 3.25, in which cracks are traced by red lines. In general, the initial cracks formed at the grout-beam interface, and then, the major cracks propagated along the interface. Small flexural cracks concentrated around the embedded connection plate and were almost perpendicular to the axis of the structural component. No visible damage was observed in other parts of the beam and column. The SBED underwent a significant plastic deformation, as shown in Fig. 3.26. The energy dissipated by the concrete is demonstrated via cracking at the material level [21], and therefore, the limited small flexural cracks observed in the precast concrete beam and column proved that only limited energy was dissipated by the precast concrete

Table 3.10 Test matrix of the precast connections

Specimen	Loading protocol	R_0^D	T_0^D (kN)	T_0^E (kN)	N_{ED}		L_{se} (mm)	Comment
					Top	Bottom		
PPED	Standard	0.35	365	367	2	2	360	Base case
PPED-R	**Repetition**	—	—	—	2	2	360	Double loadings
PPED-P	Standard	**0.45**	469	457	2	2	360	Increased design prestress
PPED-L	Standard	0.35	365	371	2	2	**240**	Varied segment total length
PPED-S	Standard	0.35	365	368	**0**	**2**	360	Varied number of SBED

Note T_0^E is the measured total initial prestressed force; N_{ED} is the number of the SBED installed at the top or the bottom surface of the beam

components and that the energy dissipation was mainly concentrated in the SBEDs. No relative perpendicular sliding between the beam and column was found during the loading histories, and the shear capacity provided by the friction forces from the prestressed force at the interface was sufficient for the precast connection.

For the first cycle, with a drift ratio of 0.25%, the initiation of a gap opening on the bottom side of the grout-beam interface was marked. Both initial cracks formed on

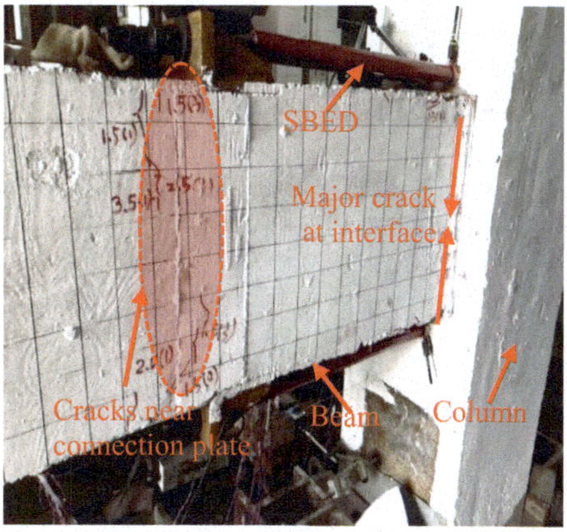

Fig. 3.25 Cracking patterns

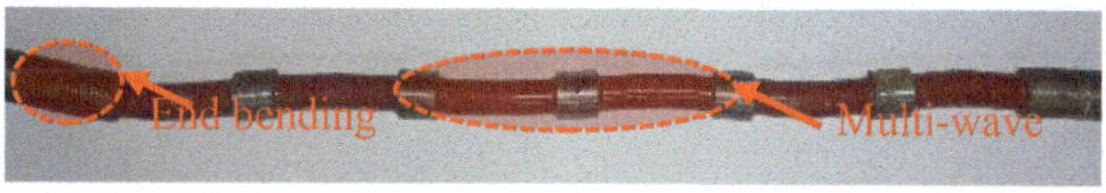

Fig. 3.26 Typical deformation pattern of the SBED

the upper and lower sides of the interface after the completion of the 0.25% drift ratio. Cracks then propagated to the middle height of the beam until the second cycle of a drift ratio of 0.5%. A major crack passed through the grout-beam interface at the first cycle of a drift ratio of 1.0%. The width of the major crack increased with the drift ratios. Crushing near the embedded steel angle was first observed at the first cycle of a 1.5% drift ratio and flexural cracks below the upper connection plate formed at the second cycle of the same drift ratio. Through the third cycle of a 2.5% drift ratio, relatively severe spalling occurred near the embedded steel angle and the connection plate at the bottom surface of the beam. When the drift ratio reached 3.5%, relatively severe crushing formed near the embedded steel angle and the connection plate at the top surface. The concrete cover near the embedded sleeve was slightly pulled out, as shown in Fig. 3.27, but the embedded sleeve still worked normally. All of the flexural cracks in the beam closed after evacuation of the load.

No failure was found in the SBEDs at the end of the test, even in specimen PPED-R, which showed that the employed SBED had a reliable low-cycle fatigue behavior and worked effectively under repeated loadings. As shown in Fig. 3.28, bending deformation was observed at the end of the SBED near the connection plate. Considering that the SBED could be assumed to be fixed to the precast connection, a rotational tendency of the SBED existed with the precast beam moving upward and downward. As shown in Fig. 3.29a, a two-point contact state was observed in the point contact area near the end of the SBED due to the connection rotation, and a similar contact state was discussed in a previous study. Outside the point contact area, the end bending of the SBED was caused by the coexistence of the axial force, shear force and end moment at the bolt connection. Furthermore, the actual displacement at the end of the SBED near the embedded connection plate could be decomposed into the horizontal and vertical directions, as depicted in Fig. 3.29b.

(2) Load versus displacement/drift ratio response

The force versus displacement, or drift ratio hysteresis loops, of the test specimens subjected to cyclic loading is illustrated in Fig. 3.30, as well as the envelop curves obtained from the forces and displacements or drift ratios at the positive (up) and negative (down) peak amplitudes of each cycle. Both the forces and displacements

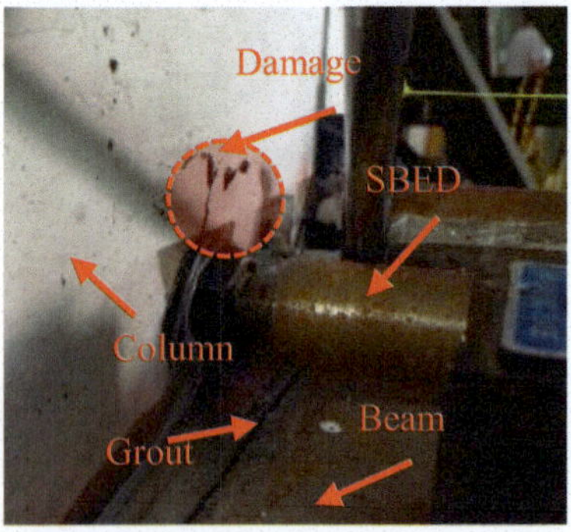

Fig. 3.27 Concrete cover damage

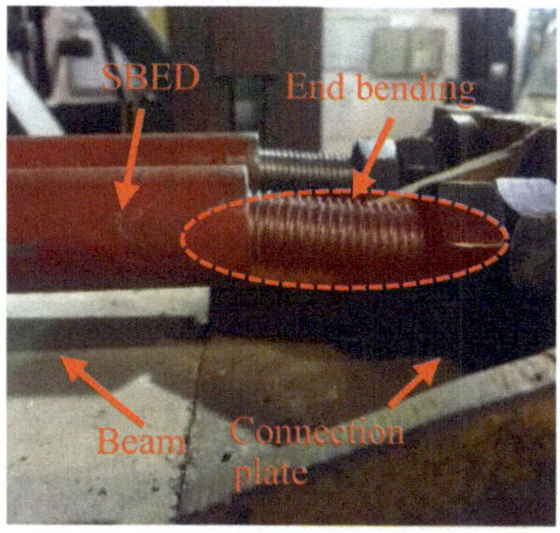

Fig. 3.28 End bending of the SBED

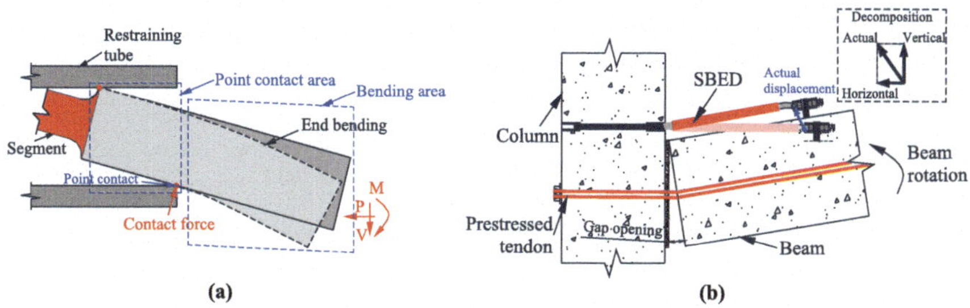

Fig. 3.29 **a** Local analysis and **b** displacement decomposition for end bending of the SBEDs

were measured at the loading points on the beam. The experimental results were adopted to investigate the strength capacity, equivalent viscous damping and self-centering capabilities of the precast connection.

According to Fig. 3.30, no strength degradation was observed, even at a maximum drift ratio of 3.5%. In specimen PPED, marked as the base case (refer to Fig. 3.30a), the average maximum tensile and compressive forces, defined as the average value of the maximum tensile or compressive forces obtained from three repeated cycles, were 86.0 kN and 79.4 kN, respectively, at a drift ratio of 3.5%. Compared to specimen PPED, specimen PPED-R, tested under a repetition loading mode (refer to Fig. 3.30b), had almost the same average maximum tensile and compressive forces at the same drift ratio of 3.5%, which demonstrated that the precast connection could sustain double loadings without any loss of strength. As shown in Fig. 3.30c, the average maximum tensile and compressive forces of specimen PPED-P, with an increased design prestress level, improved by 21.0% and 20.4%, respectively, compared to the values from specimen PPED. It was found that an increase in the initial prestress force significantly improved the strength.

For specimen PPED-L (see Fig. 3.30d), the average maximum tensile and compressive forces increased by 14.7% and 8.2%, respectively, compared to the values of specimen PPED. This phenomenon can be explained as follows: a larger average strain of the SBED was obtained in specimen PPED-L with a smaller total

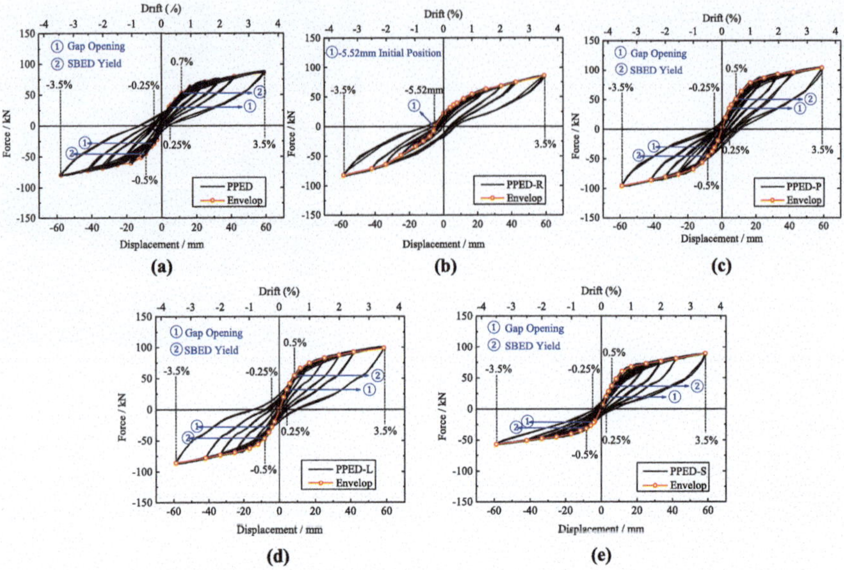

Fig. 3.30 Load–displacement/drift ratio curves

length of the segments compared to that in specimen PPED under the same drift ratio, which resulted in a larger axial force on the SBED in specimen PPED-L. Based on force equilibrium, larger tensile and compressive forces were thus observed in specimen PPED-L. An asymmetric hysteretic behavior was observed in specimen PPED-S, as depicted in Fig. 3.30e, and the average maximum tensile force was 1.6 times larger than the average maximum compressive force. Compared with specimen PPED, specimen PPED-S had a similar average maximum tensile force, but had a relatively lower average maximum compressive force. The SBEDs installed at the bottom surface of the beam in specimen PPED-S participated in the load transfer when the beam moved upward in the positive direction, but contributed little load capacity for the precast connection in the negative loading direction. This characteristic accounted for the larger tensile force compared to the compressive force in specimen PPED-S and the similar tensile force, but lower compressive force, in specimen PPED-S compared with specimen PPED.

(3) Prestressed force-drift ratio performance

The curves of the prestressed force versus the drift ratio for specimens PPED and PPED-S are depicted in Fig. 3.31. It was found that the prestressed force remained approximately constant before the gap opening, corresponding to a drift ratio below 0.25%. The small increase in the prestressed force before gap opening was due to the elastic deformation of the beam. Then, the prestressed force increased almost linearly with the drift ratios after gap opening, which was attributed to the elongation of the prestressed tendons with an increased gap. The maximum increase of the prestressed force was approximately 37.6–48.5% of the measured total initial prestressed force

T_0^E. Slim hysteretic loops were observed, demonstrating different prestressed forces during the loading and unloading phases. The reasons for this phenomenon can be because different forces during loading and unloading were obtained in the SBED at the same drift ratio.

The results of the analysis of the prestressed forces in specimens PPED, PPED-R, PPED-L and PPED-S is presented in Table 3.11. The maximum prestressed force, $T_{\max}$, was observed at approximately half of the total ultimate prestressed force T_u, which showed that the prestressed tendons stayed in the elastic state during throughout the loading history. In specimens PPED, PPED-L and PPED-S, the loss of the prestressed forces during the tests ranged from 5.2% to 8.2% of T_0^E, which was caused by the tendon and anchorage seating. In addition, in specimen PPED-R, which was loaded without retensioning of the prestressed tendons, further loss of the prestressed force was limited. The prestressed forces decreased by approximately 0.8% in the precast concrete connections loaded with the standard loading pattern.

(4) Self-centering capability

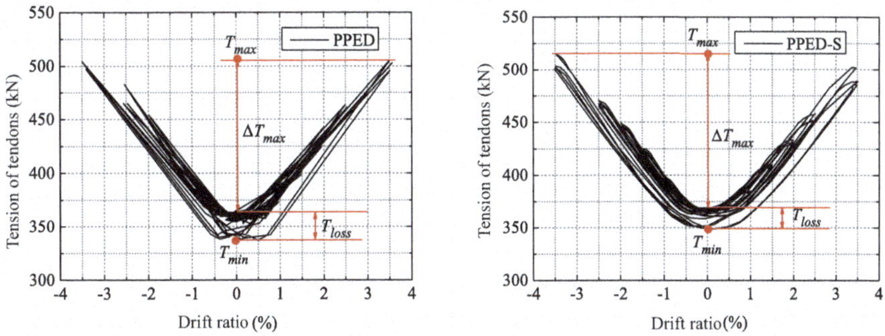

Fig. 3.31 Tension of tendons-drift ratio response

Table 3.11 Analysis of the prestressed force

Force (kN)	Specimen			
	PPED	PPED-R	PPED-L	PPED-S
T_0^E	367 (35%)	357 (34%)	371 (36%)	368 (35%)
$T_{\min}$	337 (32.4%)	349 (33.5%)	350 (33.6%)	349 (33.5%)
T_{loss}	30 (2.9%)	8 (0.8%)	21 (2.0%)	19 (1.8%)
$T_{\max}$	506 (48.5%)	530 (50.9%)	510 (49.0%)	516 (49.5%)
$T_{\max}$	506	530	510	516

Note $T_{\min}$ is the minimum prestressed force; T_{loss} is the loss of the prestressed force calculated as the difference between T_0^E and $T_{\min}$. The value in brackets is the ratio of T_{loss} to T_0^E. The prestressed forces in specimen PPED-P were unavailable due to the accidental failure of the 1000 kN force transducer during the test

The self-centering capability of the tested precast concrete connections was quantified by the relative self-centering efficiency (RSE) [22] defined as Eq. (3.12):

$$RSE = 1 - \frac{u_{res}^{+} - u_{res}^{-}}{u_{\max}^{+} - u_{\max}^{-}} \tag{3.12}$$

where u_{res}^{+} and u_{res}^{-} are the positive and the negative residual displacements, respectively. The residual displacement is defined as the displacement of the loading point at the zero force per loading cycle. An RSE value corresponding to a perfectly self-centering structure with zero residual deformation is equal to one, while an RSE value of zero corresponds to a system without any re-centering capacity.

The relationship between the relative self-centering efficiency versus the drift ratio for all of the tested precast connections is provided in Fig. 3.32. In all test specimens, the RSE decreased with the amplitudes of the drift ratios for the drift ratios exceeding 1%. This phenomenon can be attributed to the increasing plastic deformation in the SBEDs with the drift ratios, the prestress loss during loading histories and the slight indentation of the grout layer. Specimen PPED-R, reloaded without any repair, had slightly smaller RSE values than the base case specimen PPED, which stated that the proposed precast connection lost little self-centering efficiency even if it experienced double loadings without any repair. Compared to the base case, a larger RSE value was observed in specimen PPED-P at the end of the test, and specimen PPED-S, which was designed with only two SBEDs, had larger RSE values for drift ratios above 1%. It was determined that a fewer number of SBEDs and an increasing prestressed force had a positive influence on the self-centering capacity of the precast connection. A larger restoring force, caused by a larger prestress level and a smaller resistance force resulting from a fewer number of SBEDs, led to a better self-centering capacity. Specimen PPED-L, with a smaller L_{se}, experienced smaller RSE values than specimen PPED for drift ratios exceeding 1% due to the larger unrecoverable plastic deformation that existed in specimen PPED-L. For all specimens, the RSE values were above 75% and a good self-centering capability was achieved.

In addition to the RSE values of the precast connections discussed above, the average residual displacements (the mean absolute value of the difference between u_{res}^{+} and u_{res}^{-}) of the precast connections at the end of tests were analyzed. The average residual displacements for specimens PPED, PPED-R, PPED-P, PPED-L and PPED-S were 8.00 mm, 8.65 mm, 7.01 mm, 12.04 mm and 5.16 mm, respectively, and the corresponding average residual drift ratios were 0.48%, 0.52%, 0.42%, 0.72% and 0.31%, respectively.

(5) Strain analysis

To determine the state of the precast concrete beam and column when loaded and to evaluate the strain distributions in the steel rebars. The maximum strains of BT2, BT5 and C6 for each drift ratio in the PPED are shown in Fig. 3.33. All of the strains in the rebars were less than the nominal yield strain of the rebar, which was assumed

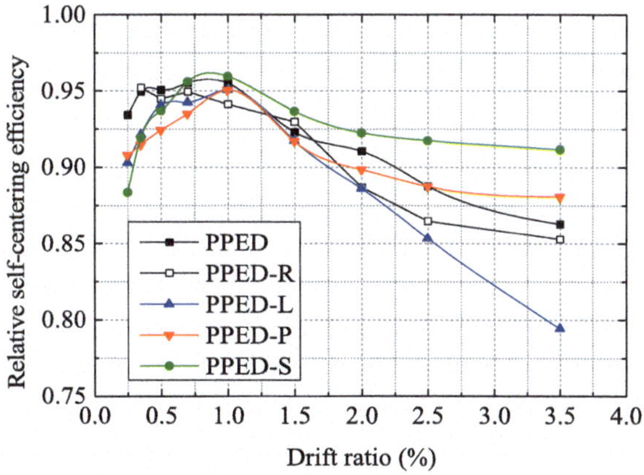

Fig. 3.32 Relative self-centering efficiency of precast concrete connections

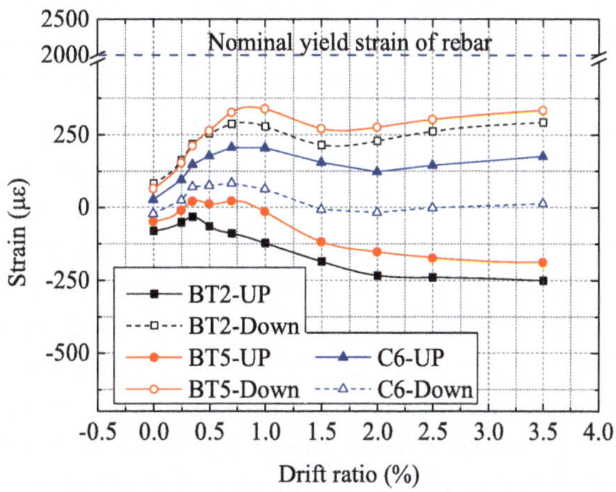

Fig. 3.33 Strains of steel rebars versus drift ratios

to be 2000 $\mu\varepsilon$, which meant that most of the parts of the beam and column stayed in the elastic state.

3.4 Conclusions

This chapter systematically introduces the latest research progress of non-emulative prestressed prefabricated concrete frame structure system in terms of the proposed external replaceable fuse-type dissipaters and prestressed precast concrete connections. In the research, analytical, numerical and experimental methods are adopted. The main highlights are listed as follows.

(a) The new all-steel bamboo-shaped dissipaters consist of elastic slubs and yielding portions which are short enough to avoid plastic buckling. Additionally, drawbacks induced by the grouting process are eliminated, such as the high weight and difficulty in controlling the grouting quality. The new dissipating device can thus be applied as replaceable structural fuses in precast concrete frames.
(b) The new partially restrained dissipater is characterized by no welding, no grouting, high utilization ratio of material and ease of construction, the efficiency of which is validated by analytical, numerical and experimental methods. Furthermore, the low-cycle fatigue life decreases with increasing the gap width and the length of the yielding portion.
(c) A prestressed precast concrete connection with external dissipaters is proposed, where the dissipaters provide energy absorption and the self-centering capacity is achieved by the post-tensioned tendons. The structural damage is expected to be concentrated on the dissipaters under the earthquake. And the elastic post-tensioned tendons provide post-event resilience. What is more, the cyclic loading tests demonstrate that the precast concrete connection sustains a similar seismic performance subjected to repeated loading.

References

1. Wang C-L, Liu Y, Zhou L, Zhou G, Lu Z (2018) Concept and performance testing of an aluminum alloy bamboo-shaped energy dissipater. Struct Design Tall Spec Build 27(4):e1444
2. Liu Y, Wang C-L, Wu J (2018) Development of a new partially restrained energy dissipater: experimental and numerical analyses. J Constr Steel Res 147:367–379
3. Wang C-L, Liu Y, Zhou L (2018) Experimental and numerical studies on hysteretic behavior of all-steel bamboo-shaped energy dissipaters. Eng Struct 165:38–49
4. Wang C-L, Liu Y, Zheng X, Wu J (2019) Experimental investigation of a precast concrete connection with all-steel bamboo-shaped energy dissipaters. Eng Struct 178:298–308
5. Usami T, Wang C-L, Funayama J (2012) Developing high-performance aluminum alloy buckling-restrained braces based on series of low-cycle fatigue tests. Earthq Eng Struct Dynam 41(4):643–661
6. American Institute of Steel Construction (2016) ANSI/AISC 341-16 seismic provisions for structural steel buildings. Chicago, IL
7. Wang C-L, Usami T, Funayama J (2012) Improving low-cycle fatigue performance of high-performance buckling-restrained braces by toe-finished method. J Earthquake Eng 16(8):1248–1268
8. Black CJ, Makris N, Aiken ID (2004) Component testing, seismic evaluation and characterization of buckling-restrained braces. J Struct Eng 130(6):880–894
9. Wang C-L, Usami T, Funayama J, Imase F (2013) Low-cycle fatigue testing of extruded aluminium alloy buckling-restrained braces. Eng Struct 46:294–301
10. Manual AS (2005) Seismic provisions for structural steel buildings. In: ANSI/AISC, pp 341-05
11. Chen Q, Wang C-L, Meng S, Zeng B (2016) Effect of the unbonding materials on the mechanic behavior of all-steel buckling-restrained braces. Eng Struct 111:478–493
12. Wang C-L, Chen Q, Zeng B, Meng S (2017) A novel brace with partial buckling restraint: An experimental and numerical investigation. Eng Struct 150:190–202
13. China Planning Press B (2003) Code for design of steel structure. (in Chinese)

14. Chen J, Chen H-J (2011) Stability of steel structures theory and design, 5th ed. Science Press, Beijing
15. Hoveidae N, Rafezy B (2012) Overall buckling behavior of all-steel buckling restrained braces. J Constr Steel Res 79:151–158
16. Sun X, Fang X, Guan L (2009) Mechanics of Materials (in Chinese). Higher education press, Beijing
17. Dehghani M, Tremblay R (2017) Design and full-scale experimental evaluation of a seismically endurant steel buckling-restrained brace system. Earthq Eng Struct Dyn
18. National Standard of the People's Republic of China GB 50010-2010 (2010) Code for design of concrete structures. Ministry of Housing and Urban-Rural Development of the People's Republic of China. (in Chinese).
19. ACI T12-03 special hybrid moment frames composed of discretely jointed precast and post-tensioned concrete members. ACI innovation task group 1 and collaborators, 2003
20. Song LL, Guo T, Chen C (2014) Experimental and numerical study of a self-centering prestressed concrete moment resisting frame connection with bolted web friction devices. Earthquake Eng Struct Dynam 43(4):529–545
21. Parastesh H, Hajirasouliha I, Ramezani R (2014) A new ductile moment-resisting connection for precast concrete frames in seismic regions: an experimental investigation. Eng Struct 70:144–157
22. Sideris P, Aref AJ, Filiatrault A (2014) Quasi-static cyclic testing of a large-scale hybrid sliding-rocking segmental column with slip-dominant joints. J Bridg Eng 19(10):04014036

Chapter 4
Friction Damped Self-Centering Precast Concrete Frame

Abstract In this chapter, the seismic performance of self-centering precast concrete structures is discussed at the node and structure levels. A novel energy dissipator employing non-asbestos organic material which provides stable energy dissipation capacity has recently been developed for controlled ductile structures, and the effects of different loading history, loading rating and bolt torque on the energy dissipator are analyzed from quasi-static axial tests. The seismic response evaluation of self-centering prestressed concrete frames with infill walls (SCPC-IW frames) is presented, while three types of infill walls with different load capacity are selected to be installed in a bare self-centering prestressed concrete frame (SCPC Bare frame). Beam web friction devices are included in the beam-column connections to provide energy dissipation capacity. Numerical models of the SCPC frame, and the three SCPC-IW frames are established and analyzed in a finite element software, OpenSees. The nonlinear static analyses and nonlinear dynamic analyses under 44 ground motions are performed. The incremental dynamic analyses (IDAs), seismic fragility analyses, and collapse resistance capacity evaluation have also been conducted. The results show that the SCPC system has superior seismic resilience.

Keywords Self-centering precast concrete structure · Dissipator employing non-asbestos organic material · Friction devices · Quasi-static axial test · Prestressed frames with infill walls · Incremental dynamic analyses

4.1 Introduction

Friction damped self-centering precast concrete frames is a fully assembled structure in which precast beams and columns are assembled through post-tensioned prestressed tendons, and friction devices are included in the beam-column connections to provide energy dissipation capacity. This structural form not only maintains the advantages of fast production speed, stable quality, and saving formwork in the traditional assembled reinforced concrete frame, but also greatly improves the strength and ductility of the assembling joints through the stable performance of the load bearing-energy dissipation dual-function friction energy dissipaters. In addition, post-tensioned prestressing tendons can also provide stable elastic restoring force for

G. Wu et al., *Novel Precast Concrete Structure Systems*, Springer Tracts in Civil Engineering, https://doi.org/10.1007/978-981-19-6821-1_4

the structure, so that the Friction damped self-centering precast concrete frame with dry connection joints has excellent assembly construction performance, structural seismic performance and functional recoverability. This chapter will analyze the five aspects of fabricated structure construction mechanism, joints damper test, analytical models, seismic performance and vulnerability analysis, and provide theoretical basis and technical support for the promotion and application of friction damped self-centering precast concrete frame system.

4.2 Structure and Mechanism of Friction Damped Self-Centering Precast Concrete Frames

According to the position of the friction energy dissipater, the Friction damped frames can be divided into Top and Bottom Friction-Damped (TBFD) fabricated frame and Web Friction-Damped (WFD) fabricated frame. The principle is that beams and columns are assembled by post-tensioning unbonded prestressing tendons, and friction energy dissipaters are added at the beam-column interface to improve the energy dissipation capacity of the structure.

4.2.1 Top and Bottom Friction-Damped Fabricated Frame

Figure 4.1a shows the structure of a Top and Bottom Friction-Damped (TBFD) fabricated frame. Among them, beams, columns, friction energy dissipaters, etc. can be prefabricated in the factory. As shown in Fig. 4.1b, during the prefabrication process, the corresponding prestressed duct and bolt holes (divided into: friction bolts and fixing bolts) are reserved at the corresponding positions of the beams, columns and the energy dissipaters. At the same time, in the beam prefabrication process, the internal friction device (mainly including the inner friction steel plate and the end plate that fix the inner friction steel plate as a whole) is pre-embedded at the top and bottom positions of the beam end. In the field assembly process, first, the exterior friction device (including the exterior friction steel plate and the exterior friction fixing plate) is fixed on the column by fixing bolts, so that the two become a whole. Then, the prestressed steel strands are passed through the holes reserved in the beams and columns and tensioned. The tensioned prestressed steel strand is beneficial to the assembly of beams and columns in the construction stage, and can bear the bending moment of the beam end in the use stage. After an earthquake, the structure can return to the original position under the self-centering forces of the prestress, reducing or eliminating the residual deformation of the structure. After the earthquake, the exterior friction device can be easily replaced by loosening the fixing bolts and then disassembling the exterior friction device. In the joint structure, the

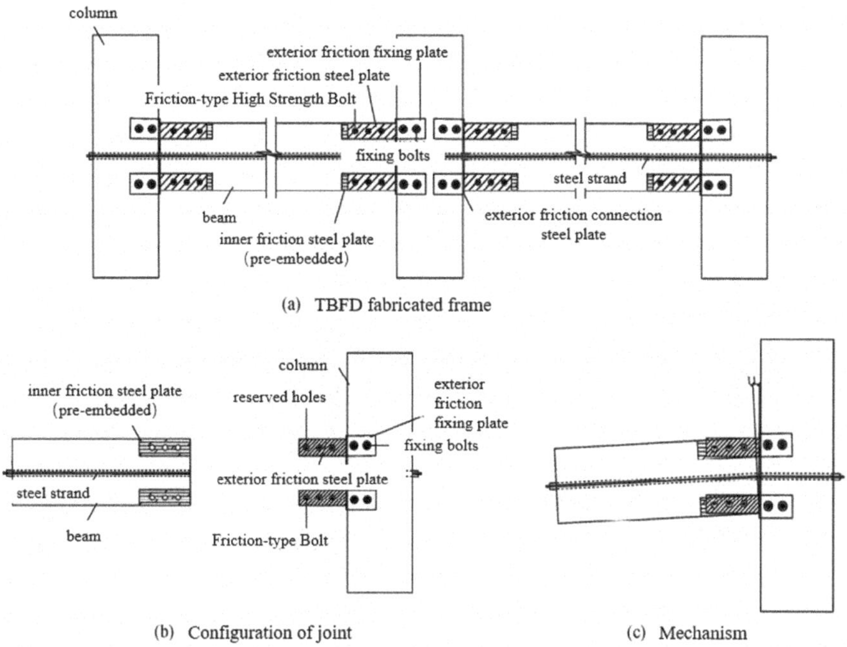

Fig. 4.1 Proposed FD-SCPC beam-column connections

external friction device and the column are connected by bolts for easy disassembly, which can facilitate the replacement of the friction plate after the earthquake.

The energy dissipation mechanism of the joint is shown in Fig. 4.1c. The external friction device and the column are a whole, and the internal friction device and the beam are a whole. The relative rotation of the beams and columns under the action of external loads such as earthquakes causes the internal and external friction steel plates to move mutually to realize energy dissipation of the joints. The friction material is placed between the internal friction device and the external friction device to obtain stable energy dissipation. In order to ensure that the friction plate is integrated with the external friction device during the rotation of the joint, a groove with the same size as the friction plate is opened on the inner surface of the external friction device to fix the friction plate. The detailed structure of the energy dissipater is very important to the energy dissipation effect of the joint. In order to ensure that the beam and column can produce relatively large relative rotation, the diameter of the reserved holes for the split friction bolt at the beam end must be significantly larger than the diameter of the friction bolt. With the gradual increase of the beam-column gap θ, the internal force in the steel strand gradually increases, providing a gradually increasing restoring force for the joint. The restoring force restores the gap formed by the structure to the initial position after the earthquake, reduces the residual displacement of the structure, and realizes the repair ability of the structure. The key to the resetting of the structure is that the moment of the initial prestress to the rotation point is not less than the moment of the friction to the rotation point.

In addition, in order to improve the vertical shear resistance of the structure, the "friction energy dissipation", "prestressed assembling" and " Hidden Corbel shear resistance" can be closely integrated to form a top and bottom friction damped assembled frame with hidden corbels (Fig. 4.2a), its joint structure is shown in Fig. 4.2b. The introduction of hidden corbels provides sufficient vertical shear resistance for the joints, and the precast beams can be placed during the construction stage to Improve the assembly efficiency of the structure; at the same time, the friction energy dissipator is reliably connected to the beam and column at the top and bottom positions of the side of the beam by means of embedded parts or bolt connection without affecting the layout of the floor, and the efficiency of frictional energy dissipation can be maximized. Figure 4.2c shows the joint rotation deformation diagram of the top and bottom friction-damped (TBFD) fabricated frame with hidden corbels. When the beam-column joints rotate relative to each other, the extrusion at the rotation point occurs between the end plates of the inner friction steel plate and the exterior friction steel plate of the damper, the local extrusion of the concrete is avoided, and the restorability of the structure is further improved. When the joints are assembled, the external friction steel plate is fixed to the column by the bearing-type high strength bolts as a whole, and the internal friction steel plate is fixed to the beam by the high strength bolts with bearing type as a whole, the part where the exterior friction steel plate and the inner friction steel plate overlap are embedded in a friction plate (such as a brass plate, etc.), and the inner friction plate is fixed in a groove reserved for the exterior friction steel plate.

When the structure deforms under the action of an earthquake (Fig. 4.2c), the beam-column connection produces a relative displacement, which drives the internal friction device and the external friction device to misalign each other, and the friction energy dissipater starts to dissipate energy. At the same time, the mutual misalignment between the beams and columns also causes the prestressed tendons (steel strands) to elongate and generate restoring force. Compared with a joint without a corbel, the lower rotation point of the joint is at the contact part of the corbel and the beam, and the upper rotation point is at the contact part of the beam-column. This will result in a slight difference in the distance between the node's up and down rotation point and the energy dissipator, and ultimately lead to a certain asymmetry in the hysteretic behavior of the joint.

4.2.2 *Web Friction Damped Precast Frames*

Figure 4.3 shows the web friction damped precast frame and its working mechanism. Among them, the frame beams and columns are prefabricated in the factory. After hoisting in place on site, the prestressed steel strands are passed through the holes reserved in the beams and columns, and then the prestressed steel strands are tensioned. The post-tensioned unbonded prestressing steel strand is not only an

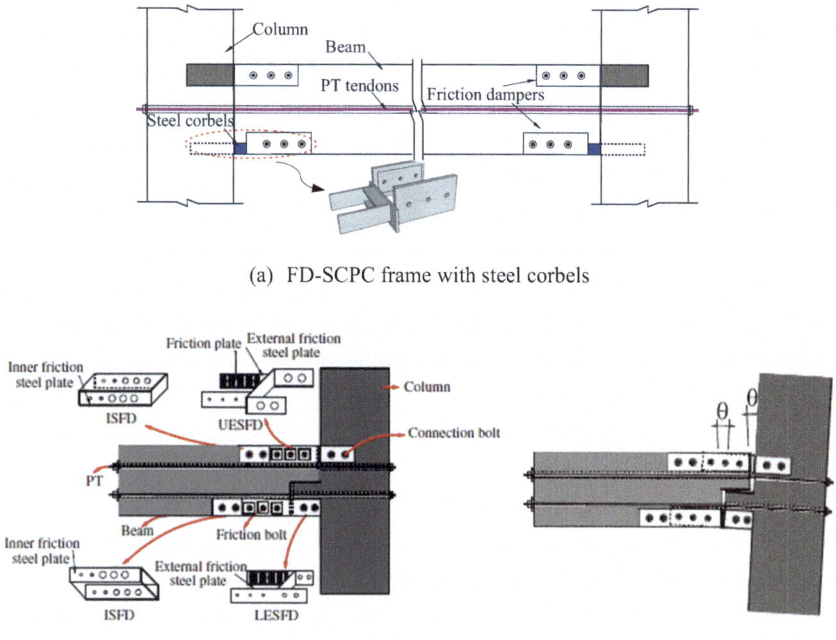

(a) FD-SCPC frame with steel corbels

(b) the configuration of HC-FD-SCPC connections

(c) rotation mechanism of HC-FD-SCPC connections

Fig. 4.2 Proposed HC-FD-SCPC beam-column connections

assembly method in the construction stage, but also bears the beam end bending moment in the use stage. Different from the traditional fabricated structure, the connection interface of the beam and column of the self-centering frame is no longer subjected to post-cast concrete or grouting treatment, but mainly depends on the friction on the connection interface of the beam and column to bear the shear force (if necessary, shear connector, connecting angle steel or corbel can also be added). Under the action of an earthquake, when the bending moment of the beam end exceeds the critical moment of the beam-column contact surface, the joint opens and the stress of the steel strand increases. After the earthquake, the frame is restored to the original vertical center position under the action of the prestressed steel strands, thereby eliminating (or greatly reducing) the residual deformation of the structure under the earthquake, and the deformation of the main structure such as beams and columns can be basically controlled within the elastic range (no loss).

In order to ensure that the local bearing failure of the concrete at the connection interface of the beam and column does not occur during the deformation process, the concrete is restrained and the overall joint work of steel plates and concrete is strengthened by setting beam end steel sleeves, beam-column spiral stirrups, stud shear connector, column embedded steel plates and other measures. At the same time, friction energy dissipaters are installed at the beam end webs. The energy dissipater

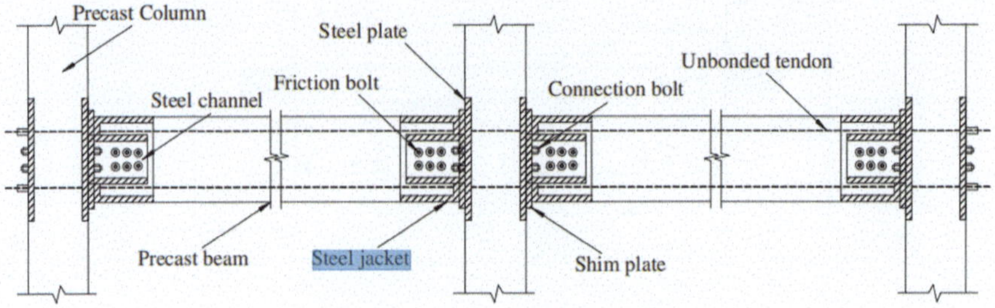

(a) SCPC-moment resisting frame (elevation)

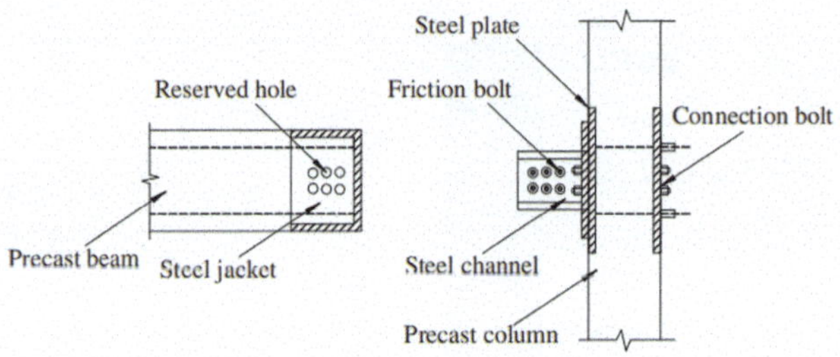

(b) Free-body diagram of SCPC connections

Fig. 4.3 Web friction-damped (WFD) fabricated frame

can be composed of a pre-embedded steel sleeve and a channel steel connected to the frame column, and the pressure perpendicular to the friction surface is provided through a pre-stressed high-strength bolts (split bolts). The contact surface between the steel sleeve and the channel steel is provided with a friction plate. The diameter of the split bolt hole reserved at the beam end is obviously larger than the diameter of the prestressed bolts, so as to ensure that the beam does not touch the bolt rod when a certain turning angle occurs.

4.3 Performance Test of Friction Energy Dissipaters

The joint friction energy dissipater is the key to the overall performance guarantee of the fabricated friction damped frame. Generally, a long-hole bolt sliding friction structure is adopted. The friction materials mainly include traditional ordinary steel plate, brass plate, and new non-asbestos organic (NAO) that have been researched and applied in recent years.

4.3.1 Structure of Friction Energy Dissipaters

The basic principle of the friction energy dissipater is to generate friction to dissipate energy through the relative movement of the friction plates. For example, in the self-centering beam-to-column joint (Fig. 4.4a), the exterior friction plate is connected to the column, and the inner friction plate is connected to the beam. As the beam-column connection gap opens, the friction plate will move relative to achieve friction dissipation; In the self-centering brace (Fig. 4.4b), the exterior friction plate is connected to the outer sleeve, and the inner friction plate is connected to the inner sleeve. The deformation of the brace drives the inner and outer sleeves to move relative to each other to achieve frictional energy dissipation.

The structure of the "sandwich" friction energy dissipater used in the friction damped self-centering precast frame is shown in Figs. 4.5a–c. The friction damper is mainly composed of an inner friction plate and two exterior friction plates, end fixing plates, friction materials and high-strength bolts. The inner and exterior friction plates are assembled together by friction-type high strength bolts, and the inner friction plate and the end fixing plate are sandwiched between the two exterior friction plates. Both ends extend 60 mm to fix the end when loading. The materials are all Q235 steel. The parts are shown in Fig. 4.5c. The stainless-steel plate is directly bonded to the interior friction plate by epoxy resin to form a whole, and is point welded around it to strengthen. Two grooves with a depth of 3 mm are reserved on both sides of the duct of the exterior friction plate, and the friction material is fixed in the groove with epoxy resin. This prevents the friction material from peeling off from the exterior friction plate during the sliding process. Under the action of the large clamping force, friction material can be constrained by the groove of the pretension force to be in a multiaxial compression state. During the sliding process of the energy dissipator, the stainless steel on the inner friction plate and the friction material on the exterior friction plate slide relative to each other to dissipate energy. The assembled schematic diagram and component dimensions are shown in Fig. 4.5a, b. It should be noted that the end fixing plate and steel plate used in this test 2 is only used to fix the ends when loading, and is not required when installing to the structure.

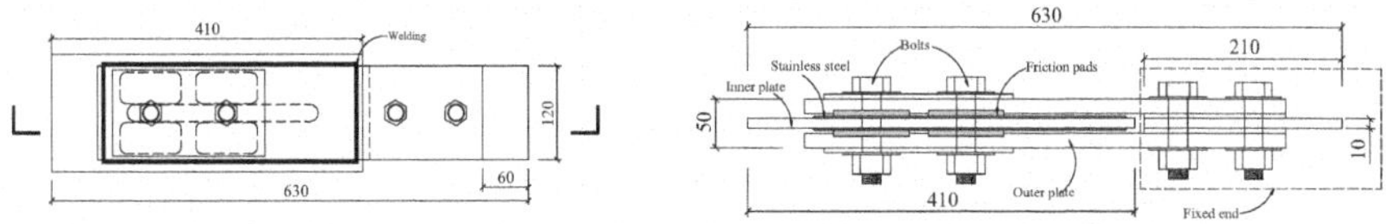

(a) Plan view of FD (mm)

(b) Section view of FD (mm)

Fig. 4.4 Details of the bolted FD

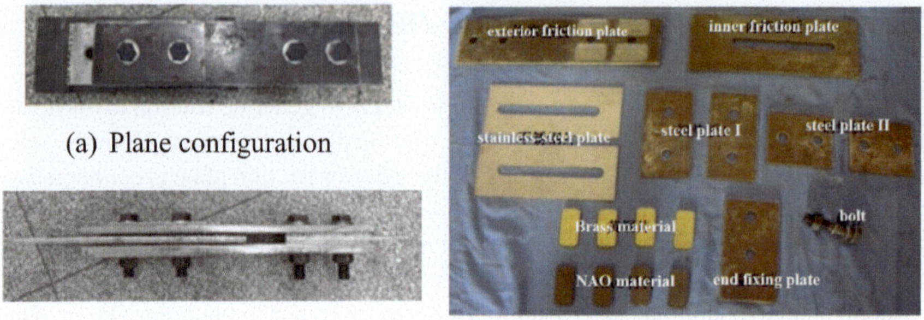

(a) Plane configuration

(b) Side view

(c) Members

Fig. 4.5 Configuration of friction energy dissipater

4.3.2 Test Scheme

The friction materials were compared with non-asbestos organic (NAO), brass and ordinary steel. The low-cyclic reciprocating load test of the friction energy dissipaters was completed in the materials laboratory of the Jiulonghu campus of Southeast University. The test used the MTS810 fatigue test machine with a measuring range of 25 t with a ±75 mm maximum stroke. The calibration test of friction-type high-strength bolts is to transmit the bolt preload of the bolt to the TDS-530 data acquisition instrument through a force sensor, and read the bolt preload data through the TDS-530 data acquisition instrument. In addition, the calibration test also used auxiliary tools such as a torque wrench with a range of 800 N·m, an ordinary wrench, two 8 mm thick steel plates with ducts, and a circular steel tube.

Figure 4.6 shows the structure of the test device and the physical diagram of the test piece. As shown in Fig. 4.6a, when the test piece is installed, the fixture selected in advance by MTS are used to connect with the end of the inner friction plate at the bottom of the test piece. The length of the end of the inner friction plate is about 5 cm and has been scratched to make it better clamp with the fixture; On the upper part of the test piece, MTS is also used to connect the pre-selected fixture to the end plate and the end plate is also scratched within 5 cm. During the test installation, clamp the upper end plate first, and then clamp the lower end position, adjust the end fixing button on the MTS, until the upper and lower ends of the clamping force is far greater than the sliding friction force of the friction energy dissipater. Figure 4.6b, c are the physical diagrams of the upper and lower parts of the friction energy dissipater test device.

The bolt calibration test uses a torque wrench to control the bolt preload value, and establishes the numerical relationship between the bolt preload of a single bolt and the applied torque. Figure 4.6d is the loading device for the bolt calibration test. Two steel plates with bolt ducts are used to sandwich the sensor. The bolts that need to be calibrated pass through the steel plate and the force sensor is fixed with a nut on the outside of the other steel plate. To ensure the accuracy of the calibration, the

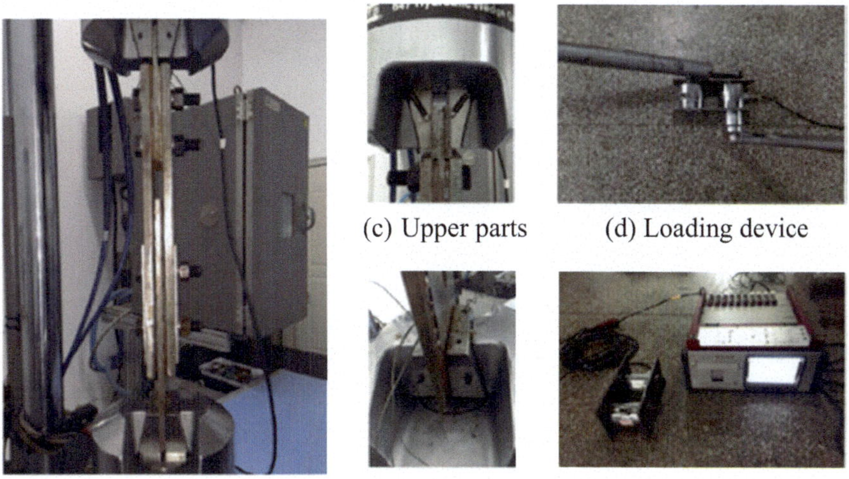

(a) Specimen (b) Lower parts (c) Upper parts (d) Loading device (e) Measurement device

Fig. 4.6 The structure of the test device and the physical diagram of the test piece

sensor is also placed on the other side of the steel plate, and also put the same type of high-strength bolts. After everything is installed in place, connect the force sensor of the bolt that needs to be calibrated to the TDS-530 data acquisition instrument. Figure 4.6e is the bolt calibration test measurement device. After loading the target torque by the torque wrench, the force sensor will synchronously transmit the force information to the data acquisition instrument TDS-530, and then record the force on the data acquisition instrument in time. The test carried out two elastic stage torque tests on single high-strength bolt of 10.9 and 12.9 respectively. The minimum torque value applied in the test is 200 N·m, and the maximum torque value is 350 N·m. The relationship between the preload of grade 10.9 bolts and the applied torque is shown in Table 4.1, the relationship between the preload of grade 12.9 bolts and the applied torque is shown in Table 4.2. Linear regression of the actual measured mean values of bolt preload and applied torque obtained by statistics in Tables 4.1 and 4.2 (Fig. 4.7), the relationship between torque and preload in the elastic stage of the bolt can be obtained (Eqs. 4.1 and 4.2).

Table 4.1 The applied torque of the preload of grade 10.9 bolts (kN)

Number	Bolt torque (N·m)			
	200	250	300	350
1	27.66	34.34	43.41	50.26
2	26.42	32.68	41.57	49.36
Average	27.04	33.51	42.49	49.81

Table 4.2 The applied torque of the preload of grade 12.9 bolts (kN)

Number	Bolt torque (N·m)			
	200	250	300	350
1	40.39	48.65	60.89	69.36
2	41.05	50.45	62.21	71.44
Average	40.72	49.55	61.55	70.40

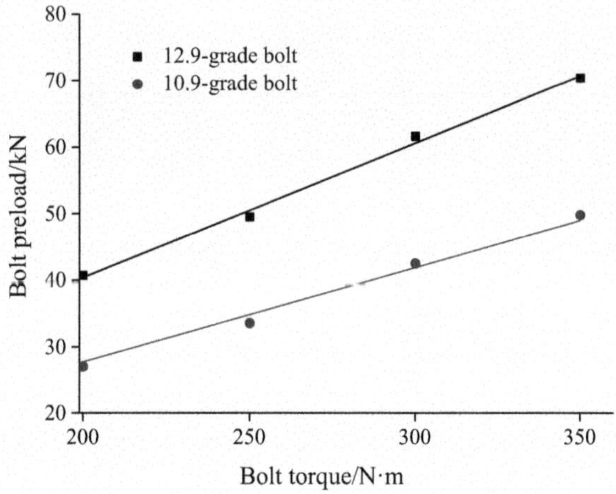

Fig. 4.7 Linear regression curve between bolt torque and bolt preload

The relationship between the pre-tightening force and the applied torque of the 10.9-grade bolts:

$$F_i = 0.142\mathrm{T}_i - 0.59 \approx 0.142\mathrm{T}_i \quad (4.1)$$

The relationship between the pre-tightening force and the applied torque of the 12.9-grade bolts:

$$F_i = 0.2\mathrm{T}_i - 0.006 \approx 0.2\mathrm{T}_i \quad (4.2)$$

NAO materials are different according to the content of their synthetic materials, low temperature (<200 °C) wear resistance is good, weight is lighter (1.9–2.0 g/cm^2), but its compressive strength is lower, about 108–184 MPa. As a metal friction material, brass plate is easy to rust after long-term use, but it has the characteristics of high strength and high heat resistance temperature. Because the two materials are not easy to show high temperature when they are used in the friction energy dissipater, the friction performance of the two materials under different compressive strengths when the bolt preload is changed is analyzed. In addition, in order to restore the force process of the friction energy dissipater under the action of an earthquake as much as possible, the effect of the sliding displacement amplitude, slip rate, loading

Table 4.3 Loading conditions of friction energy dissipaters

Number	Type	Bolt torque T (N·m)	Loading rate v ($mm \cdot s^{-1}$)	Amplitude A (mm)	Loading systems
1	Test	200	0.1	5	Program 1
2	Amplitude varies	300	1	15, 30, 45	Program 2
3	Loading rate varies	300	1, 2, 3	15, 30, 45	Program 2
4	Constant amplitude	300	1	30	Program 3
5	Pretightening force varies	200, 300, 400	1	45	Program 1

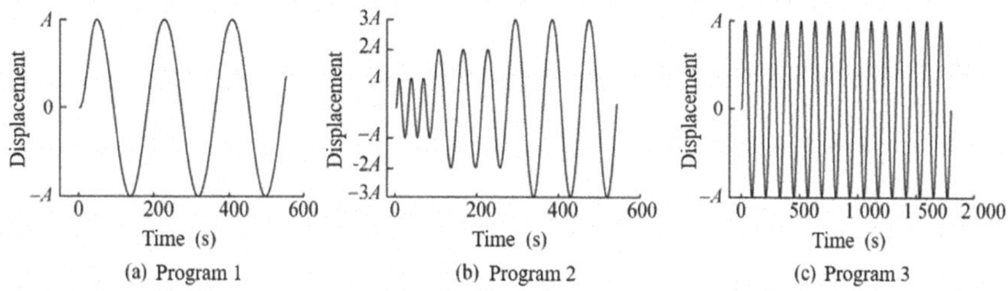

Fig. 4.8 Programs of loading

cycle number and loading system on the hysteretic behavior of the friction energy dissipater is also studied.

Table 4.3 shows the bolt torque and loading schemes used under 5 working conditions. The design parameters of each working condition are bolt torque, loading rate, amplitude and loading system. The three loading systems are shown in Fig. 4.8. The first type of loading system keeps the amplitude unchanged and loops for 3 cycles, the second type of loading system changes the amplitude every 3 cycles, and the third type of loading system keeps the amplitude unchanged and loops for 15 cycles.

4.3.3 *NAO Test Results*

First, use NAO as the friction material to test, and the test results are shown in Fig. 4.9. In working condition 1, the displacement amplitude and loading rate of the test piece are both small, mainly to estimate the difference between the actual friction and the theoretically estimated friction force, and the difference between the maximum static friction and the sliding friction. The test results are shown in Fig. 4.9a, the maximum static friction is about 54.3 kN, and the average sliding friction is 42.8 kN.

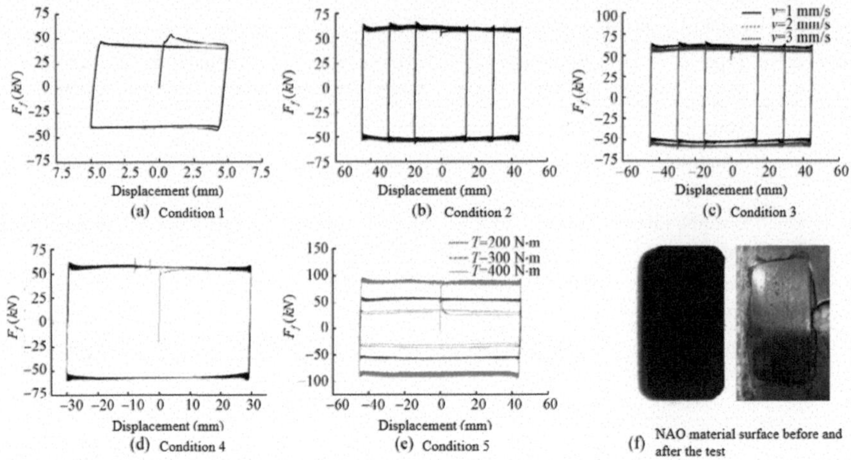

Fig. 4.9 Test results of NAO material

In working condition 2, the sliding friction remains basically unchanged during variable amplitude loading, and the hysteresis performance in the two directions of tension and compression is relatively symmetric. In working condition 3, the friction decreases slightly after the loading rate increases, and its influence can be ignored within a certain range of loading rate.

In working condition 4, the test hysteresis curve is relatively stable under constant amplitude loading, the material performance is stable, and the energy dissipation effect is ideal. It can be seen from Working Condition 5 that changing the bolt preload has a greater impact on the test results. With the gradual increase of the bolt preload, the normal force received by the NAO material continues to increase, but the test curve remains relatively stable. And the average friction is approximately 2:3:5. This is mainly because the NAO material is embedded in the groove of the exterior friction plate, and it is in a multiaxial stress state when it is compressed, so there is no premature failure.

During the entire test, the noise generated by the friction energy dissipater is very small. After the test, the inner and exterior friction plates are taken out. It can be seen from Fig. 4.9f that there are only a few scratches on the surface of the NAO material, and there is no obvious flaking phenomenon.

4.3.4 Comparison of Experimental Results Between NAO and Brass Plates

After replacing the friction material with brass plates, the test is performed again, and the loading conditions in Table 4.3 are still used, the hysteresis curve obtained

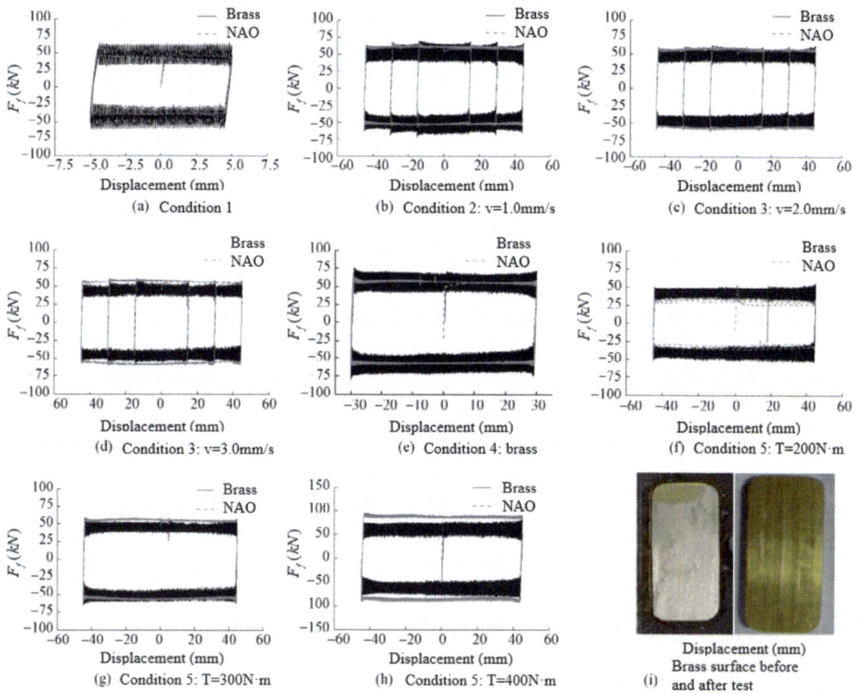

Fig. 4.10 Comparison of test results of two friction materials

is shown in Fig. 4.10. The test results of the NAO material are kept in the figure to facilitate the comparison of the two materials.

In working condition 1, the hysteresis curve when using brass plate has a large up-and-down jitter, and it is accompanied by large metal friction noise from the sensory point of view, while NAO material is always quieter during the test friction process. In working condition 2, under variable amplitude loading, the curve of the brass plate still shows large jitter, but as the amplitude increases, the frictional jitter amplitude will be slightly reduced.

In working condition 3, the frictional jitter is still much larger than that of NAO material when using brass plate. But with the gradual increase of the loading rate, the frictional jitter of the two materials decreased. Under different loading rates, the friction when using NAO material is basically close to the upper limit of the friction when using brass plate. When the strain rate is 3 mm/s, the friction of NAO materials exceeds the friction when using brass plates.

In working condition 4, the two materials did not show serious friction degradation under constant amplitude continuous circulation. The ratio of the difference between the average value of the upper and lower limits of the friction of the brass plate material to the average value of the lower limit is 28%, while that of the NAO material is 2%. In working condition 5, changing the bolt preload has a greater impact on the brass test results. The friction obtained by using NAO material increases with the bolt torque at a faster rate than when using brass plates.

During the test, the friction material of the entire brass plate is accompanied by a lot of friction noise. After the end of the test, a large number of flaking brass fragments were found under the specimen. From Fig. 4.10i, it can be seen that there are many nicks on the surface of the brass plate, and the ground stainless steel is embedded in the nicks. It can be seen that the wear resistance of brass plate is not as good as that of NAO material.

Figure 4.11 shows the relationship between the average dynamic friction and the loading rate of energy dissipators using different materials. Because the friction of the two materials fluctuates during loading, the upper limit and the lower limit of each step of the two materials are averaged. It can be seen from the figure that the difference between the upper and lower limits of the friction produced by the two materials decreases with the increase of the loading rate. When the loading rate increases, the ratio of the difference between the average value of the upper and lower limits of the friction of the brass plate material to the average value of the lower limit decreases from 28.2 to 19.5%, and the average value of the friction of the brass plate material decreases by about 7.1%. At different loading rates, the friction of NAO material is relatively stable.

The relationship between the average dynamic friction and bolt torque of energy dissipaters using different materials is shown in Fig. 4.12. The friction generated by the two materials is approximately linear with the bolt torque, but as the torque increases, the ratio of the difference between the average value of the upper and lower limits of the friction and the average value of the lower limit when the brass sheet is used remains between 20.7 and 28.9%. The difference between the upper and lower limits of the friction of NAO materials decreases with the bolt torque. When the bolt torque is 400 N·m, the difference is only 3.5%. In addition, when the bolt torque increases from 200 to 400 N·m, the average friction of the NAO material increases

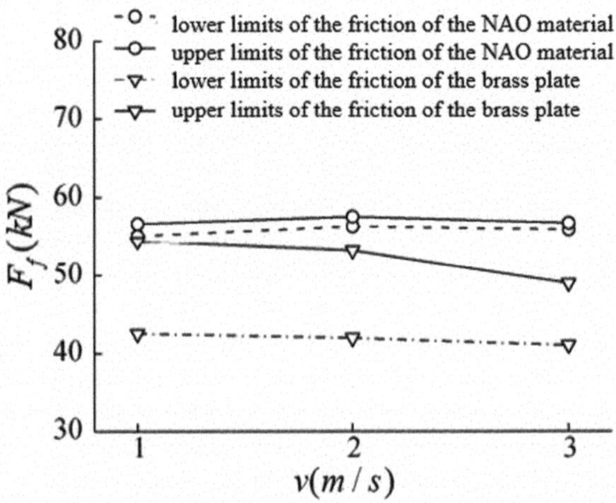

Fig. 4.11 Friction force-load rate diagram

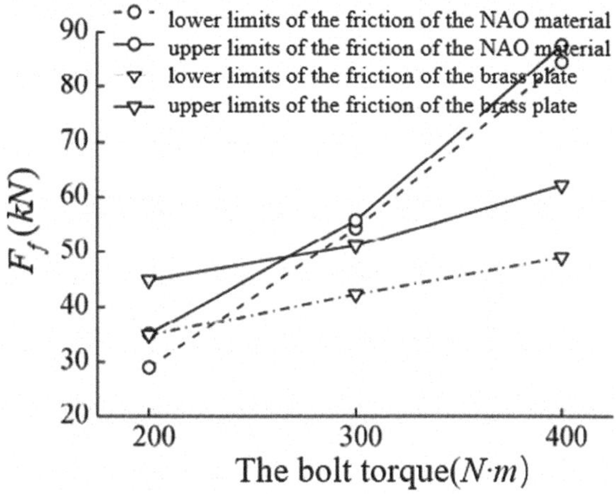

Fig. 4.12 Friction-torque diagram

with the torque by about 71.8%, while the brass plate material only increases by about 37.1%. This may be because the Poisson's ratio of the NAO material is greater than that of the brass plate, and its lateral deformation is constrained by the groove under the action of the preload, which produces greater friction than the brass plate material under the same torque.

In summary, when using NAO friction materials in civil engineering, attention should be paid to: (1) Due to the low compressive strength of NAO materials, NAO materials should be used in friction dampers; (2) NAO materials have low heat resistance. It is not suitable for use under extreme high temperature above 200 °C; (3) When epoxy resin is used to fix NAO material on the friction plate, grooves should be set to prevent it from slipping. Attention should be paid to the use of brass materials: (1) Due to the relatively high compressive strength of brass materials, brass materials are used in the friction protection device; (2) in order to reduce the jitter generated by brass materials during friction, try to keep the surface of the brass friction plate as flat as possible.

4.3.5 Analysis of NAO and Ordinary Steel Test Structure

Finally, the Q235 steel plate was used as the friction material for the test, and the loading system was unchanged. The noise during the sliding process of the friction energy dissipater was very small. The comparison between the test results and the NAO friction plate is shown in Fig. 4.13. It can be seen from Fig. 4.13 that there is a small range of jitter in the hysteresis curve of ordinary Q235 steel under various working conditions. From Fig. 4.13b–d), it can be found that the friction of ordinary steel changes significantly with the loading rate, while the NAO basically does not change, which further shows that ordinary steel is not as stable as NAO; From Fig. 4.13e, it can be seen that the hysteresis curve of ordinary steel jitters more

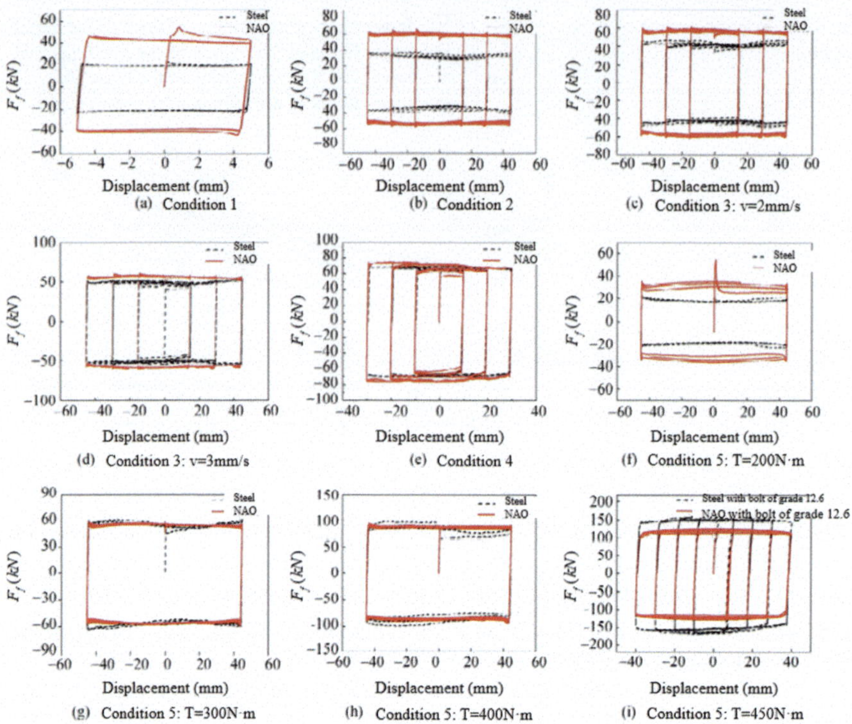

Fig. 4.13 Comparison of test results

severely, and the friction is slightly degraded, while the friction of NAO is basically not degraded; from Fig. 4.13f–i, it can be found that the friction of ordinary steel relative to NAO material has a larger variation, and its value is more susceptible to external conditions.

In summary: NAO materials are more stable than ordinary Q235 steel materials under various conditions, and there is basically no noise during the test; the change of loading rate basically has no effect on NAO materials, and the change of loading rate has a great impact on ordinary Q235 steel materials in many cases; Under a variety of working conditions, the average friction of NAO material is greater than that of ordinary steel, and its energy dissipation capacity is stronger. Ordinary Q235 steel material is greatly affected by external conditions such as loading rate, and its performance stability is not as good as that of NAO material.

4.4 Analytical Models of Self-Centering Prestressed Concrete (SCPC) Frames

4.4.1 Prototype Structure

In this study, a 6-story 6-bay SCPC moment resisting frame is selected as the prototype structure for analyses. The plan view of the structure is shown in Fig. 4.14a, in which the lateral force resisting systems are four SCPC frames at the perimeter, two in each primary direction. The lateral earthquake action is mainly resisted by the two SCPC frames along this direction when the structure is under one-directional earthquake ground motion, while the other frames only resist gravity load and their lateral resistance capacity could be neglected. The structure height is 22.2 m, with a 4.2 m floor-to-floor height for the first story and 3.6 m floor-to-floor height for other stories. The center-to-center space of columns is 6 m. The design base shear of one single SCPC frame is 1494 kN.

Table 4.4 presents the geometry and reinforcement details of beams and columns. Where, ρ_b and ρ_c represent the total reinforcement ratio of beam and column, respectively. P_0 represents the total initial prestressing force, A_{PT} represents total prestressed steel bars area of beam and F_f represents the friction force applied to web friction devices. The design unconfined concrete strength is $f_c = 19.1$ MPa, the yield strength and ultimate strength of PT tendons are $f_{py} = 1675$ MPa, and $f_{pu} = 1860$ MPa, respectively.

The beam-column joints were adopted the same as Sect. 4.2.2. Considering the infill walls strength may have an influence on the global lateral resistance capacity, three types of infill walls with different strength designed based on GB50003-2011 are installed in this SCPC frame, as shown in Table 4.5. Parameters are studied to change the strength of infill walls. In the following text, the SCPC-IW frames with Infill-1, Infill-2, and Infill-3 are referred as SCPC-IW1, SCPC-IW2, SCPCIW3 frames, respectively, and the SCPC frame without infill walls is referred

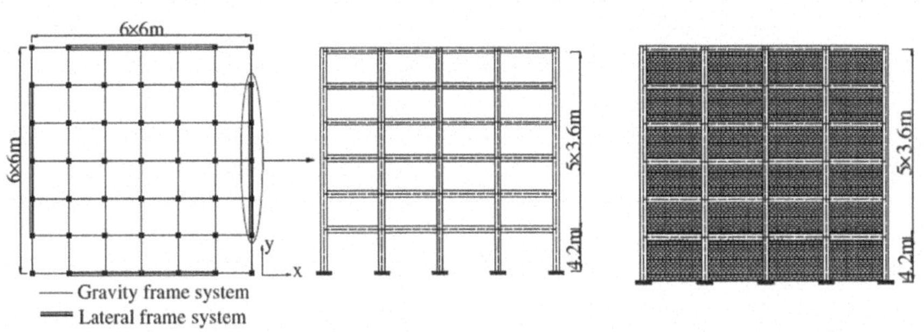

(a) Prototype structure

(b) Arrangement of in walls

Fig. 4.14 Prototype SCPC-IW frame

Table 4.4 Design parameters of SCPC moment resisting frames

Building floor	Beam area	Column area	ρ_b	ρ_c	A_{pt}	P_0	F_f
	mm	mm	%	%	mm^2	mm^2	kN
6	400 × 600	550 × 550	0.76	1.61	834	413	138
5	400 × 600	550 × 550	0.76	1.94	834	718	239
4	400 × 600	550 × 550	0.76	2.27	973	980	327
3	400 × 650	600 × 600	0.71	1.91	1112	1131	377
2	400 × 650	600 × 600	0.71	2.18	1112	1209	403
1	400 × 700	650 × 650	0.66	1.86	1251	1202	401

Table 4.5 The relative parameters of infill walls

Sequence numbers	Strength grade of flock	Strength grade of mortar	f_w (MPa)	F_{IW}/F_{FD}
Infill-1	MU20	Mb20	6.30	1.32
Infill-2	MU15	Mb15	4.02	0.79
Infill-3	MU7.5	Mb5	1.71	0.37

as SCPCBare. Because infill walls and friction devices are the main energy dissipation elements in such SCPC-IW frames, the ratio of the load capacity of all infill walls to the friction forces of all friction devices (FIW/FFD) is selected as the index to quantify the influence of infill walls on SCPC-IW frames. Where the load capacity of an infill walls (FIW) is defined by the product of the equivalent area and design strength of infill walls.

4.4.2 *Numerical Simulation of SCPC Frame with WFDs*

The numerical model of the beam-column connection is shown in Fig. 4.15, where the precast beam and column are modeled using the dispBeamColumn element that captures distributed plasticity by fiber cross sections, in which the steel fiber and concrete fiber are assigned with the Steel02 material and Concrete02 material, respectively. The Steel02 material is based on a Giuffre-MenegottoPinto constitutive model. This material has a bilinear backbone curve with a post-yield stiffness expressed as a proportion of the initial modulus of elasticity of the steel. The material model accounts for the Bauschinger effect, which contributes to the gradual stiffness degradation of steel members under cyclic loading, and has an isotropic hardening option for tension and compression portions of the hysteresis. The Concrete02 material, which is a modification of the typical Kent-Scott-Park constitutive model, effectively considered tensile stress of concrete. In order to account for the confinement effect of transverse reinforcement, the concrete of column and beam section were divided into cover concrete and core concrete, and each was defined by the corresponding constitutive

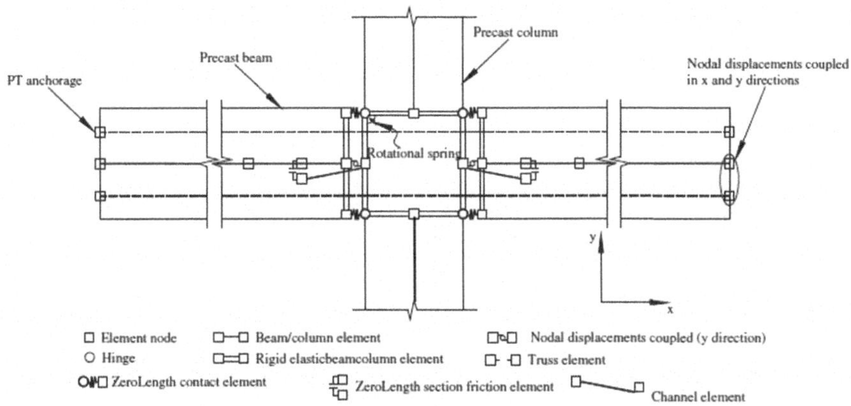

Fig. 4.15 Numerical model of SCPC beam-column connection

parameters of Concrete02 constitutive model. In order to simulate the gap opening and closing behaviour of the connections, the beam-column interface is modeled by two ZeroLength contact elements, which are assigned with the ElasticPPGap material. The definition of ElasticPPGap material with a compression-only force–deformation relationship mainly includes elasticity modulus, which determines the initial tangent stiffness of the contact elements.

The dispBeamColumn element is selected to model the two steel channels, in which the Steel02 material is selected to simulate the axial-flexure deformation and shear deformation, respectively. The beam-column panel zone is simulated by eight stiff elasticBeamColumn elements at four sides of the panel zone and a rotational spring, which is assigned with a bilinear-elastic moment relationship material, is located at one of the four corners to represent the panel zone shear deformation. One node of steel channels is connected with the panel zone, the other node of the steel channel is connected with a node on the beam element by a Zero-length section friction element, which is defined by Steel02 material to model the energy dissipation developed by web friction devices. The truss element defined by Steel02 and ElasticPP material, is used to conduct the post-tensioning tendons and their rupture characteristics. The definition of ElasticPP material mainly includes a maximum and minimum value to determine the failure state of other materials. So, the ElasticPP material is used in conjunction with Steel02 material to specify an upper bound of tendon strains. Once the strain exceeds the predefined upper bound, the steel02 material stops working, representing the rupture of tendons.

A lean-on column is modeled by elasticBeamColumn elements, in which axial rigid stiffness is defined by a large value and flexural stiffness is defined by a small value so as to exactly model the P-Δ effect and the mass of the building. The connection between the lean-on column and SCPC frame is established by a truss element assigned with elastic material by large axial regality stiffness.

4.4.3 *Numerical Simulation of Infill Walls at the SCPC-IW Frames*

The infill wall panel is modeled by two equivalent struts suggested in literature as shown in Fig. 4.16a. The equivalent strut is considered compression-only, and the tension is neglected. The stress–strain law of the strut is defined by Fig. 4.16b. It was proved in the literature that this model can effectively model the influence of the in-plane strength and stiffness of infill walls on the failure modes of infilled RC frames. The two equivalent struts are implemented by the truss element in the OpenSees, in which the force–displacement behavior relationship is modeled by the Hysteretic material.

The geometry of the equivalent strut is described as follows:

The thickness t_{inf} of the strut equals to the actual infill walls thickness, the width of the strut a is calculated by Eqs. (4.3) and (4.4)

$$a = 0.175(\lambda_1 h_{\mathrm{col}})^{-0.4} L_{\mathrm{diag}} \tag{4.3}$$

$$\lambda_1 = \left(\frac{E_{\mathrm{inf}} t_{\mathrm{inf}} \sin 2\theta}{4 E_{\mathrm{f}} I_{\mathrm{col}} h_{\mathrm{inf}}}\right)^{\frac{1}{4}} \tag{4.4}$$

where h_{col} is the height of column adjacent to infill walls, h_{inf} is the height of infill walls, L_{diag} is the diagonal length of infill walls, E_f and E_{inf} are the Young's modulus of elasticity of the frame and infill walls materials, respectively, I_{col} is the moment of inertia of the column cross section, θ is the angle whose tangent value equal to the infill height-to-length aspect ratio.

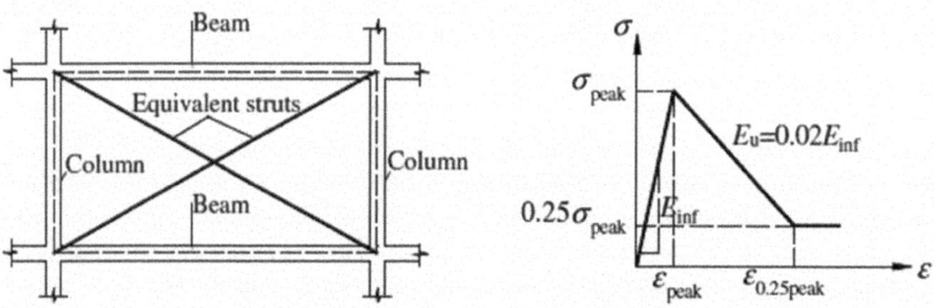

(a) Equivalent diagonal struts (b) The stress-strain relationship

Fig. 4.16 The numerical model of infill walls

4.4.4 Model Validation and Calibration of SCPC Beam-Column Connection and Infill Walls Model

In order to apply above analytical modeling approach for beam-column connections and the model of infill walls, it is necessary to validate the two analytical models and calibrate relative parameters through comparisons between numerical and experimental results.

The low cycle loading test of a SCPC beam-column connection conducted by Song et al. [1] is selected to validate the SCPC beam-column connection model. Comparisons between the experimental and numerical results of the SCPC beam-column connection model are shown in Fig. 4.17a. It can be seen that the modeling result matches well with the experimental result. It is indicated that the numerical model for the SCPC beam-column connection is valid. It is observed that the modeling of the gap opening and closing performance is crucial in achieving the Flag-shape hysteretic curve. The experimental and analytical results can match each other as shown in Fig. 4.17a only when the initial tangent stiffness of Zero Length contact element shown in Fig. 4.15 is sufficient large. Besides, the harden ratio of Steel02 material should be defined as zero to accurately model the characteristic of friction forces in defining the Zero-length friction section element.

The low-cycle loading test of a one-story one-bay infilled RC frame conducted by Essa et al. [2] is selected to validate the numerical model proposed by Al-Chaar and calibrate several parameters, as shown in Fig. 4.17b. The hysteretic behavior is analyzed using the hysteretic material in the OpenSees software, in which the residual strength equal to 25% of the peak strength and the softening branch slop equal to 2% of the initial stiffness. The strength degradation and the loading and unloading characteristics are determined by two parameters relating to pinching factor (pinch x and pinch y) for strain (or deformation) and stress (or force) during reloading, two damage factor (damage1 and damage2) due to ductility and energy, and a coefficient

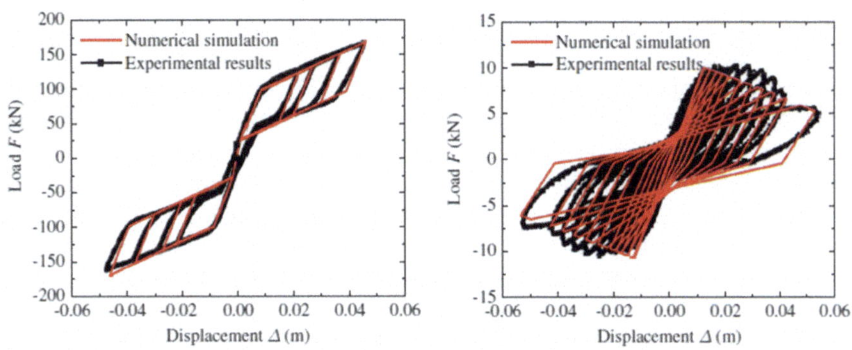

(a) SCPC beam-column connection (b) Infill walls

Fig. 4.17 Comparisons between numerical and experimental results factor (beta) used to determine the degraded unloading stiffness based on ductility. Figure 4.17b presents comparison between the experimental and numerical results of infill walls model. It is observed that when pinch x and pinch y equal to 0.4, damage1 and damage2 equal to 0.01 and beta equal to 0.4, there is good agreement of stiffness, strength, energy dissipation and degradation characteristics between numerical and experimental results.

4.5 Nonlinear Static Analyses

Using the numerical model in Sect. 4.4, the nonlinear static performance of the SCPC frame under repeated low-cycle loading and unidirectional pushover loading are studied respectively.

4.5.1 Low Reversed Cycle Loading Analyses

Figure 4.18 shows the hysteretic behavior of the four models which is reflected by the relationship between nonlinear shear force V and roof drift ratio Δ. Displacement control loading is carried out on nonlinear static analysis by scaling Δ to four maximum values, namely, 0.5, 1.0, 1.5, 2.0%, with increments of 0.1 mm per analytical step. The roof drift ratio Δ is calculated as lateral displacement at roof divided by total height of the building.

For calculating the stiffness of structures, the "yield" point, in which the base shear strength is corresponding to a significant change in the structure lateral stiffness, is provided. The "yield" point is the position in which gap opening at the interface of beam and column occurs and the friction devices begin to dissipate energy. For SCPC-IW frame, the plastic deformation even the development of cracks on the surface of infill walls begin to occur. Before "yield" point occurs, the stiffness of the SCPC-Bare frame is mainly decided by initial posttensioning force and friction force. While the stiffness of the SCPC-IW frame mainly depends on the combination of posttensioning force, friction force and the stiffness of infill walls. After yield point occurs, the stiffness of SCPC-Bare frame is mainly decided by the stiffness of posttensioning tendons. Whereas the stiffness of SCPC-IW frames mainly depends on the stiffness of posttensioning tendons as well as infill walls. So the stiffness of structure is calculated by the ratio of displacement to base force for the "yield" point shown in Fig. 4.18. The stiffness of SCPC-Bare frame is 1.16×10^4 kN/m, while are 2.53×10^4 kN/m, 2.20×10^4 kN/m and 1.77×10^4 kN/m for the SCPC-IW1, SCPC-IW2 and SCPC-IW3 frames, respectively, which are obviously increased in comparison with SCPC-Bare frame. In addition, it is obviously observed that the hysteresis loop area of the SCPC-IW1, SCPC-IW2 and SCPC-IW3 frames are larger than those of the SCPC-Bare frame, indicating that all the three SCPC-IW frames have better energy dissipation capability compared with the SCPC-Bare frame. It is

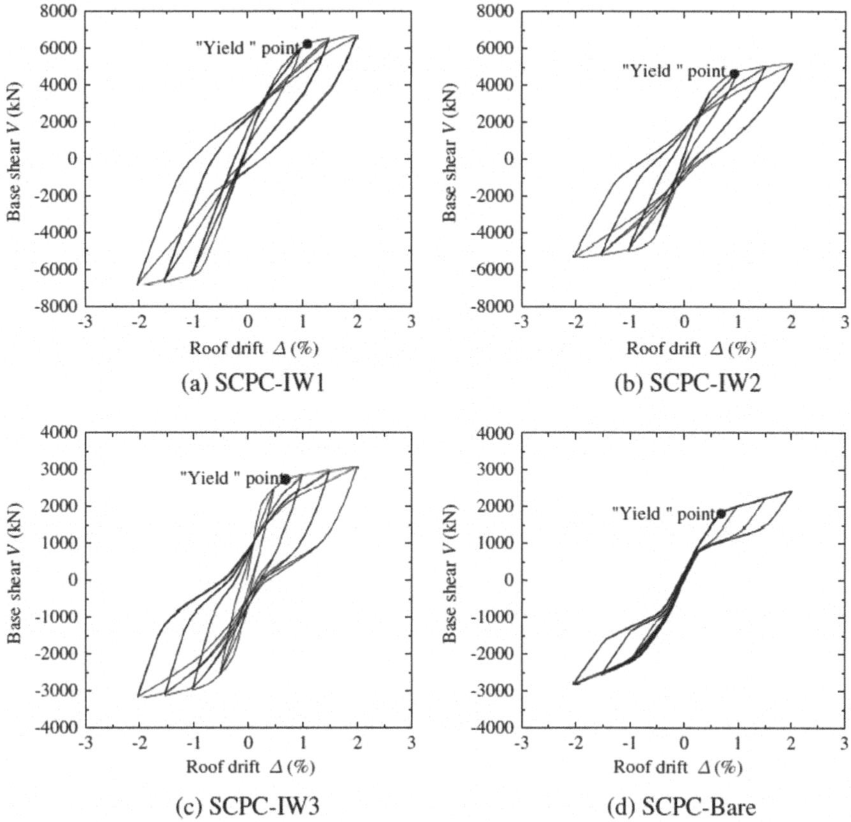

Fig. 4.18 Comparison of hysteretic behavior of four structures

indicated that the infill walls generate considerable contribution to the stiffness and energy dissipation of the SCPC-IW frames.

It is observed that although the SCPC-IW2 and SCPC-IW3 frames exhibit apparent flag shape characteristic, they exhibit some residual displacement. While the energy dissipation capacity of SCPC-IW1 frame is apparently larger than that of the SCPC-Bare frame, its hysteretic curve almost has no flag shape, indicating the self-centering capability almost diminished. Therefore, the SCPC-IW1 frame possesses the largest amount of energy dissipation but the smallest amount of self-centering capability.

By comparing the hysteretic responses of the SCPC frame and three SCPC-IW frames, it can be concluded that the self-centering capability gradually diminished and the energy dissipation capability increase significantly with the raising of infill walls strength. The SCPC-IW2 and SCPC-IW3 frames have similar flag-shape hysteretic curve with SCPC-Bare frame and are provided a good agreement between strength, stiffness, energy dissipation and self-centering capability.

4.5.2 *Pushover Analyses*

Previous research showed that the presence of infill walls may result in the increase of column axial compression ratio of infilled RC frames so as to cause brittle failure of column, so the Pushover analyses are carried out on the four numerical models to obtain the column axial compression ratio, which is needed for further exploring the influence of infill walls on structural components. The load distribution is enforcing the first mode response in analyses. Table 4.6 shows different limit values corresponding to four aseismic grades of RC frames proposed by GB50011-2010 to guarantee the column has sufficient ductility, deformation ability and collapse resistance capability as well as avoid the structural seismic short column failure under earthquakes. The inter-story drift ratio verse axial compression ratio relationship curves for side column of each floor are shown in Fig. 4.19. The axial compression ratio is defined as the ratio of compression force to the product of column cross section area A and concrete design strength f_c.

The axial compression ratio of SCPC-IW frame is increased with the raising of infill walls strength, and as the increase of floor level the axial compression ratio is decreased, indicating that SCPC-IW frames are more likely present soft story failure than SCPC-Bare frames and the soft story may more likely present in its lower floor. It is observed that as the increasing of inter-story drift ratio, axial compression ratio of SCPC-IW1 frame exceed the limit values of all the four seismic grades at the first floor and the first seismic grade at the second floor; axial compression ratio of SCPC-IW2 frame exceed the limit value of the first seismic grade at the first floor; axial compression ratio of SCPC-IW3 and SCPC-Bare frames always less than all the limit values of the four seismic grades. Previous research has shown that column axial compression ratio exceeds corresponding limit values, the brittle failure, short column failure and soft story failure even collapse of structures may be occurred due to the lack of ductile behavior. So SCPC-IW1 frame may present higher collapse probability under severe earthquake.

The above results obtained from low reversed cycle loading analyses and pushover analyses indicate that high strength infill walls can result in larger residual deformation and increase axial compression ratio of structures, while the low strength infill walls can provide a good balance between the self-centering capability and energy dissipation ability. The SCPC-IW frames whose capacity load ranges from 37 to 79% of web friction force are successful in satisfying the expected stiffness and energy dissipation requirement, axial compression ratio as well as self-centering capability.

Table 4.6 Limit values of axial compression ratio for reinforced concrete column

Seismic grade	One	Two	Three	Four
Limit values of the axial compression ratio	0.65	0.75	0.85	0.90

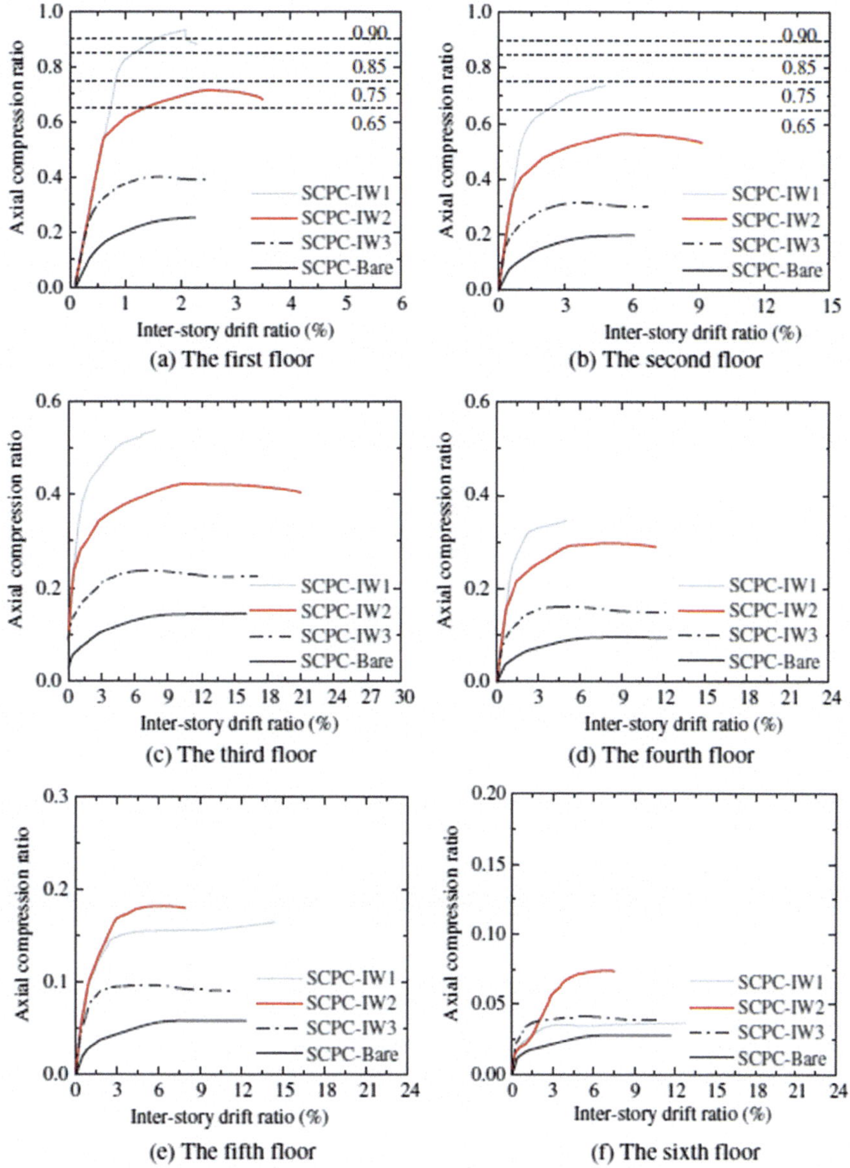

Fig. 4.19 Inter-story drift ratio verse axial compression ratio relationship curves

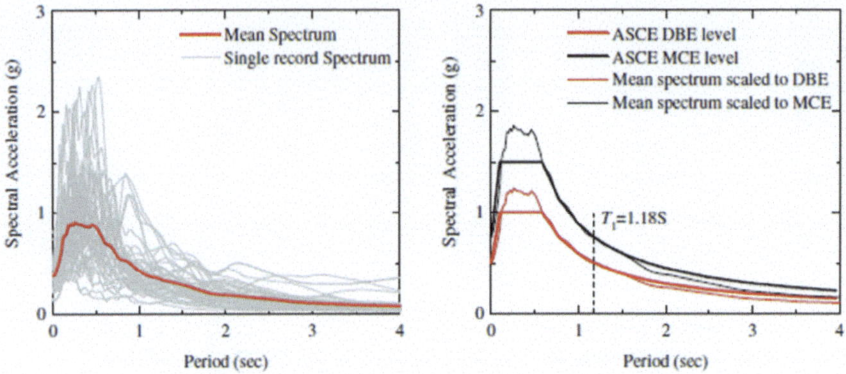

(a) Spectrum of the selected ground motion records
(b) Mean Spectrum scaled to the design spectrum

Fig. 4.20 Pseudo acceleration spectra of selected ground motion records and the design spectra

4.6 Nonlinear Dynamic Time History Analysis

4.6.1 Selection of Ground Motions Records

To investigate the seismic influence of infill walls on SCPC frames, 44 ground motion records shown in Fig. 4.20a, which are two horizontal components of 22 far-field ground motion ensembles applied by FEMA P695, are selected as excitation to input the structures in dynamic analysis. Because the intensities of the ground motions are different, such as peak ground accelerations (PGA) and spectral acceleration S_a(T1, 5%), the epicentral distance and durations and so on, they can describe the uncertainty of earthquakes.

4.6.2 Comparison of Seismic Response Under the DBE and MCE Seismic Hazard Levels

Nonlinear dynamic time history analyses are carried out using above 44 ground motion records to investigate the seismic behavior of SCPC-IW frames. It can be concluded from the existed research that the spectral acceleration $S_a(T_1, 5\%)$ corresponding to the basic period T_1 of the structure with damping ratio of 5% is a better intensity measure in terms of the efficiency and proficiency attributed to its smaller dispersion. So $S_a(T_1, 5\%)$ is selected as the intensity measure in seismic analysis. In this paper, the T_1 of SCPC-IW1, SCPC-IW2, SCPC-IW3 and SCPC-Bare are 0.98 s, 1.07 s, 1.18 s, 1.27 s, respectively. When $T_1 = 1.18$ s, both ground motion intensity and the spectral shape effect of the four structures are similar, which can provide a better benchmark to compare analytical results obtained from different structures. The ensemble of the 44 ground motions are linearly scaled to two seismic hazard levels that are the design basis earthquake (DBE) and the maximum considered earthquake (MCE), respectively, based on $S_a(T_1, 5\%)$, corresponding to 10% and 2% probabilities of exceedance in 50 years, respectively. Figure 4.20b shows the mean spectrum for the DBE and MCE seismic hazard levels, respectively.

In order to comprehensively investigate the seismic performance of SCPC-IW frames, the maximum inter-story drift ratio (θ_{max}) and maximum residual inter-story drift ratio ($\theta_{r,max}$) are used to investigate the seismic response and self-centering capability of structures. Figure 4.21 and Fig. 4.22 show the θ_{max} and $\theta_{r,max}$ of the four structures under DBE and MCE seismic hazard levels. The corresponding average values for the ensemble of the 44 earthquake ground motions are given in Table 4.7 and Table 4.8.

The interpretations from the Figs. 4.21, 4.22 and Tables 4.7, 4.8 are summarized as follows:

Subjected to the DBE-level ground motions, the θ_{max} of SCPC-IW1, SCPC-IW2, SCPC-IW3 frames are generally much lower than those of SCPC-Bare frame in most earthquake ground motion records (see Fig. 4.21a). However, the $\theta_{r,max}$ of SCPC-IW1, SCPC-IW2 and SCPC-IW3 frames are larger than those of SCPC-Bare frame in most cases(see Fig. 4.22a). The average values of θ_{max} and $\theta_{r,max}$ for the four

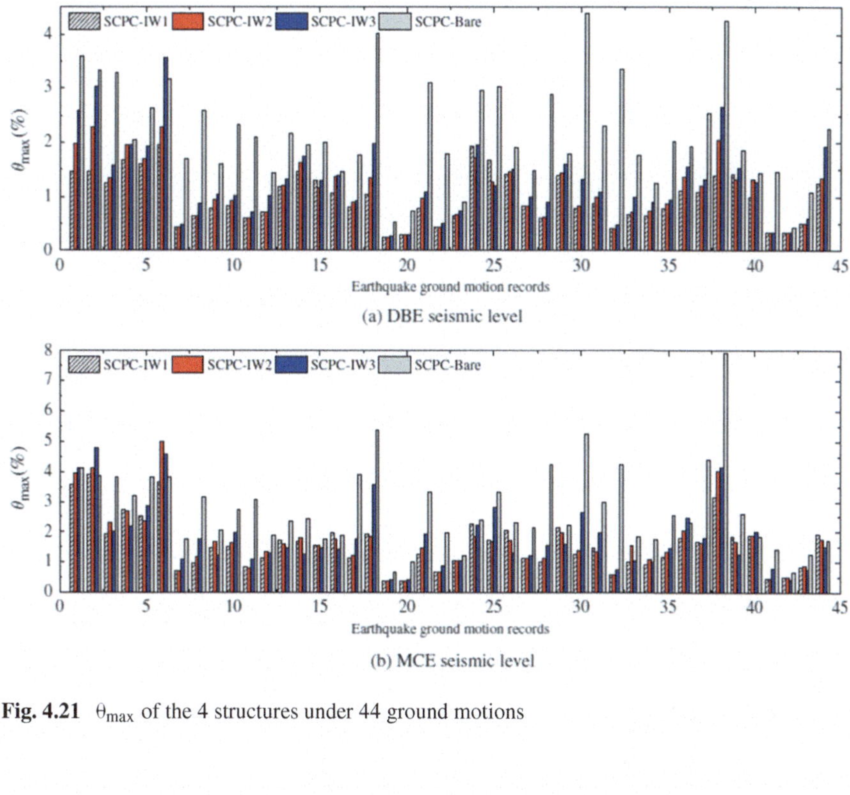

(a) DBE seismic level

(b) MCE seismic level

Fig. 4.21 θ_{max} of the 4 structures under 44 ground motions

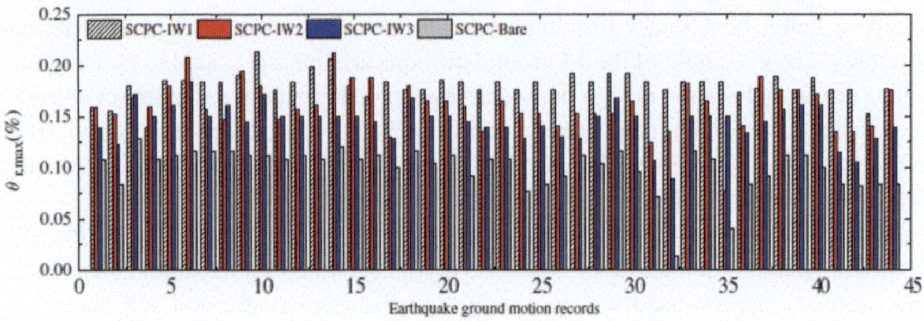

(a) DBE seismic level

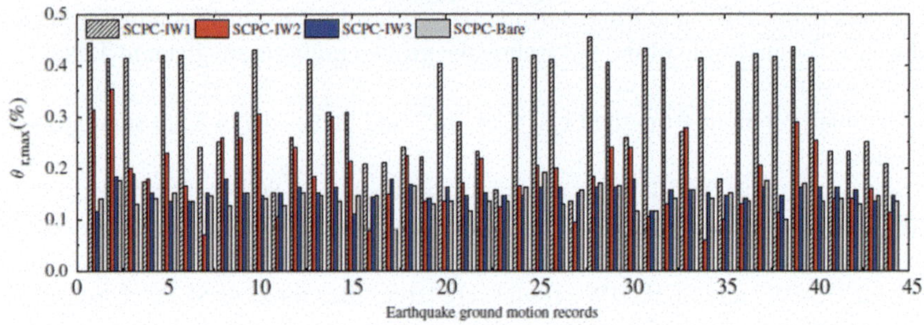

(b) MCE seismic level

Fig. 4.22 $\theta_{r,max}$ of the four structures under 44 ground motions

Table 4.7 Dynamic response of structures under DBE-level

Dynamic response	SCPC-IW1	SCPC-IW2	SCPC-IW3	SCPC-Bare
θ_{max}	0.97%	1.08%	1.28%	2.19%
$\theta_{r,max}$	0.18%	0.16%	0.14%	0.10%
D_{CF}	2.04	1.80	1.67	1.54

Table 4.8 Dynamic response of structures under MBE-level

Dynamic response	SCPC-IW1	SCPC-IW2	SCPC-IW3	SCPC-Bare
θ_{max}	1.57%	1.66%	1.80%	2.81%
$\theta_{r,max}$	0.32%	0.18%	0.15%	0.13%
D_{CF}	2.07	1.82	1.76	1.55

structures under DBE-level ground motions are shown in Table 4.7. It can be seen that the average values of θ_{max} of SCPC-IW1, SCPC-IW2, SCPC-IW3 and SCPC-Bare frames are 0.97%, 1.08%, 1.28% and 2.19% (listed in Table 4.7), respectively, which present obvious distinction, whereas the average values of $\theta_{r,max}$ for the corresponding frames are not distincted obviously, i.e. 0.18%, 0.16%, 0.14% and 0.10%, respectively (listed in Table 4.7), indicating that with the increase of infill walls strength, the θ_{max} of SCPC-IW frames is decreased while the $\theta_{r,max}$ is increased. The percentage decrease of the θ_{max} estimated by SCPC-IW1, SCPC-IW2, SCPC-IW3 frames, with respect to the SCPC-Bare frame, are 56%, 51%, 42%, respectively. These are in contrast to an increase of 80.0%, 60.1%, 40.3% for the $\theta_{r,max}$.

A similar observation obtained from DBE seismic hazard level can be found under MCE seismic hazard level ground motions. The mean values of the θ_{max} of SCPC-IW1, SCPC-IW2, SCPC-IW3 and SCPC-Bare frames are 1.57%, 1.66%, 1.80% and 2.81%, respectively, while the corresponding $\theta_{r,max}$ are 0.32%, 0.18%, 0.15% and 0.13%, respectively. The percentage decrease of the θ_{max} estimated by SCPC-IW1, SCPC-IW2, SCPC-IW3 frames, with respect to the SCPC-Bare frame are also computed, which are 44%, 41%, 36%, respectively. This are in contrast to an increase of 146%, 38%, 15% for the $\theta_{r,max}$. It is worth noting that the $\theta_{r,max}$ of the SCPC-Bare frame is obviously smaller than the three SCPC-IW frames. A recent study showed that the RC frames with sustained $\theta_{r,max}$ greater than 0.2% may suffer slight damage but can be instant occupancy with simple repair, and 0.4% may no longer be suitable for repair from an economic perspective. For SCPC-IW1 frame (Fig. 4.22), there are 21 earthquake ground motions cause the $\theta_{r,max}$ exceed 0.4%, and the average value is significantly greater than 0.2%. Therefore, the SCPC-IW1 is not suitable in design because of the excessive residual deformation.

As previously mentioned that the presence of infill walls may result in soft story failure for infilled RC frames, so as one of specific infilled RC frame, the soft story effect of SCPC-IW frames is worth investigating. The soft story effect is always evaluated by drift concentration factor (D_{CF}), in which the comprehensive factors, such as the lateral stiffness, number of story are considered. D_{CF} also reflects the uniform story drift distribution. Equation (4.5) is the definition of D_{CF}.

$$D_{CF} = \theta_{max}/(\Delta_{roof}/h_n) \tag{4.5}$$

where Δ_{roof} and h_n are the maximum top displacement and the total height of the structure, respectively.

The D_{CF} of the SCPC-Bare and three SCPC-IW frames under DBE and MCE ground motion records are shown in Fig. 4.23, respectively, and the mean D_{CF} values of the four models under 44 ground motion records are also shown in Table 4.7 and Table 4.8. The mean D_{CF} values of three SCPC-IW frames are larger than those of the SCPC-Bare frames, indicating that the SCPC-IW frames have more unevenly distributed lateral deformations over the building height and are more sensitive to the deformation of soft story in comparison to the SCPC-Bare frames; As the increase of infill walls strength, the mean DCF values of three SCPC-IW frames exhibit a descending order of SCPC-IW1 > SCPC-IW2 > SCPC-IW3, indicating that with the increase of infill walls strength, the SCPC-IW frames have more unevenly distributed lateral deformation over the building height and are more sensitive to the deformation of soft story.

The mean value of the θ_{max} and $\theta_{r,max}$ along the building elevation under the DBE and MCE seismic hazard levels are presented in Figs. 4.24 and 4.25. In the distribution

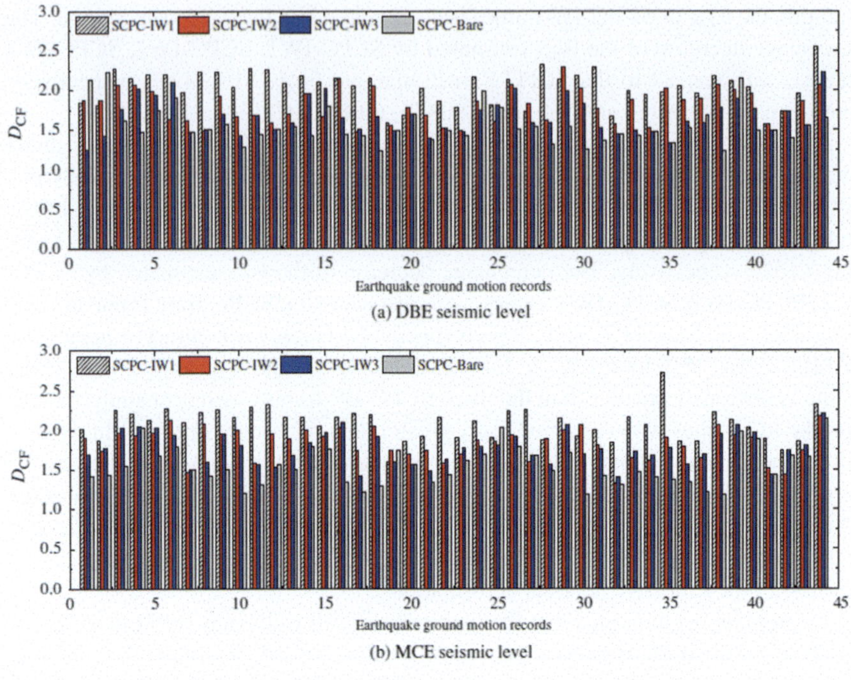

Fig. 4.23 DCF of the 4 structures under 44 ground motions

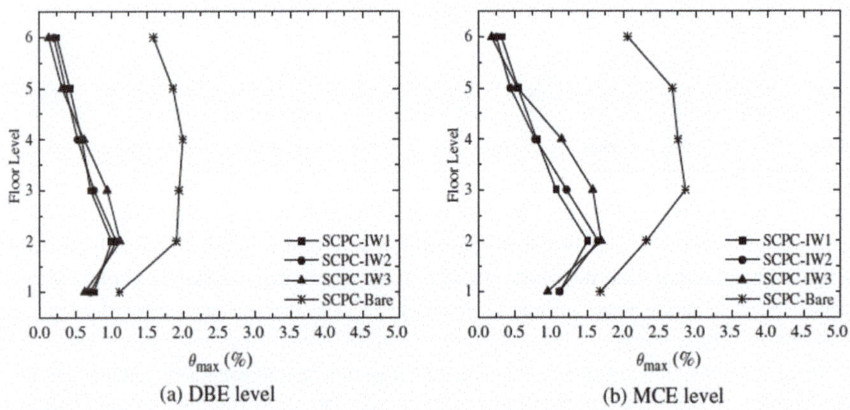

Fig. 4.24 θ_{max} along the building elevation

of the θ_{max} along the building elevation, the SCPC-Bare frame exhibit lower inter-story drift ratio at the first story as well as higher values at the upper levels, while the three SCPC-IW frames exhibit lower inter-story drift ratio at the upper floor and higher values at the second floor. The $\theta_{r,max}$ along the building elevation is similar to the results obtained from θ_{max}. The$\theta_{r,max}$ of SCPC-IW1 frame is larger than 0.2% at

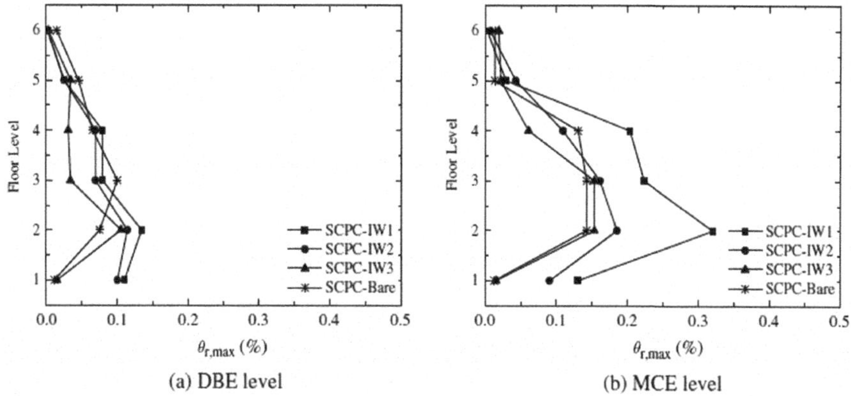

Fig. 4.25 $\theta_{r,max}$ along the building elevation

the second, third and fourth floor. It is indicated that the presence of infill walls can decrease the seismic response of SCPC frames at upper floors, but the deformation concentration effect is aggravated at lower floor of structures.

4.6.3 Comparison of Energy Dissipation Capability Under the DBE and MCE Seismic Hazard Level

The energy dissipation is one of the most crucial factors in studying the seismic behavior of SCPC frames under earthquake actions. The energy dissipation contribution of the SCPC-IW frames is mainly provided by two elements: (1) the web friction devices, in which the energy is dissipated by the friction between friction plates and steel plates; and (2) the infill walls, in which the dissipated energy depends on its plastic deformation. The ground motion record B-WSM090, which is selected from the 1987 Superstition Hills (B) Earthquake with a duration of 40 s and an epicentral displacement for 19.51 km, is found similar to the mean spectrum of the overall 44 ground motions in terms of its spectrum properties and duration. Besides, the dynamic responses of structures obtained from B-WSM090 are approximately equal to the mean results of ensemble of 44 ground motions. Therefore, the ground motion record B-WSM090 was selected and scaled to DBE and MCE seismic hazard levels to investigate the energy dissipation contribution of infill walls, which could keep a balance between the accuracy and the computational efficiency.

The hysteretic energy dissipated at each infill wall or web friction device over the ground motion record is calculated based on related force–deformation hysteresis loops. The total dissipated energy is calculated as the sum of hysteretic energy dissipated for all infill walls or web friction devices over a course of a particular ground motion time history. Total hysteretic energy is calculated as Eq. (4.6).

$$E_H = \sum_{i=1}^{N} \sum_{j=1}^{n} E_{j,i,H} \tag{4.6}$$

where $E_{j,i,H}$ is total hysteresis energy dissipated by infill walls or web friction devices in the *j*th *cycle* at the *i*th storey, n is the number of cycles in structure response and N is the total number of stories.

Defining the energy dissipation ratio of infill walls to web friction devices (denoted as $\beta = E_{IW}/E_{FD}$ in the following text) to determine the relative energy dissipation contribution provided by infill walls to the whole seismic energy dissipation.

Figure 4.26 compares β of SCPC-IW1, SCPC-IW2 and SCPC-IW3 frames along the building height under DBE and MCE seismic levels. Over all, β and θ_{max} along the building height are similar in shape. The β of three SCPC-IW frames at the second and third floor exceed 1 under DBE seismic hazard level, indicating the energy dissipation contribution of infill walls is larger than web friction devices at the second and third floor. For MCE seismic hazard level, there are also several floor whose β exceed 1 and β along the building height shows a slight difference with DBE seismic hazard level.

In DBE level, the energy dissipation contribution of infill walls is 70.9 to 105.6% of friction devices, and is 42.4 to 82.3% of friction devices for MCE level. It is indicated that the infill walls can make considerable energy dissipation contribution to SCPC-IW frames. For investigating the energy dissipation contribution of infill walls under different seismic hazard level, Fig. 4.27 plots β of three SCPC-IW frames under DBE and MCE hazard levels. It can be seen that β of three SCPC-IW frames under DBE hazard level is larger than that of MCE hazard level, indicating that the energy dissipation contribution of infill walls under DBE level is larger than MCE

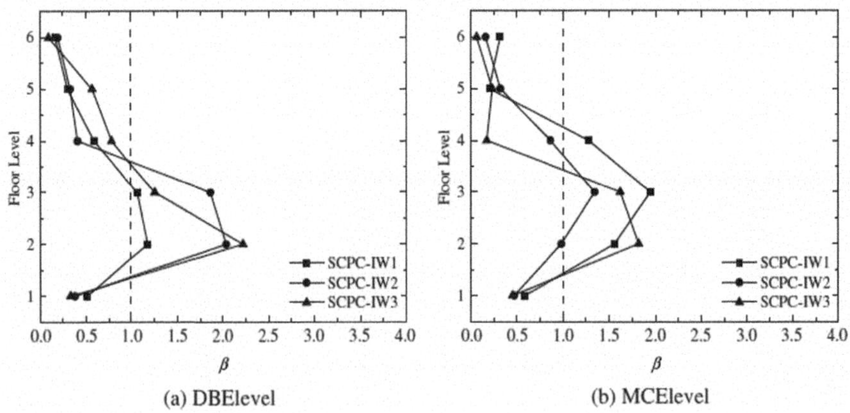

Fig. 4.26 β along the building elevation

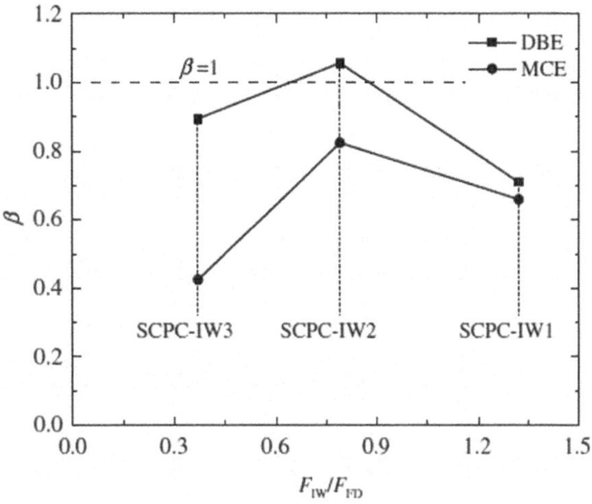

Fig. 4.27 Total energy dissipation contribution of SCPC-IW framesw

level. While the increase in level of infill walls strength does not necessarily result in a monotonic increase in energy dissipation capability. Whether DBE or MCE seismic hazard level, the most significant contribution of energy dissipation is SCPC-IW2 frame.

4.7 Seismic Fragility Analysis of Friction Damped Self-Centering Precast Frames

4.7.1 Incremental Dynamic Analysis (IDA)

To investigate the performance of SCPC-IW frames under random ground motions, incremental dynamic analyses (IDAs) are performed with the 44 ground motions (Fig. 4.20) [3]. The spectral acceleration $S_a(T_1, 5\%)$, which is proved to be a better intensity measure in terms of the efficiency and proficiency to its smaller dispersion and have well correlate with the structural response, is selected as the Intensity Measure (IM), and the maximum inter-story drift ratio (θ_{max}) and maximum residual inter-story drift ratio ($\theta_{r,max}$) are used as the Damage Measure (DM). For each ground motion record, the nonlinear history analysis is repeatedly performed, with the $Sa(T_1, 5\%)$ of the record increasing from 0.1 g with an increment of 0.1 g until the collapse occurs, where the slope of the curve is 20% of the initial slope. After the collapse point, the IDA curve is close to a horizontal line, such that further increase of ground motion intensity lead to the increase of inter-story drift ratio without bounds. All analyses are implemented with additional zero acceleration values padded to all ground motion records to allow the structure develop free vibration after exciting of ground motion to capture the residual deformation.

The results of IDA based on θ_{max} for the SCPC-Bare frame and three SCPC-IW frames are illustrated in Fig. 4.28, showing the relationship between the ground motion intensity Sa(T_1, 5%) and θ_{max} for the 44 ground motions.

For quantitatively analyze the seismic performance of structures and directly compare the IDA results, on the one hand, the 16, 50 and 84% fractile IDA curves are computed. On the other hand, four different limit states and their control objectives defined by FEMA 356 are selected in analyses, which are: Very Light damage, Light damage, Moderate damage, Severe damage. HAUZS proposed corresponding θ_{max} threshold values about the four limit states, which are 0.33%, 0.67%, 2.0% and 5.3%, respectively [4]. So the 16%, 50% and 84% fractile IDA curves of the 4 structures and four vertical lines (θ_{max} = 0.33%, 0.67%, 0.20%, 0.53%) are also marked in Fig. 4.28. The $S_a(T_1, 5\%)$ values obtained from the intersection points between three fractile IDA curves and four vertical lines are statistically summarized in Table 4.9.

The results shown in Fig. 4.28 and Table 4.9 indicate that for reaching a given seismic response θ_{max}, the three SCPC-IW frames need larger seismic excitation

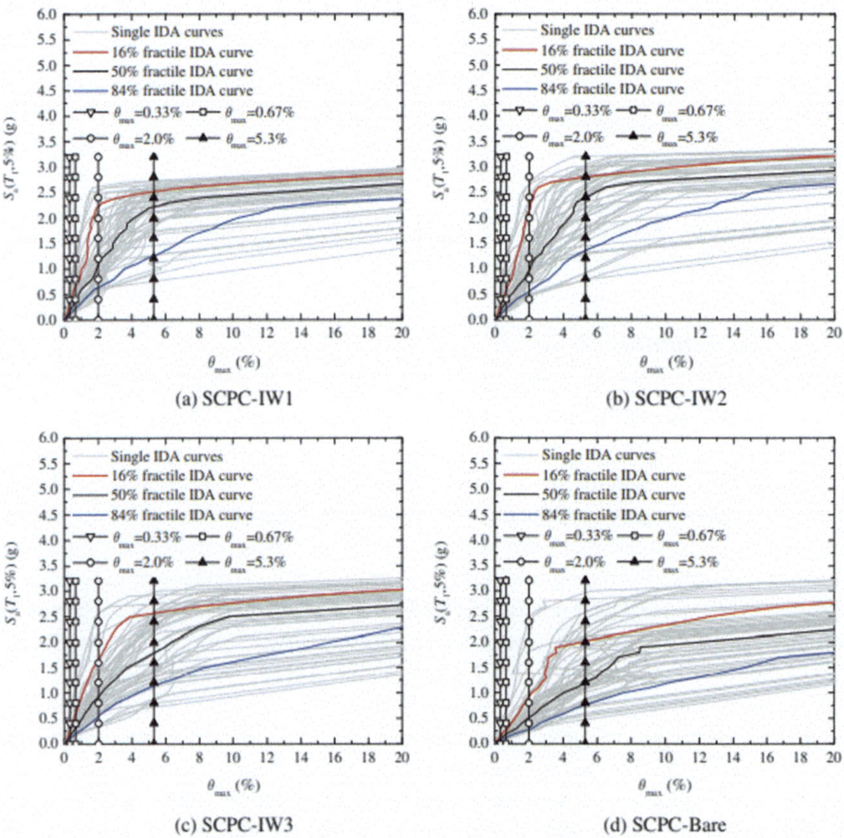

Fig. 4.28 The IDA curves based on θ_{max}

Table 4.9 Required seismic excitation $Sa(T_1, 5\%)$ to reach four limit states based on θ_{max}

	SCPC-IW1			SCPC-IW2			SCPC-IW3			SCPC-Bare		
	16%	50%	84%	16%	50%	84%	16%	50%	84%	16%	50%	84%
$\theta_{max} = 0.33\%$	0.29 g	0.17 g	0.11 g	0.31 g	0.17 g	0.11 g	0.26 g	0.16 g	0.11 g	0.15 g	0.10 g	0.07 g
$\theta_{max} = 0.67\%$	0.65 g	0.38 g	0.23 g	0.59 g	0.37 g	0.22 g	0.61 g	0.31 g	0.20 g	0.27 g	0.19 g	0.13 g
$\theta_{max} = 2.0\%$	2.24 g	1.09 g	0.64 g	2.16 g	0.96 g	0.56 g	1.67 g	0.93 g	0.54 g	0.92 g	0.51 g	0.28 g
$\theta_{max} = 5.3\%$	2.53 g	2.22 g	1.26 g	2.80 g	2.39 g	1.35 g	2.57 g	1.79 g	1.14 g	2.02 g	1.20 g	0.76 g

Sa(T1, 5%) than SCPC-Bare frame. While for a given seismic response θ_{max}, the needed seismic excitation $S_a(T_1, 5\%)$ values of SCPC-IW frame are not always increase with the increasing of infill walls strength. When the seismic response θ_{max} reach to the limit states $\theta_{max} = 0.33\%$, $\theta_{max} = 0.67\%$ and $\theta_{max} = 2.0\%$, the needed seismic excitation $Sa(T_1, 5\%)$ values with an ascending order of SCPC-IW3 < SCPC-IW2 < SCPC-IW1 (see the 50% fractile IDA curve in Fig. 4.28 and corresponding $Sa(T_1, 5\%)$ in Table 4.9), indicating that only $F_{IW}/F_{FD} < 0.79$, can the increase of infill walls strength be helpful to strengthen SCPC-IW frames at all limit states. While to reach to the limit state $\theta_{max} = 5.3\%$ (Severe damage), the ascending order is SCPC-IW3 < SCPC-IW1 < SCPC-IW2 (see the 50% fractile IDA curve in Fig. 4.28 and corresponding Sa(T1, 5%) in Table 4.9), indicating F_{IW}/F_{FD} reach to 1.32, the infill walls may result in more severe damage comparing with SCPC-IW frames with lower strength.

Similar to the IDA based on θ_{max}, the IDA curves and 16, 50, 84% fractile IDA curves based on the $\theta_{r,max}$ have been obtained to investigate the influence of infill walls on the self-centering and rehabilitation capability of SCPC-IW frames. Based on the recommendation of FEMA 356, three limit states and its control objectives are proposed by Kam et al. [5], namely, Instant Occupancy Limit State (IOLS), Repairable Limit State (RLS) and Life Safety Limit State (LSLS). The corresponding threshold values of $\theta_{r,max}$ are 0.2%, 0.4% and 1.0%, respectively. Figure 4.29 present the IDA curves and 16, 50, 84% fractile IDA curves based on $\theta_{r,max}$. Table 4.10 summarizes the intersection points between three fractile IDA curves and four vertical lines, which represent the required seismic excitation $Sa(T_1, 5\%)$ values to reach the three limit states ($\theta_{r,max} = 0.2, 0.4, 1.0\%$).

It is observed that unlike the changing tendency of θ_{max} with the increasing of infill walls strength, the results based on $\theta_{r,max}$ is increased monotonously. The seismic excitation $S_a(T_1, 5\%)$ values needed to achieve the same seismic limit states ($\theta_{r,max} = 0.2\%$, $\theta_{r,max} = 0.4\%$ and $\theta_{r,max} = 1.0\%$) with a ascending order of SCPC-IW1 < SCPC-IW2 < SCPC-IW3 < SCPC-Bare.

From above IDA results based on θ_{max} and $\theta_{r,max}$, it can be concluded that the SCPC-IW2 and SCPC-IW3 frames both exhibit superior performance in controlling the inter-story drift ratio in comparison to the SCPC-Bare frame and in controlling the residual inter-story drift ratio in comparison to the SCPC-IW1 frame. This implies that excessively high strength infill walls in SCPC-IW frames do not necessarily result in a better structural performance, even causing more severe damage during severe or mega earthquake.

4.7.2 *Seismic Fragility Analysis*

The fragility analysis, as one of the most important contents of performance-based earthquake engineering, can be used to predict the conditional probability of structural damage under different earthquake intensities. In general, the fragility analysis is based on the incremental dynamic analysis and defined as the conditional probability

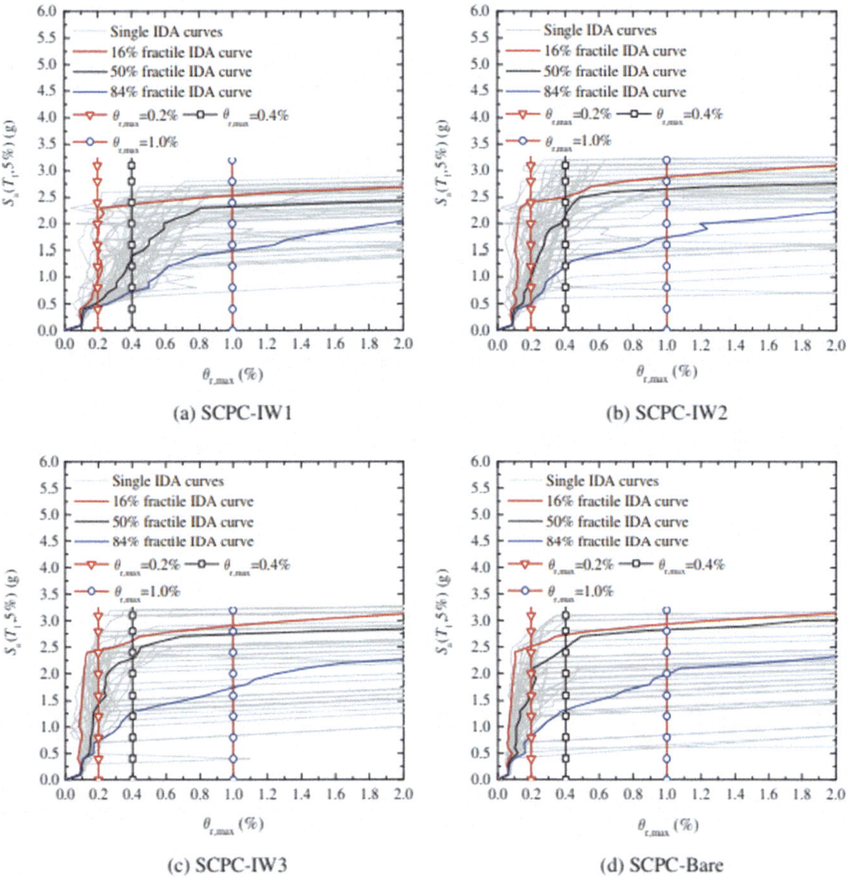

Fig. 4.29 The IDA curves based on $\theta_{r,max}$

of attaining or exceeding an expected limit state for a given set of boundary variables. The results of the fragility analysis always prescribed by a fragility curve. Given lognormal distributions for both the structural capacity (*C*) and seismic demand (*D*), the fragility function used to define the fragility curve as Eq. (4.7) [6]

$$P(D > C/IM) = \Phi\left[\frac{\ln(S_{\mathrm{d}}/S_{\mathrm{c}})}{\sqrt{\beta_{\mathrm{c}}^2 + \beta_{\mathrm{m}}^2 + \beta_{d/\mathrm{IM}}^2}}\right] \tag{4.7}$$

where $\Phi[.]$ is the standard normal distribution function; S_{d} is the median structural demand, S_{c} is the median structural capacity, associated with the limit states; β_{c} denotes the aleatoric uncertainty of structural capacity, can be assumed to be 0.25 and 0.47 for lower and higher limit states, respectively [7]; β_{m} is the modeling uncertainty, can be assumed to be 0.2 [8, 9]; $\beta_{\mathrm{d/IM}}$ is the aleatoric uncertainty (logarithmic standard

Table 4.10 Required seismic excitation Sa(T_1, 5%) to reach three limit states based on $\theta_{r,max}$

	SCPC-IW1			SCPC-IW2			SCPC-IW3			SCPC-Bare		
	16%	50%	84%	16%	50%	84%	16%	50%	84%	16%	50%	84%
$\theta_{r,max} = 0.2\%$	0.94 g	0.53 g	0.46 g	2.41 g	1.0 g	0.50 g	2.44 g	1.44 g	0.75 g	2.51 g	1.70 g	0.83 g
$\theta_{r,max} = 0.4\%$	2.28 g	1.40 g	0.73 g	2.51 g	2.14 g	1.20 g	2.62 g	2.28 g	1.27 g	2.73 g	2.48 g	1.33 g
$\theta_{r,max} = 1.0\%$	2.54 g	2.32 g	1.49 g	2.88 g	2.65 g	1.74 g	2.91 g	2.74 g	1.75 g	2.94 g	2.82 g	1.98 g

deviation) of the demand conditioned on the IM, which is determined by the linear regression of the demand-intensity measure pairs in the log-transformed space [9].

The seismic response of a structure is often directly reflected by $\theta_{\max}$, so using the four limit states as explained in the previous section as well as corresponding threshold values ($\theta_{\max} = 0.33$, 0.67, 2.0 and 5.3%) and IDA results, the fragility analysis is performed to quantify the seismic performance of the four structures. The linear regression of $\ln(\theta_{\max}) - \ln(S_a(T_1, 5\%))$ is illustrated in Fig. 4.30. Based on the regression results, the corresponding $\beta_{d/IM}$ values are calculated and also presented in Fig. 4.30.

Based on Eq. (4.7) and Fig. 4.30 the fragility curves of the four structures are shown in Fig. 4.31, in which the x-axis is the ground motion intensity $S_a(T_1, 5\%)$ and the y-axis is the exceeding probability P_f.

The fragility curves for Very Light, Light and Moderate limit states have similar graphs and only different amounts, but some anomalous regularities are found for

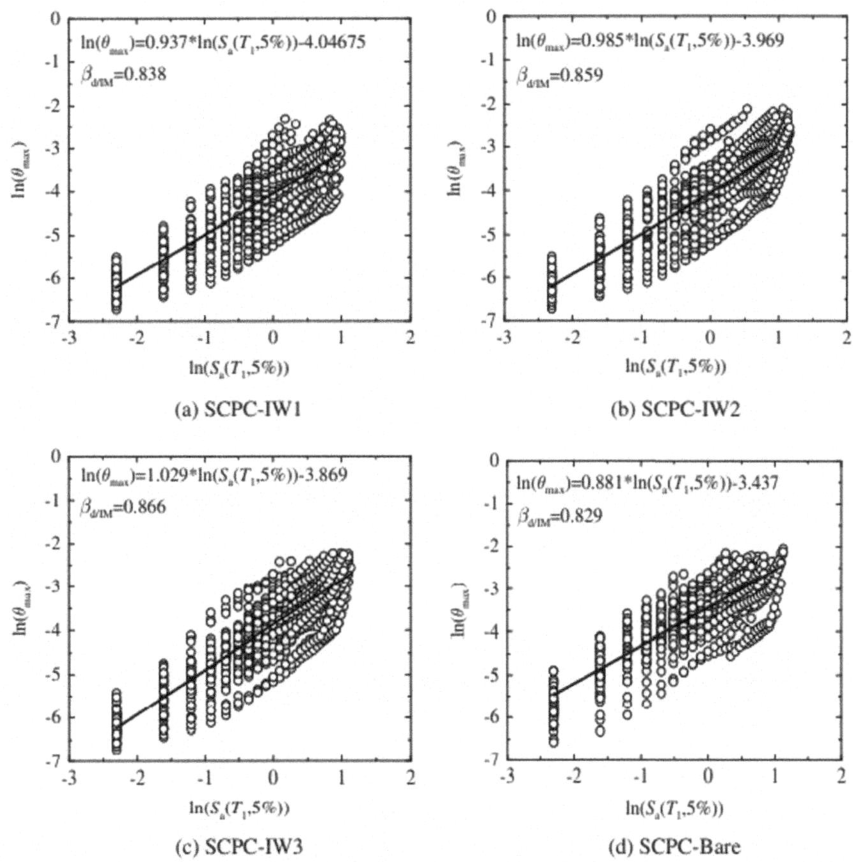

Fig. 4.30 Linear regression of $\ln(\theta_{\max}) - \ln(S_a(T_1, 5\%))$ used to approximate $\beta_{d/IM}$

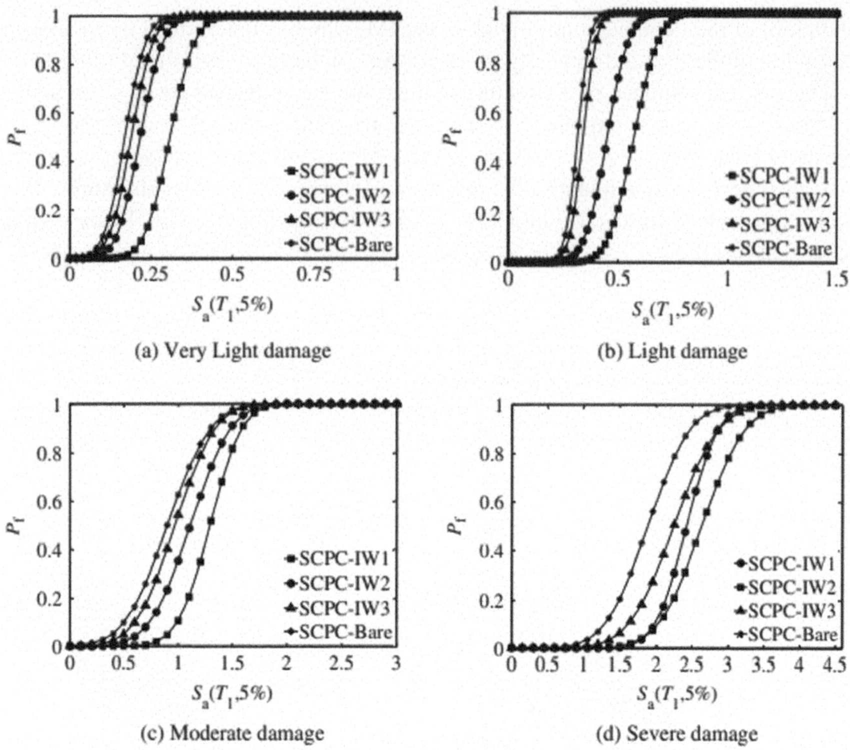

Fig. 4.31 The fragility curves based on θ_{max}

Severe limit state. The variance among fragility curves in Fig. 4.31 proves that when the damage scale does not exceed Severe limit state, the failure possibility of SCPC-IW frames may be decreased with the increasing of infill walls strength. While the damage scale reaches to Severe limit state (Fig. 4.31d), when $S_a(T_1, 5\%)$ exceed 1.74 g, the exceeding probability of SCPC-IW1 frame is larger than SCPC-IW2 frame and larger than SCPC-IW3 frame when $S_a(T_1, 5\%)$ exceed 2.71 g, indicating that the SCPC-IW frames with excessive high strength infill walls may exhibit higher exceedance probability at Severe limit state.

Recent post-earthquake functionality assessment of structures has highlighted that residual inter-story drift ratio is an important factor in the post-earthquake safety of a building and economic feasibility of repair and reconstruction [10]. This paper also carried out the fragility analyses based on $\theta_{r,max}$. Using above three limit states and corresponding $\theta_{r,max}$ threshold values ($\theta_{r,max} = 0.2\%$, 0.4%, and 1.0%), the seismic fragility analyses are also conducted. The linear regression of $\ln(\theta_{r,max}) - \ln(S_a(T_1, 5\%))$ and the seismic fragility curves based on $\theta_{r,max}$ of the four structures are derived as shown in Figs. 4.30 and 4.31. It can be clearly seen that there is a significant difference between the fragility analyses based on θ_{max} and $\theta_{r,max}$. As shown in Fig. 4.31, the SCPC-Bare frame has the lowest exceedance probability at

Table 4.11 The median values (S_d) of fragility results based on θ_{max}

Building type	Limit states			
	Very light (g)	Light (g)	Moderate (g)	Severe (g)
SCPC-IW1	0.308	0.572	1.29	2.42
SCPC-IW2	0.217	0.449	1.11	2.63
SCPC-IW3	0.191	0.341	0.97	2.24
SCPC-Bare	0.172	0.315	0.90	1.88

Table 4.12 The median values (S_d) of fragility results based on $\theta_{r,max}$

Building type	Limit states		
	IOLS (g)	RLS (g)	LSLS (g)
SCPC-IW1	0.72	1.68	2.35
SCPC-IW2	1.13	2.07	2.73
SCPC-IW3	1.56	2.13	2.89
SCPC-Bare	1.71	2.48	3.21

all limit states. At each limit state, the exceedance probability is increased as the infill walls strength increases. It is indicated that the presence of infill walls significantly decrease the possibility of structural repair. With the increase of infill walls strength, the possibility of repair for SCPC-IW frames is gradually decreased.

In order to conduct quantitative analyses for fragility curves, the median values (S_d) of fragility results based on θ_{max} are listed in Table 4.11, with a descending order of SCPC-IW1 > SCPC-IW2 > SCPC-IW3 > SCPC-Bare for Very Light, Light and Moderate limit states. While, for Severe limit state, the descending order is SCPC-IW2 > SCPC-IW1 > SCPC-IW3 > SCPC-Bare. The SCPC-IW1 frame, which is installed with the highest strength infill wall, has lower S_d comparing to SCPC-IW2 frame. The median values (S_d) of fragility results based on $\theta_{r,max}$ are also calculated as shown in Table 4.12, with a descending order of SCPC-Bare > SCPC-IW3 > SCPC-IW2 > SCPC-IW1 for all three limit states.

From this fact, it can be concluded that from the view of fragility analyses based on θ_{max}, SCPC-IW1, SCPC-IW2 and SCPC-IW3 frames perform obviously better seismic performance than SCPC-Bare frame during seismic activities, and SCPC-IW1 frame perform poorer than SCPC-IW2 frame. While from the view of fragility analyses based on $\theta_{r,max}$, SCPC-IW1, SCPC-IW2 and SCPC-IW3 frames perform obviously poorer than SCPC-Bare frame. With the increase of infill walls strength, the self-centering capability of SCPC-IW frames is gradually decreased. It is indicated that the arrangement of infill walls can decrease the exceedance probability of structural damage but increase the residual deformation of SCPC frames. Only the infill walls load capacity ranges from 37 to 79% of the web friction force, can the SCPC-IW frames achieve a good balance between self-centering capability and structural damage.

4.8 Conclusions

This chapter first introduces the structure and mechanism of friction damped self-centering precast frames. Then the stability of Nao friction material was verified through the performance test of friction energy dissipaters. Then, the influence of infilled wall on SCPC frame under earthquake is studied by numerical simulation, and the static analyses, dynamic analysis, IDA analyses and fragility analyses were carried out in this study. Based on the analyses results, the following conclusions can be obtained:

1. The arrangement of infill walls can not only enhance the stiffness but also improve energy dissipation capacity of SCPC frames. But as the infill walls strength increases, the self-centering capability of SCPC-IW frames gradually decreases or even diminishes and the axial compression ratio of column gradually increases. When the infill walls load capacity ranges from 37 to 79% of the web friction force, the examined SCPC-IW frames can achieve a good balance in terms of strength, stiffness, energy dissipation and self-centering capability.
2. Comparison of seismic response indicates that all SCPC-IW frames have achieved good results in decreasing maximum inter-story drift ratio at both DBE and MCE. But the maximum residual inter-story drift ratio increases as the infill walls strength increases, and the corresponding values of SCPC-IW1 frame have almost exceed the threshold of Instant Occupancy Limit State even likely beyond Repairable Limit State under MCE.
3. Regarding the fragility analyses based on maximum inter-story drift ratio and maximum residual inter-story drift ratio, it can be concluded that for the former case, the SCPC-IW1, SCPC-IW2 and SCPC-IW3 frames perform better than SCPC -Bare frame in earthquakes, but SCPC-IW1 frame perform poorer than SCPC-IW2 in Severe limit state. While, in the latter case, SCPC-IW1, SCPC-IW2 and SCPC-IW3 frames perform poorer than SCPC-Bare after earthquakes and with the increase of infill walls' strength, the self-centering capability of SCPC-IW frames gradually decrease. Therefore, in the design of SCPC-IW frames, the load capacity of infill walls and web friction force should be controlled in a reasonable range to guarantee structures repairable under earthquake.
4. The collapse resistance capacity of fully-infilled SCPC-IW frames is larger than SCPC-Bare frames. However, the collapse resistance capacity of fully-infilled SCPC-IW frames is not necessarily enhanced with the increase of infill walls' strength. If excessively high strength infill walls are used in the design of SCPC-IW frames, it is likely to produce an uneconomical over-designed structure and the collapse resistance capability of structures may decrease.

References

1. Song LL, Guo T, Chen C (2014) Experimental and numerical study of a self-centering prestressed concrete moment resisting frame connection with bolted web friction devices. Earthq Eng Struct Dynam 43(4):529–545
2. Tawfik Essa ASA, Kotp Badr MR, Elzanaty AH (2014) Effect of infill wall on the ductility and behavior of high strength reinforced concrete frames. Hbrc J 10(3):258–264
3. Huang L, Zhou Z , Zhang Z et al (2008) Seismic performance and fragility analyses of self-centering prestressed concrete frames with infill walls. J Earthq Eng:1–31
4. FEMA (2003) HAZUS-MH MR4 technical manual, earthquake model, Federal Emergency Management Agency, Washington, D.C
5. Kam WY, Pampanin S, Palermo A, Carr A (2008) Design procedure and behaviour of advanced flag-shape (AFS) MDOF systems, University of Canterbury Civil and Natural Resources Engineering
6. Jamnani HH, Abdollahzadeh G, Faghihmaleki H (2017) Seismic fragility analysis of improved RC frames using different types of bracing. J Eng Sci Technol 12(4):913–934
7. Celik OC, Ellingwood BR (2009) Seismic risk assessment of gravity load designed reinforced concrete frames subjected to mid-America ground motions. J Struct Eng 135(4):414–424
8. Celik OC, Ellingwood BR (2010) Seismic fragilities for non-ductile reinforced concrete frames—role of aleatoric and epistemic uncertainties. Struct Saf 32(1):1–12
9. Jeon JS, DesRoches R, Brilakis I, Lowes LN (2012) Modeling and fragility analysis of non-ductile reinforced concrete buildings in low-to-moderate seismic zones. Struct Cong 2012:2199–2210
10. Solomon T, Katsuichiro G (2015) Seismic performance evaluation framework considering maximum and residual inter-story drift ratios: application to non-code conforming reinforced concrete buildings in Victoria, BC, Canada. Front Built Environ 1

Chapter 5
Cast-In-Place Frame-Prefabricated Sub-Frame System

Abstract This chapter studies the cast-in-place frame-prefabricated sub-frame system. Seismic isolation, energy dissipation technology, and primary-secondary structure system are adopted to reduce the seismic demand of prefabricated secondary frames, improve the energy dissipation capacity of the whole structure, and ensure the seismic safety of the structure.

Keywords Cast-in-place frame-prefabricated sub-frame system · Primary-secondary structure system · Mega frame · Seismic isolation · Energy-dissipation hinge point connection · Skateboard corbel

5.1 Introduction

The primary and secondary structural system is a new type of structural system for super high-rise buildings. Such system has reasonable arrangement and structural design with clear mechanics, the main frame bears all the vertical load and lateral load, and the secondary frame bears a small part of the lateral load and transfers the vertical load to the main frame [1]. The construction sequence of the primary and secondary frame structure is different from the ordinary frame structure. The main frame can be constructed first, and the secondary frame structure can then be constructed at the same time. Thanks to this special construction sequence, the construction period can be greatly shortened. If a prefabricated assembly of secondary structures can be realized, this primary and secondary frame structure will be the first choice for prefabricated buildings. In other words, scale housing construction and industrialization will be within reach. At present, prefabricated buildings are being vigorously promoted in China, which has the advantages of improving production efficiency, shortening construction period, improving building quality, convenient management, environmental protection and so on [2–4]. However, it is difficult to realize prefabricated assembly in high seismic hazard zone. There are all kinds of difficulties in structure design and construction of traditional prefabricated concrete structure in high seismic hazard zone, which constrains the advantages and great protentials of prefabricated concrete structure, such

G. Wu et al., *Novel Precast Concrete Structure Systems*, Springer Tracts in Civil Engineering, https://doi.org/10.1007/978-981-19-6821-1_5

as low cost, fast construction and environmental protection. Furthermore, safety concern also needs to be considered [5–10].

Therefore, this chapter studies the cast-in-place frame-prefabricated sub-frame system. Seismic isolation technology, energy dissipation technology and primary-secondary structure system are adopted to reduce the seismic demand of prefabricated secondary frame, to improve the energy dissipation capacity of the whole structure, and to ensure the seismic safety of the structure.

5.2 Primary and Secondary Frame System Based on Dissipation Hinge Joints

The internal force distribution of the primary and secondary frame has significant secondary force characteristics. The main frame should bear not only its own load but also the load transferred by the secondary frame, so the beam-column internal force of the main frame is much larger than that of the secondary frame. The yield mechanism of mega-frame under earthquake is as follows: successively yield of sub-frame beam, sub-frame column, mega-beam and mega-column [11]; the sub-frame is seriously damaged under a rare and sever earthquake. In order to realize the prefabricated assembly in high seismic intensity area, the seismic damage and ductility requirements of sub-frame structural members must be reduced.

This section uses hinge joints that can assemble and dissipate energy [12] to realize the assembly connection between the main frame and the sub-frame, and eccentrically BRB is used to further improve the rotation of the hinge joints, so that the plastic deformation of the structure is concentrated in the hinge joints. Other members of the sub-frame are in the elastic or slight damage stage. The damage and energy distribution of the structural system are studied by elastoplastic time-history analysis. The parametric analysis of hinge joints is carried out.

5.2.1 Hinge Joints Between Primary and Secondary Frames

A steel hinge is pre-embedded in each of the sub-frame beam and the mega column, a pin is employed to connect steel hinges, and the steel hinge is bolted with two groove steel plates around the steel hinge. The pin bears the vertical load, the groove steel plate provides anti-rotation force so that the energy consumption is concentrated on the steel plate. The design of the joint is to design the anti-rotation ability of the pin and the surrounding groove steel plate. The connection of the energy dissipation hinge joint is shown in Fig. 5.1.

In this system, a joint of energy dissipation hinge is adopted to connect the sub-frame beam and the main frame column, and the connection of the other sub-frame beam-column is considered as equivalent to cast-in-place. This section specifically

Fig. 5.1 Energy dissipation hinge point connection

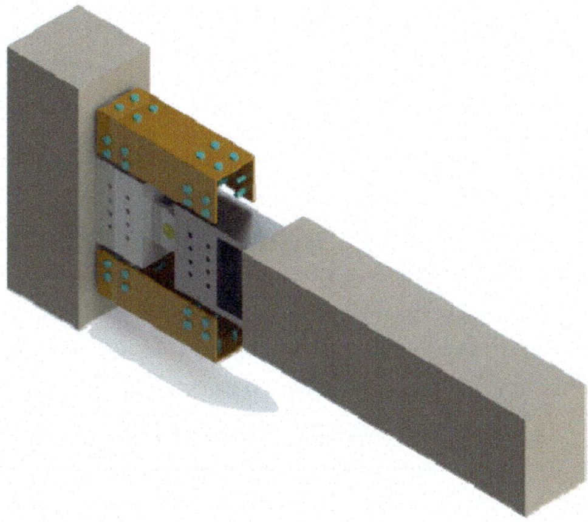

introduces the constitutive and parameter setting of the energy dissipation hinge joint. The pin of the energy dissipation hinge joint mainly bears the shear force, and the bending moment is borne by the surrounding groove cover plate. Since the shear capacity of the pin is sufficient, only the bending capacity of the groove cover is considered in design.

The moment–curvature relationship of the groove cover plate is calculated based on the buckling theory of unidirectional uniformly compressed thin plates. It is assumed that the four sides of the plate are simply supported, and the upper and lower two groove cover plates of the energy dissipation hinge joints are made of Q235 steel to form a closed box section. In the design of energy dissipation hinge joints, the section size of the groove cover plate can be preliminarily selected according to the section size of the sub-frame beam. The local correlation buckling coefficient of the web is calculated according to formula (5.1), and the critical buckling stress is calculated according to formula (5.2). The buckling moment and buckling curvature of trough cover plate are calculated according to formula (5.3).

$$\chi_{\mathrm{w}} = \frac{1}{k}\left(\frac{2}{\sqrt{1+15(b/h)^3}} + \frac{2+4.8b/h}{1+15(b/h)^3}\right) \tag{5.1}$$

$$\sigma_{\mathrm{cr}} = \frac{\chi_w k \pi^2 E}{12(1-\nu^2)(b/t)^2} \tag{5.2}$$

$$M_y = \frac{\sigma_{cr} I}{b}, \theta_y = \frac{M_y}{EI} \tag{5.3}$$

where K—Buckling coefficient, Desirable 4; χ_w—Local dependent buckling coefficient; E—Elastic modulus of steel; υ—Poisson's ratio of steel.

5.2.2 An Example Model of Primary and Secondary Frame Structures

On one hand, the main frame column and the secondary frame beam are connected by prefabricated energy dissipation hinge joints, and an anti-buckling eccentric brace is used to improve the seismic performance of the structure. On the other hand, the deformation is concentrated on the energy dissipation hinge joints to reduce the seismic damage of the internal sub-frame structure. The sub-frame of the energy dissipation hinge joint structure is prefabricated and assembled, in which the connection between the sub-frame beam and the mega-column is connected by the prefabricated energy dissipation hinge joint, and the prefabricated assembly of the other sub-frame beams and columns is equal to cast-in-place. In order to compare the seismic performance of energy dissipation hinged joint structure with that of unbraced structure, steel braced structure and BRB structure, the primary and secondary frames are cast-in-place, as shown in Fig. 5.2.

The example is a 30-story super high-rise reinforced concrete mega-frame structure with a height of 123 m, an ordinary story height of 4 m, a mega-beam layer-by-layer height of 5 m, a plane size of 4 spans on each side, and an aspect ratio of 3.9. The reinforced concrete mega-columns are located at four corners of the structural plane under earthquake action: according to the current Code for Seismic Design of buildings (GB 50, 011–2010), when the seismic fortification intensity is 8 degrees, the basic acceleration of seismic design is 0.20 g. For type II site soil, the design earthquake is divided into the first group, and the seismic grade of the frame is first class. The secondary frame column is disconnected in the lower layer of the mega beam, forming a large bay floor. The bottom of the structure is fixed with the foundation, and the mega beam and mega column are rigidly connected. The giant beams are arranged on the 10th floor, the 20th floor and the 30th floor respectively. The main frame adopts C60 grade concrete, and the main frame structure has three

a) Unsupported structure b) Steel braced structure c) BRB structure d) Energy dissipation hinge

Fig. 5.2 Example structural model

Table 5.1 Component parameter table

Structural member		Size (mm)	Steel bar type
Main frame	Column	3000 × 3000	HRB400
	Main beam	3000 × 1500	
	Secondary beam	3000 × 1500	
	Plate	200	
Sub-frame	Column	900 × 900	
	Beam	600 × 300	
	Plate	100	

mega-stories of 41 m each, which are the 1st–10th, 11th–20th and 21st–30th floors respectively, the height of the secondary frame is 4 m, the height of the mega-beam layer is 5 m, the column distance of the secondary frame is 7.8 m, and the secondary frame adopts C40 grade concrete. The structural member parameters of the primary and secondary frame are shown in Table 5.1.

A Perform-3D finite element analysis software is used to analyze the elastic–plastic dynamic time-history of primary and secondary reinforced concrete frames. The beam-column element adopts the plastic hinge model, and the XTRACT section calculation software is used to calculate the moment–curvature curve and axial force-moment curve of reinforced concrete members. The beam-column element is composed of elastic region and plastic region with column element P-Δ effect considered. The floor is considered as a rigid floor with infinite in-plane stiffness. The lateral stiffness of braces in steel braced structures is 2.4 times that of frame columns at this level. The BRB of BRB structure and energy dissipation hinge joint structure reflect the design principle of lateral stiffness equal to steel braces. In order to make BRB in energy dissipation hinge joint structure yield first, the yield axial force of braces is 0.6 times that of BRB braces. The basic periods of steel braced structure, BRB structure and energy dissipation hinge structure are the same, and the constitutive relation of bracing element is shown in Fig. 5.3.

The time history curve of the selected ground motion record is shown in Fig. 5.4.

5.2.3 *Yield State and Structural Energy Dissipation Distribution of Unbraced Structural Members*

This section studies the yield state and energy dissipation distribution of structural members in reinforced concrete primary and secondary frame. The structures under the action of small earthquake, moderate earthquake, large earthquake and super large earthquake is analyzed. The time history envelope value of moment–curvature plastic hinge at the beam end of the secondary frame, the ductility demand of the primary and secondary frame and the energy dissipation percentage of the primary and secondary frame are also analyzed.

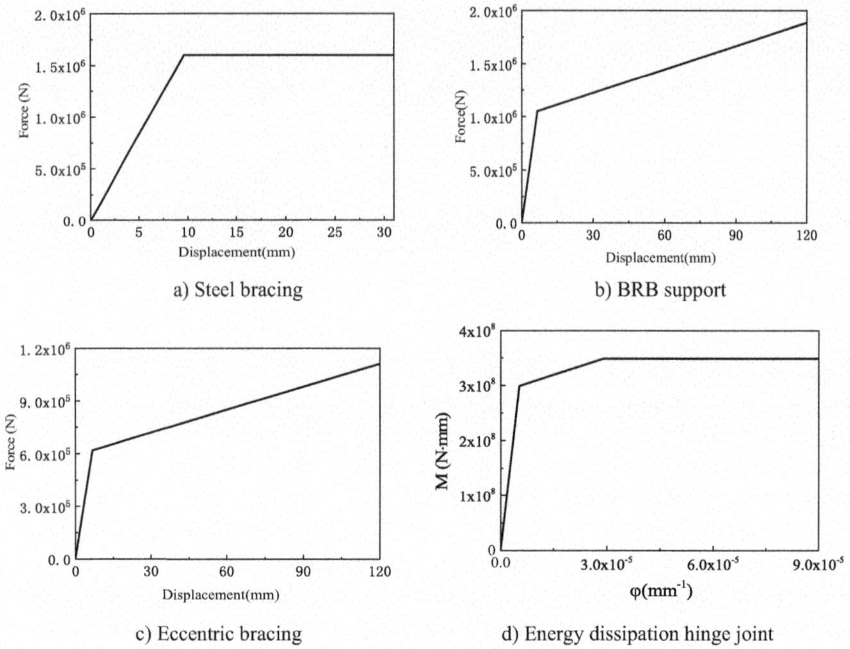

a) Steel bracing

b) BRB support

c) Eccentric bracing

d) Energy dissipation hinge joint

Fig. 5.3 Constitutive relation of support element in perfrom-3D

When the inter-story displacement angle of the structure is maximum, it is considered that the bending moment and the plastic hinge at the beam end of the hierarchical frame are the largest, that is, the plastic hinge at the beam end of the secondary frame connected to the main frame column on the 15th floor is shown in Fig. 5.5a. The plastic hinge at the beam end of the secondary frame unconnected by the main frame column is shown as shown in Fig. 5.5b. It can be seen from the figure that the yield at the beam end of the secondary frame is not serious, and the ductility coefficient is less than 3 under the action of super earthquake. Compared with the unconnected sub-frame beam end of the main frame column, the yield of the sub-frame beam end connected to the main frame column is lighter.

In order to analyze the yield quantity of secondary frame beams, the results of statistical analysis of the ductility coefficient of plastic hinges at the end of beams are greater than 2 (that is, $\mu > 2$, the yellow line in Fig. 5.5) are collected. The percentage of sub-frame beams reaching $\mu > 2$ under large and super earthquakes is shown in Table 5.2.

As can be seen from Table 5.2, the percentage of secondary frame beams $\mu > 2$ reached 81.67% and 91.23% respectively under large and super earthquakes, and a large number of secondary frame beams yielded. The percentage of energy dissipation of secondary frame beams connected to main frame columns and unconnected secondary frame beams to the total energy dissipation of secondary frame beams is shown in Tables 5.3 and 5.4.

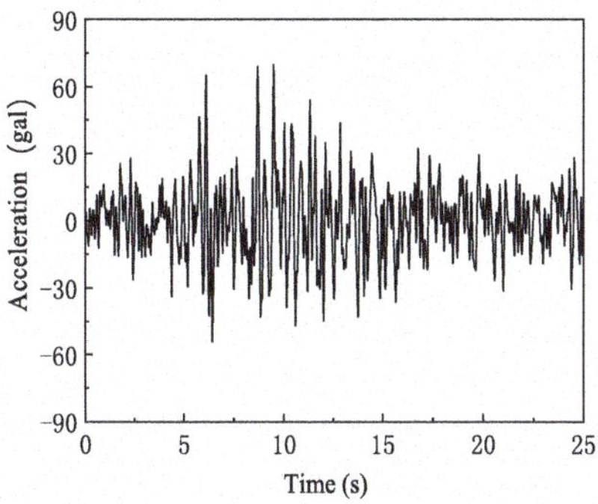

a) Seismic record GM1

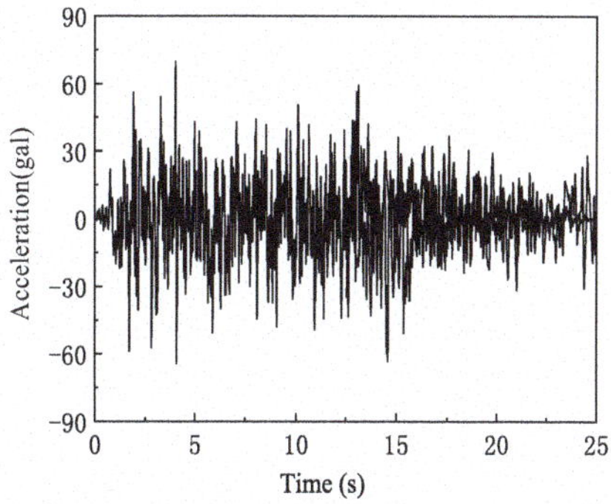

b) Seismic record GM2

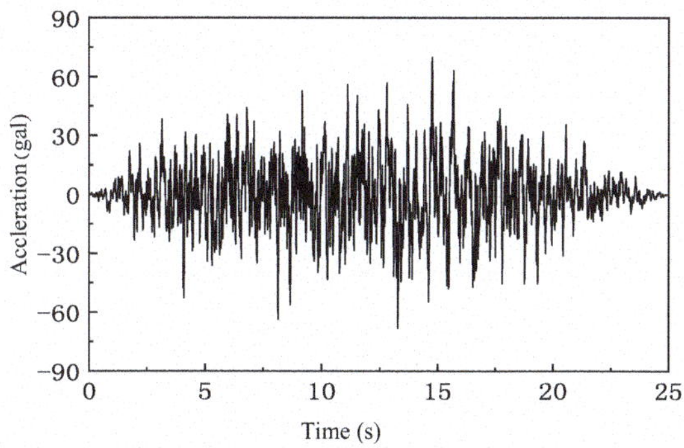

c) Seismic record GM3

Fig. 5.4 Ground motion acceleration time history curve

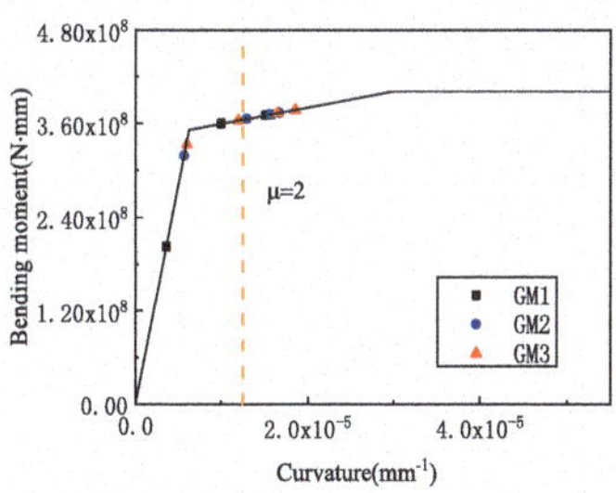

a) The main frame column is connected to the beam end of the secondary frame

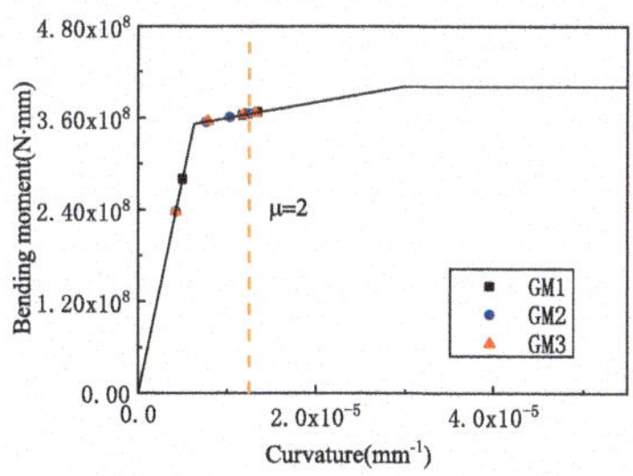

b) Unconnected beam end of secondary frame with main frame column

Fig. 5.5 Plastic hinge M-φ curve at the end of secondary frame beam under super-large earthquake

Table 5.2 Percentage of the number of frame beams $\mu > 2$

Ductility coefficient	PGA (gal)	GM1 (%)	GM2 (%)	GM3 (%)	Average (%)
$\mu > 2$	400	82.41	77.96	84.63	81.67
	510	92.04	87.04	94.63	91.23

Table 5.3 Percentage of energy dissipation of secondary frame beams at PGA = 400 gal

Relationship between secondary frame beam and main frame column	GM1 (%)	GM2 (%)	GM3 (%)	Average (%)
Connected	32.61	31.51	34.14	32.75
Disconnected	67.39	68.49	65.86	67.25

Table 5.4 Percentage of energy dissipation of secondary frame beams at PGA = 510 gal

Relationship between secondary frame beam and main frame column	GM1 (%)	GM2 (%)	GM3 (%)	Average (%)
Connected	34.36	33.52	35.15	34.34
Disconnected	65.64	66.48	64.85	65.66

From Tables 5.3 and 5.4, it can be seen that under the action of large and super-large earthquakes, the energy dissipation of the sub-frame beam connected to the main frame column is less than that of the unconnected sub-frame beam, and the damage of the internal sub-frame beam is more serious.

The main and secondary frames begin to enter plasticity during moderate earthquakes. Under the action of moderate, large and super earthquakes, the energy consumption percentages of main frames and secondary frames are shown in Tables 5.5, 5.6 and 5.7.

Table 5.5 Percentage of energy consumption of primary and secondary frame structures at PGA = 220 gal

Structure	GM1 (%)	GM2 (%)	GM3 (%)	Average (%)
Sub-frame	91.97	89.83	93.04	91.62
Main frame	8.03	10.17	6.96	8.38

Table 5.6 Percentage of energy consumption of primary and secondary frame structures at PGA = 400 gal

Structure	GM1 (%)	GM2 (%)	GM3 (%)	Average (%)
Sub-frame	88.28	87.56	89.74	88.53
Main frame	11.72	12.44	10.26	11.47

Table 5.7 Percentage of energy consumption of primary and secondary frame structures at PGA = 510 gal

Structure	GM1 (%)	GM2 (%)	GM3 (%)	Average (%)
Sub-frame	82.46	83.52	83.13	83.04
Main frame	17.54	16.48	16.87	16.96

It can be seen that under the action of moderate earthquake, large earthquake and super earthquake, the energy dissipation percentage of the secondary frame structure is 91.62%, 88.53% and 83.04% respectively. The secondary frame structure is the main energy dissipation structure of the system. With the increase of the peak value of ground motion, the energy dissipation percentage of the main frame also increases, but the secondary frame structure members are still the main energy dissipation members. It can be seen that the damage to the next frame structure under earthquake is very serious, which is consistent with the analysis result of the yield percentage of the secondary frame beam $\mu > 2$.

5.2.4 *Damping Performance of Mega-Frame Structure Based on Energy Dissipation Hinge*

The following shows the yield limit state of energy dissipation hinge structure and unbraced structure, steel braced structure and BRB structure when the ductility coefficient $\mu = 2$ under the action of seismic record GM1, and the yield condition of four kinds of structures, as shown in Figs. 5.6, 5.7 and 5.8.

As can be seen from Figs. 5.6, 5.7 and 5.8, under the action of moderate earthquake, some sub-frame beams and bottom-level frame columns of the second mega-floor of

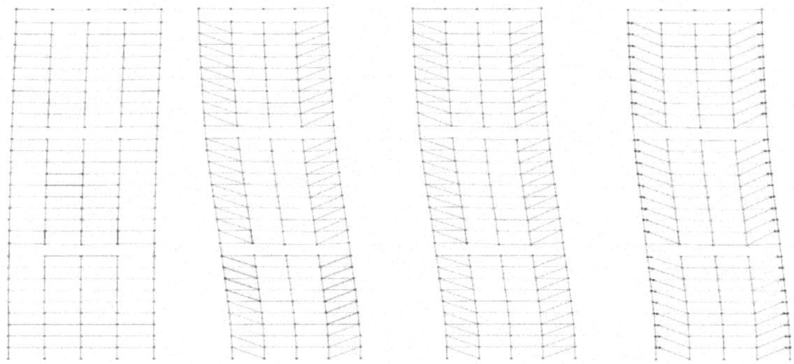

a) Without support   b) Steel bracing   c) BRB   d) Energy dissipation hinge

Fig. 5.6 Yield limit state of structure (PGA = 220 gal)

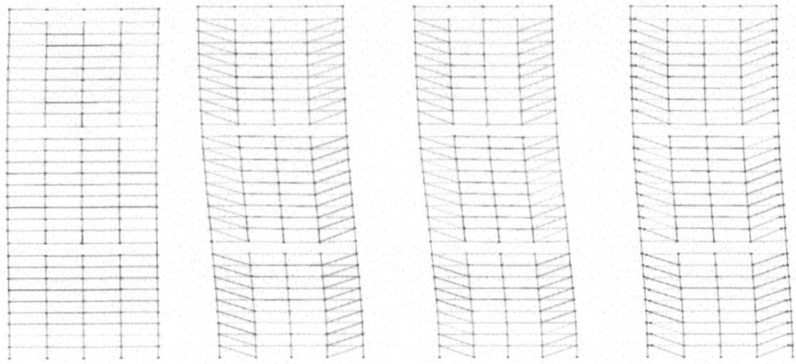

a) Without support b) Steel bracing c)BRB d) Energy dissipation hinge

Fig. 5.7 Yield limit state of structures (PGA = 400 gal)

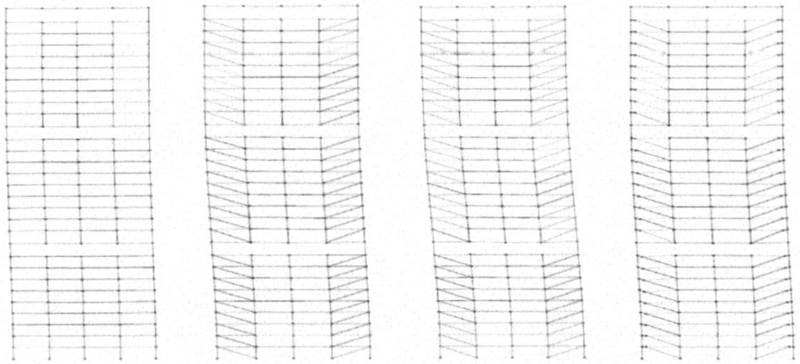

a) Without support b) Steel bracing c) BRB d) Energy dissipation hinge

Fig. 5.8 Yield limit state (PGA = 510 gal) of structures

unbraced structures begin to yield, while the secondary frame beams of steel braced structures, BRB structures and energy dissipation hinge joints do not yield, but the bottom-level frame columns of the second mega-floor yield. Under the action of large earthquake, a large number of sub-frame beams of unbraced structures yield, the sub-frame columns of the third mega-floor begin to yield, the sub-frame beams of steel braced structures, BRB structures and energy dissipation hinge joints gradually begin to yield, but the non-energy dissipation beams of energy dissipation hinge joints basically do not yield. Under the action of super-large earthquake, the sub-frame beams of unbraced structures yield almost completely, the sub-frame beams of steel braced structures and BRB structures yield in large numbers, and the energy dissipation hinged joints also yield a large number of sub-frame beams except the energy dissipation beam. Thus it can be seen that the sub-frame bracing can concentrate the energy dissipation on the brace and effectively reduce the yield degree of the sub-frame beam-column under the earthquake, in which the energy dissipation hinge joint structure focuses on the energy dissipation hinge joint and eccentric bracing, reducing the yield of the secondary frame structure under the earthquake to a greater extent.

In order to analyze the yield of secondary frame beams and columns, the energy dissipation values of secondary frame members under moderate, large and super earthquakes are listed separately in bar Figs. 5.9, 5.10 and 5.11, so as to compare the yield behavior of secondary frame beams and secondary frame columns under earthquake.

As can be seen from Figs. 5.9, 5.10 and 5.11, compared with unbraced structures, the subframe braces of steel braced structure, BRB structure and energy dissipation hinge joint structure effectively reduce the energy dissipation of secondary frame structure, especially the energy dissipation of secondary frame beam. Among them, the energy dissipation hinge joint can effectively reduce the yield degree of the non-energy dissipation beam, the energy dissipation is concentrated in the energy dissipation beam hinge joint and eccentric BRB, and most of the members are in the

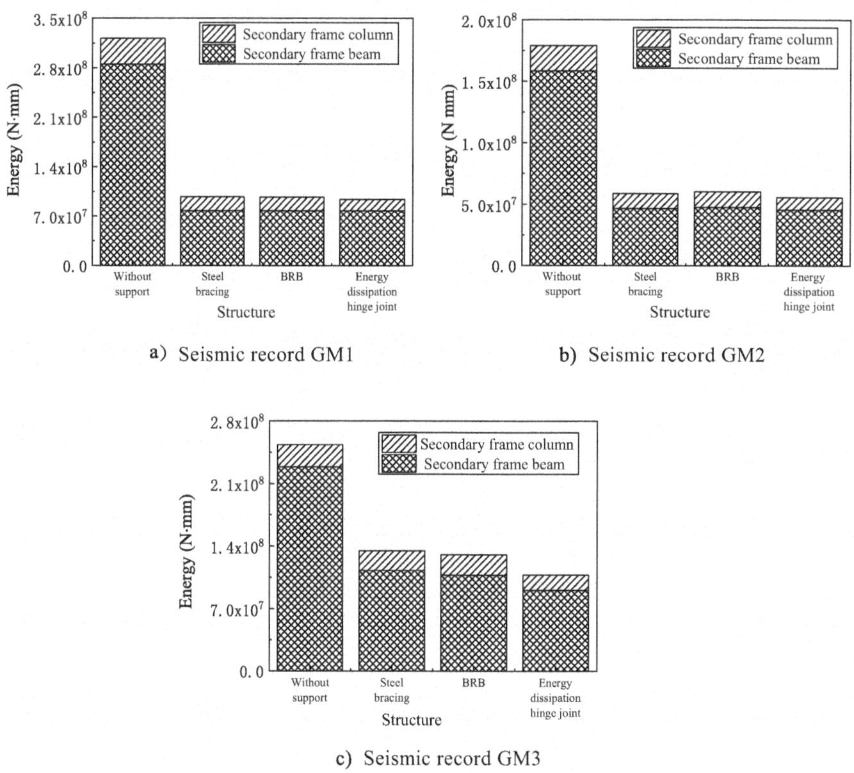

Fig. 5.9 Comparison of sub-frame energy consumption of the structure (PGA = 220 gal)

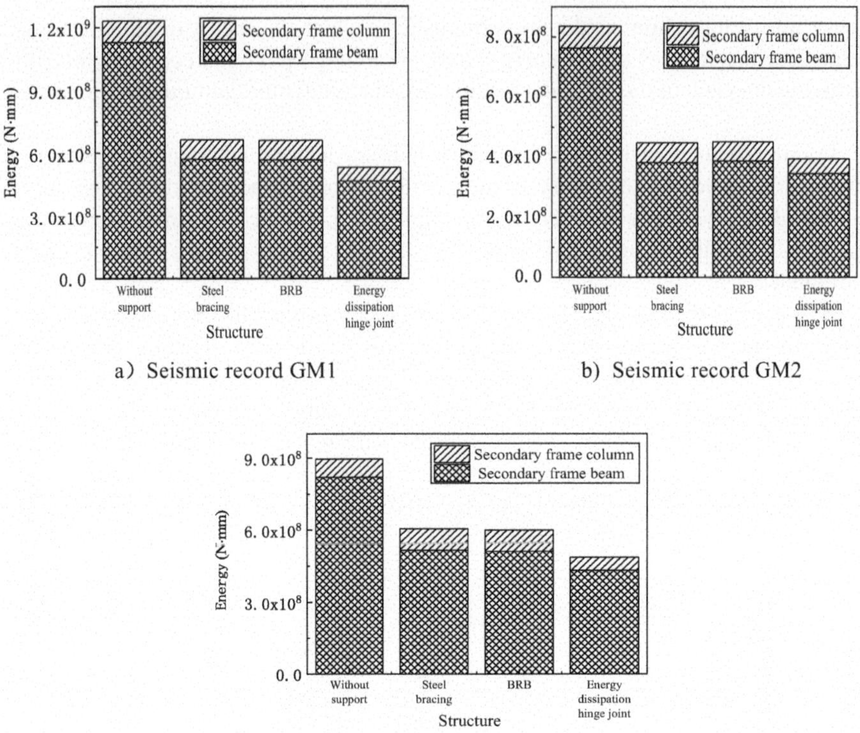

c) Seismic recordGM3

Fig. 5.10 Energy consumption comparison of subframes of four structures (PGA = 400 gal)

elastic stage. The direct force transfer action of steel bracing and BRB, which transfers part of the force of the main frame to the secondary frame column, thus increasing the energy dissipation of the secondary frame column, but the eccentric brace is not directly connected with the main frame column, and the force transfer effect of the brace is not obvious. With the increase of earthquake PGA, the energy dissipation of secondary frame column of steel braced structure and BRB structure is greater than that of unbraced structure and energy dissipation hinge joint structure. Therefore, the energy consumption of the sub-frame column of the energy dissipation hinge joint structure is the minimum. It can be seen that the energy dissipation control effect of sub-frame structure under earthquake is as follows: energy dissipation hinged joint structure > BRB structure > steel braced structure.

Tables 5.8 and 5.9 show the percentage of curvature ductility coefficient $\mu > 2$ of secondary frame beams under large and super-large earthquakes. It can be seen from the table that the percentage control effects of secondary frame beams $\mu > 2$ of steel braced structures, BRB structures and energy dissipation hinge joint structures under large earthquakes are 54.49%, 58.43% and 62.36%, respectively. Under the action of super-large earthquake, the percentage control effect of sub-frame beam $\mu > 2$ of

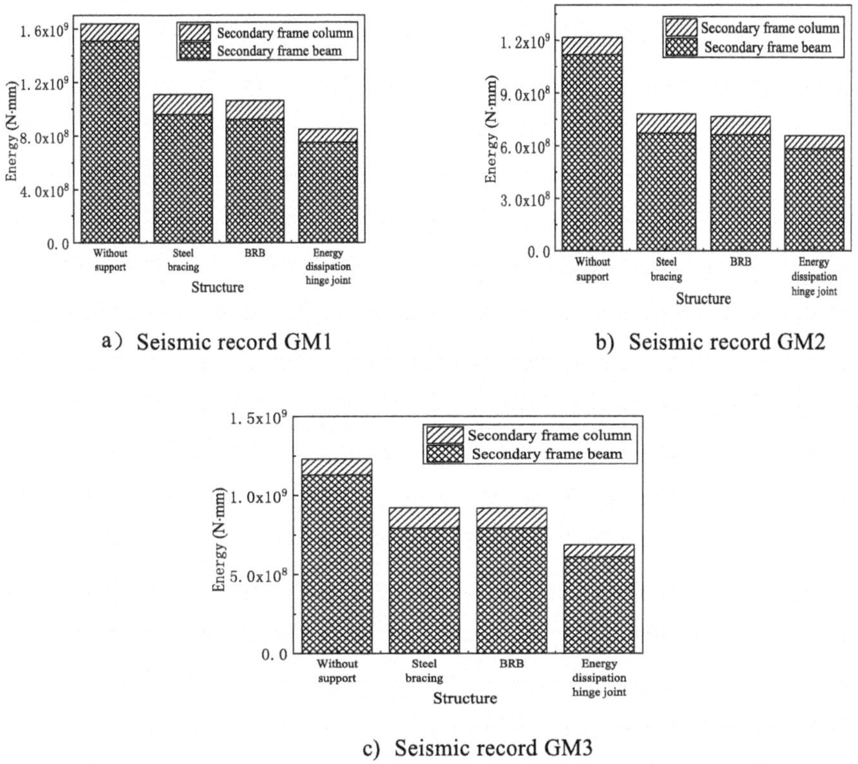

Fig. 5.11 Energy consumption comparisonof subframes of four structures (PGA = 510 gal)

steel braced structure, BRB structure and energy dissipation hinge joint structure are 29.61%, 32.62% and 43.35%, respectively. It can be seen that with the increase of earthquake PGA, the control effect of brace on the percentage of sub-frame beams $\mu > 2$ decreases gradually, and the control effect is as follows: energy dissipation hinged joint structure > BRB structure > steel braced structure.

Table 5.8 Percentage of secondary frame beams $\mu > 2$ at PGA = 400 gal

Ductility coefficient	Structure	GM1 (%)	GM2 (%)	GM3 (%)	Average value (%)	Control effect (%)
$\mu > 2$	Without support	62.04	40.74	62.04	54.94	–
	Steel bracing	29.63	12.04	33.33	25.00	54.49
	BRB	27.78	11.11	29.63	22.84	58.43
	Energy dissipation hinge	25.93	9.26	26.85	20.68	62.36

Table 5.9 Percentage of secondary frame beams $\mu > 2$ at PGA = 510 gal

Ductility coefficient	Structure	GM1 (%)	GM2 (%)	GM3 (%)	Average value (%)	Control effect (%)
$\mu > 2$	Without support	72.22	69.44	74.07	71.91	–
	Steel bracing	50.00	44.44	57.41	50.62	29.61
	BRB	48.15	43.52	53.70	48.46	32.62
	Energy dissipation hinge	42.59	36.11	43.52	40.74	43.35

5.2.5 *Parameter Analysis of Energy Dissipation Hinge Joint*

From the analysis results of the previous section, it can be seen that the energy dissipation hinge joint structure is very effective in reducing the yield degree of the subframe structure under earthquake. This section further studies the influence of the energy dissipation hinge joint on reducing the yield degree of the subframe with constitutive matching relationship between the energy dissipation hinge joint and the eccentric BRB brace considered. By reasonably designing the parameter M-φ curve of energy dissipation hinge joints and the constitutive structure of eccentric BRB braces, this paper tries to figure out a better energy dissipation hinge joint structure, so as to greatly reduce the yield degree of sub-frame structure under earthquake, reduce the design standard of sub-frame structure, and realize the prefabricated assembly of mega-frame sub-frame structure in high intensity area. In order to clarify the role of eccentric BRB brace in energy dissipation, this section adds elastic bracing energy dissipation hinge joint structure. The eccentric BRB brace in 5.2.4 energy dissipation hinge joint structure is replaced by elastic brace with other parameters unchanged. The elastic brace is designed according to the principle of equal lateral stiffness. An elastic bar element in Perform-3D is selected, but not participating in the energy dissipation of the structure.

Based on 5.2.4 energy-saving hinge joint structure, the M-φ curve of energy dissipation hinge joint and the constitutive model of eccentric BRB bracing are changed. In this section, five kinds of relatively weakened energy dissipation hinge joint structures (1.0 M_y, 0.9 M_y, 0.8 M_y, 0.7 M_y, 0.6 M_y) are designed, in which M_y is the yield moment at the beam end of the sub-frame, and the yield moment M_y and limit moment M_u in the M-φ curve of energy dissipation hinge joints are reduced proportionally. The initial stiffness K_0 and the post-yield stiffness K_H remain unchanged. The yield force F_y of BRB constitutive force also decreases in sequence, and K_0 remains unchanged. Figure 5.12 shows the M-φ curve of energy dissipation hinge joints and the constitutive diagram of eccentric BRB braces.

In order to understand the yield of secondary frame beams and columns under earthquake, the energy dissipation values of sub-frame beams, columns and energy

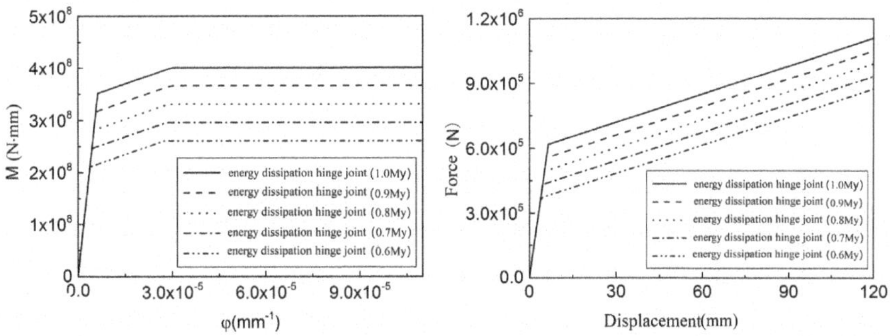

a) M-φ curve of energy dissipation hinge joint

b) BRB constitutive relation

Fig. 5.12 Energy dissipation hinge joint and eccentric brace parameter curve

dissipation hinge joints of seven kinds of structures under moderate, large and super earthquakes are listed in bar diagram Figs. 5.13, 5.14 and 5.15 respectively. The yield of sub-frame beams, columns and energy dissipation hinge joints under earthquake are compared.

As can be seen from Figs. 5.13, 5.14 and 5.15, compared with the unbraced structure, the energy dissipation of the energy dissipation hinge joint structure effectively reduces the energy dissipation of the sub-frame structure, especially the energy dissipation of the sub-frame beam; with the weakening of the M-φ curve parameters of the energy dissipation hinge joint, the energy dissipation of the sub-frame beam-column decreases gradually; the energy dissipation of the elastic bracing structure is more notable than that of the unbraced structure, but the control effect of the elastic brace is not as good as that of BRB. With the increase of seismic PGA, the control effect of M-φ curve parameters of weakening energy dissipation hinge joints on the energy dissipation of subframe beams and columns decreases gradually. Thus, it can be seen that weakening the parameters of the M-φ curve of energy dissipation hinge joints helps further reduce the energy consumption of subframe beams and columns under earthquake. Specific control effect needs to be further studied.

Tables 5.10 and 5.11 show the percentage of curvature ductility coefficient $\mu > 2$ of sub-frame beams under large and super-large earthquakes. It demonstrates that under the action of large earthquakes, with the weakening of M-φ curve of energy dissipation hinge joints, the percentage of secondary frame beams $\mu > 2$ decreases, and the control effect is 62.36% and 66.58%. The percentage control effect of secondary frame beams $\mu > 2$ of elastically braced structures is 50.0%. Under the action of super-large earthquake, with the weakening of the M-φ curve of energy dissipation

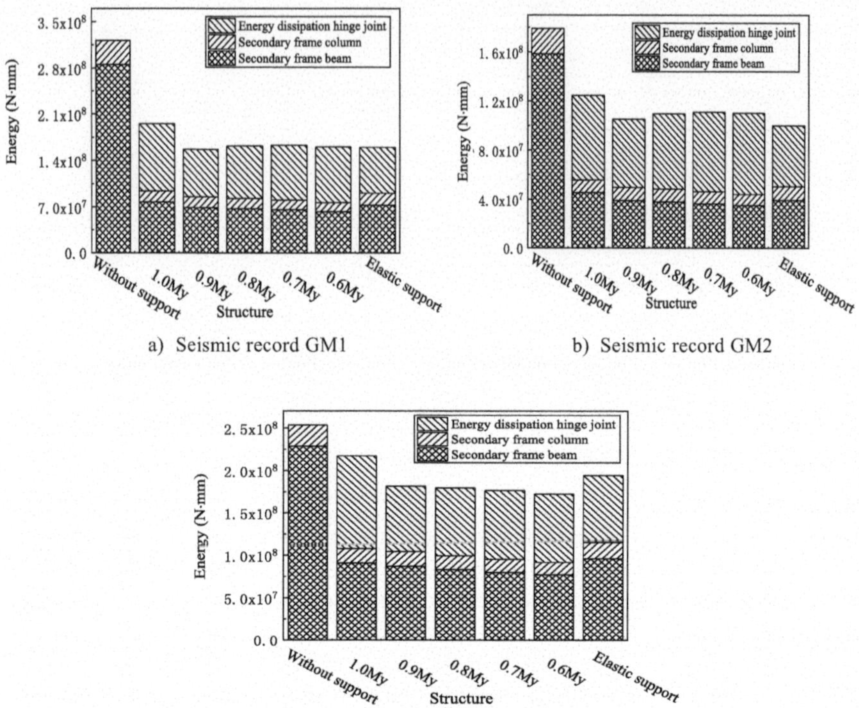

a) Seismic record GM1

b) Seismic record GM2

c) Seismic record GM3

Fig. 5.13 Sub-frame energy dissipation bar diagram of 7 kinds of structures (PGA = 220 gal)

hinge joints, the percentage of sub-frame beams μ > 2 decreases, and the control effect is 43.35% and 47.21%. The percentage control effect of the number of sub-frame beams μ > 2 of elastic braced structures is 39.06%. With the increase of earthquake PGA, the control effect of the M-φ curve of weakened energy dissipation hinge joints on the percentage of sub-frame beams μ > 2 gradually decreases, and the control effect of energy dissipation hinge joints (0.6 M_y) always works the best, and the control effect of elastic bracing structures is no match for that of energy dissipation hinge joints, so it can be seen that the energy dissipation effect of BRB prevails.

To sum up, under large and super-large earthquakes, the maximum inter-story displacement angle of the energy dissipation hinge joint structure is reduced by about 26%, and the number of moderate damage sub-frame beams is reduced by about 62% and 43%. When the yield moment of the energy dissipation hinge joint is 0.6 times the yield moment at the beam end of the sub-frame, the control effect of the structure works the best, better to implement prefabricated assembly of the sub-frame.

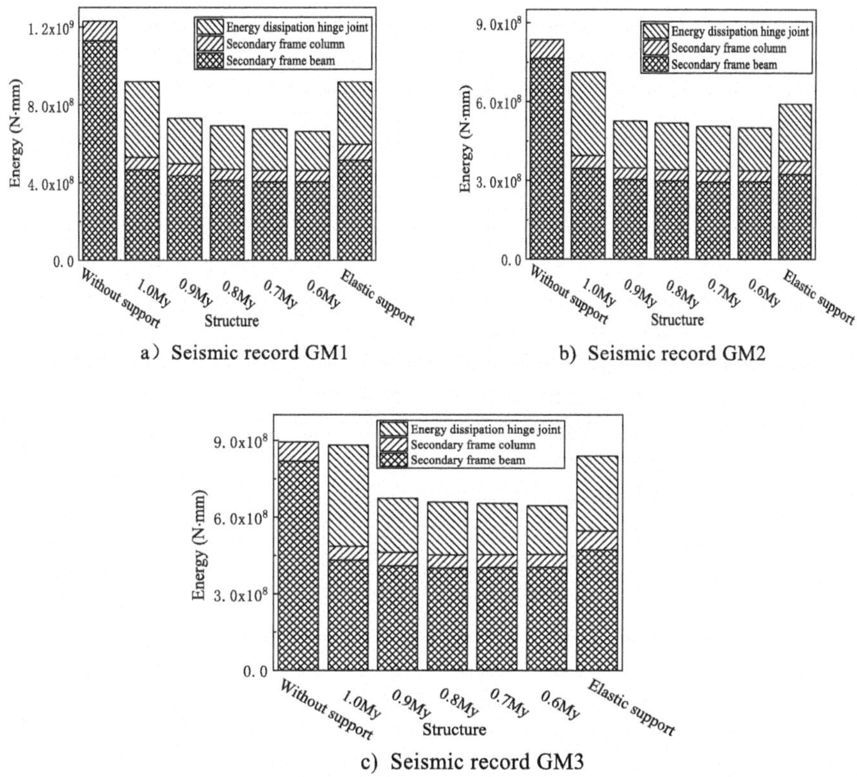

Fig. 5.14 Sub-frame energy dissipation bar diagram of 7 kinds of structures (PGA = 400 gal)

5.3 Cast-In-Place Main Frame-Prefabricated Assembly Tuned Secondary Frame System

The primary and secondary frame is the main anti-lateral force system. Changing the connection mode of the primary and secondary frame has little influence on the internal force distribution, deformation and other structural performance of the whole frame structure, however, the internal force of the secondary frame beam hinged with the main frame column decreases. The study on the seismic performance of the primary and secondary frame structure system based on the energy dissipation hinge joint shows that the energy dissipation hinge joint can't avoid but reduce the damage of the sub-frame. Such a result is subject to the coordination of lateral deformation between the secondary frame and the main frame. Based on the above, the structural system that changes the coordination condition of lateral deformation between sub-frame and main frame is further explored, and a structural system connecting main frame and sub-frame beam by skateboard corbels and adjustable liquid damper is put forward in order to adjust the lateral deformation relationship between the subframe and the main frame.

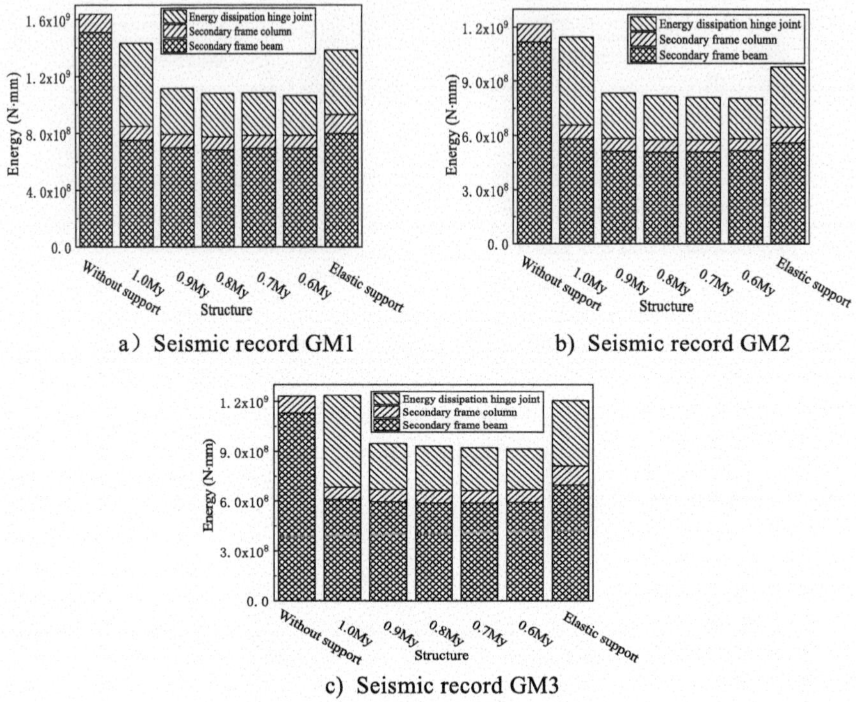

Fig. 5.15 Sub-frame energy dissipation bar diagram of 7 kinds of structures (PGA = 510 gal)

Table 5.10 Percentage of secondary frame beams μ > 2 at PGA = 400 gal

Ductility coefficient	Structure	GM1 (%)	GM2 (%)	GM3 (%)	Average value (%)	Control effect (%)
$\mu > 2$	Without support	62.04	40.74	62.04	54.94	–
	1.0 M_y	25.93	9.26	26.85	20.68	62.36
	0.9 M_y	22.22	9.26	26.85	19.44	64.61
	0.8 M_y	21.30	9.26	26.85	19.14	65.17
	0.7 M_y	21.30	9.26	25.93	18.83	65.73
	0.6 M_y	20.37	8.33	25.93	18.21	66.85
	Elastic support	35.19	10.19	37.04	27.47	50.00

The cast-in-place main frame-prefabricated assembly tuned secondary frame system connects the ground floor column of the secondary frame with the floor of the main frame through the isolation bearing to form a tuned secondary frame structure. The main frame column is connected with the secondary frame side beam through the corbel and skateboard supports, a viscous damper is installed between

Table 5.11 Percentage of secondary frame beams μ > 2 at PGA = 510 gal

Ductility coefficient	Structure	GM1 (%)	GM2 (%)	GM3 (%)	Average value (%)	Control effect (%)
$\mu > 2$	Without support	72.22	69.44	74.07	71.91	–
	1.0 M_y	42.59	36.11	43.52	40.74	43.35
	0.9 M_y	42.59	35.19	43.52	40.43	43.78
	0.8 M_y	40.74	34.26	43.52	39.51	45.06
	0.7 M_y	38.89	34.26	43.52	38.89	45.92
	0.6 M_y	37.96	33.33	42.59	37.96	47.21
	Elastic support	44.44	41.67	45.37	43.83	39.06

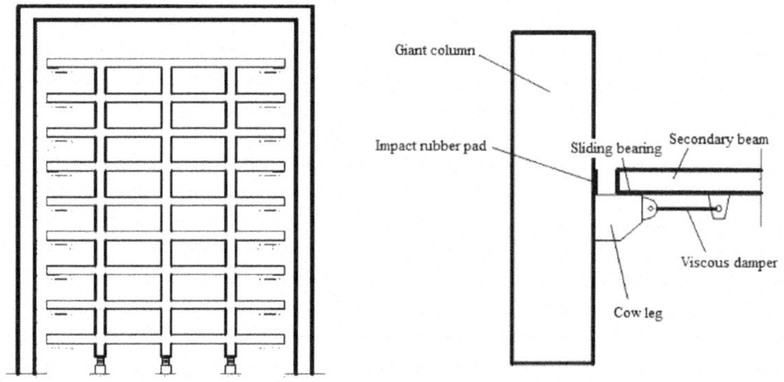

a) Single primary and secondary frame

b) Primary and secondary frame node connection

Fig. 5.16 Schematic diagram of tuned isolation subframe system

the corbel and the secondary frame beam, and an anti-collision device is installed on the main frame column, which can realize the prefabricated assembly of super high-rise structures in high-intensity areas, as shown in Fig. 5.16.

5.3.1 Simplified Calculation Model of Sub-Frame Isolated Mega-Frame

The simplified calculation model of sub-frame isolation mega-frame is shown in Fig. 5.17. After seismic isolation, the upper sub-frame basically translates as a whole, so it can be simplified to a single degree of freedom, while the bottom sub-frame

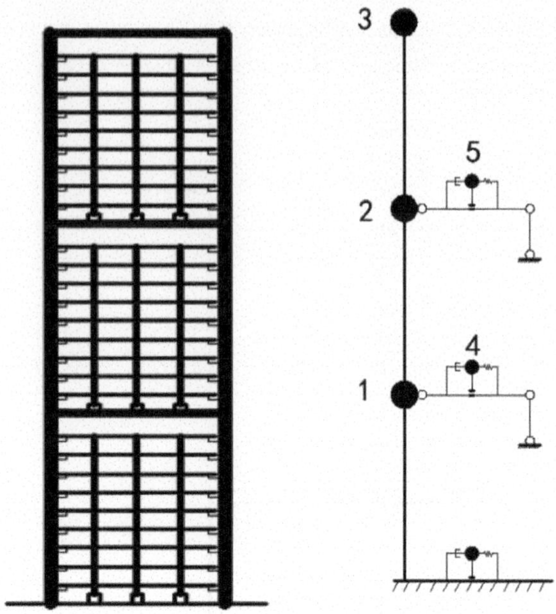

Fig. 5.17 Simplified calculation model of frame-isolated mega-frame structure

has no interaction with the main frame because it is completely isolated from the main frame, that is, it is a separate multi-story frame isolation system. Therefore, the influence of the bottom sub-frame is not considered in the simplified calculation model. Because of the large size and stiffness of the beam and column, the main frame only has translation motion, so the interstory shear calculation model can be used. The rods hinged at both ends to the equivalent particles of the main frame layer are equivalent to the floor of the main frame.

The motion equation of cast-in-place main frame-isolation tuned sub-frame is:

$$[\boldsymbol{M}]\{\ddot{X}\} + [\boldsymbol{C}]\{\dot{X}\} + [\boldsymbol{K}]\{X\} = -[\boldsymbol{M}]\{I\}\ddot{x}_g \tag{5.4}$$

where $\{\mathbf{X}\}$, $\{\dot{X}\}$ and $\{\ddot{X}\}$ are the displacement, velocity and acceleration vectors of each degree of freedom of the primary and secondary frame relative to the ground, respectively; [**M**], [**C**] and [**K**] are mass matrix, damping matrix and stiffness matrix of the system, respectively., {**I**} is a column vector with a unit of 1, $\ddot{x}_g$ is the acceleration of ground motion. The mass matrix and stiffness matrix are given respectively.

$$[\boldsymbol{M}] = \begin{bmatrix} m_{p1} & & & & \\ & m_{p2} & & & \\ & & m_{p3} & & \\ & & & m_{s1} & \\ & & & & m_{s2} \end{bmatrix}$$

$$[\boldsymbol{K}] = \begin{bmatrix} k_{p1} + k_{p2} + k_{s1} & -k_{p2} & & -k_{s1} & \\ -k_{p2} & k_{p2} + k_{p3} + k_{s2} & -k_{p3} & & -k_{s2} \\ & -k_{p3} & k_{p3} & & \\ -k_{s1} & & & k_{s1} & \\ & -k_{s2} & & & k_{s2} \end{bmatrix}$$

The damping matrix of the main frame structure adopts Rayleigh damping, and the damping of the isolated secondary frame is determined according to the damping coefficient of the isolation layer.

5.3.2 Damping Performance of Primary and Secondary Frame Structures with Dampers

To limit the response of the mega-frame under seismic excitation and to protect the sub-frame, this section will first analyze the disconnection between the side beam of the sub-frame and the mega-column of the main frame, and these two are connected by dampers. The damper dissipates energy under the earthquake, so that the damage of the secondary frame can be reduced, and the primary and secondary frame can be protected. Figures 5.18 and 5.19 show the interstory displacement angles of primary and secondary frames and pure frame structures with dampers under the action of small and super earthquakes of seismic records GM1, GM2 and GM3.

As can be seen from Figs. 5.18 and 5.19, in small earthquakes and super-large earthquakes, the inter-story displacement angle of primary and secondary frame structures with viscous dampers decreases more than that of structures without viscous dampers. The viscous damper is a velocity damper that dissipates energy under the action of small, moderate, large and super earthquakes, which means it plays an obvious damping effect under all kinds of earthquakes. The parameter design of viscous dampers in this section is based on the design of super large earthquakes, so the dampers can not come into full play under the action of small earthquakes. The average damping rates under small, moderate, large and super earthquakes are 4.85%, 10.16%, 16.31% and 13.28%, respectively. Besides, for the primary and secondary frame structures with viscous dampers, the displacement angles between the upper and lower adjacent layers of the 10th and 20th floors of the transfer floor without secondary frame columns are significantly reduced, however, the inter-story displacement angles of the transfer floor are enlarged. In addition, after the installation of viscous dampers, the maximum inter-story displacement angle between primary and secondary frames appears on the 13th floor, which is lower than that of pure frame structures without dampers.

Tables 5.12, 5.13 and 5.14 show the yield percentage of sub-frame beams of two kinds of structures under different seismic record PGA strength. The average control effects of viscous dampers on the yield percentage of structural sub-frame beam members under moderate, large and super-large earthquakes are 80.15%, 41.66% and

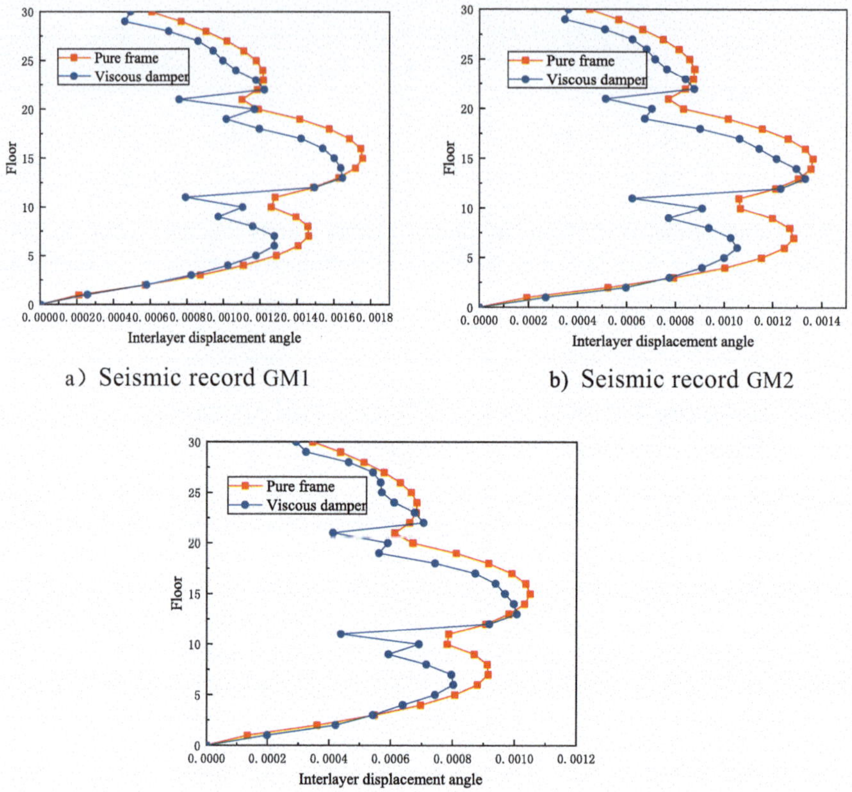

a) Seismic record GM1

b) Seismic record GM2

c) Seismic record GM3

Fig. 5.18 Comparison of interlayer displacement angles (PGA = 70 gal)

26.33%, respectively. It can be seen that with the increase of earthquake action, the control effect of viscous damper on the yield of subframe beam members decreases gradually, this is because the velocity index of viscous damper is 0.5. With the increase of the peak value of ground motion, the growth rate of the damping force of the damper slows down.

Under the action of an earthquake, the internal sub-frame beam connected with the sub-frame column reaches yield before the sub-frame beam connected to the main frame mega-column through the damper. Even under the action of super-large earthquake, the secondary frame beam members connected to the top main frame floor and the damper still do not yield, which fully illustrates the protective effect of viscous dampers on the secondary frame members.

With the increasing earthquake action, the secondary frame beam of the middle main frame layer that is not connected with the damper and the secondary frame beam of the bottom main frame floor yield first, and the hierarchical frame beam of the top main frame yields later. At the same time, the protective effect of viscous damper on

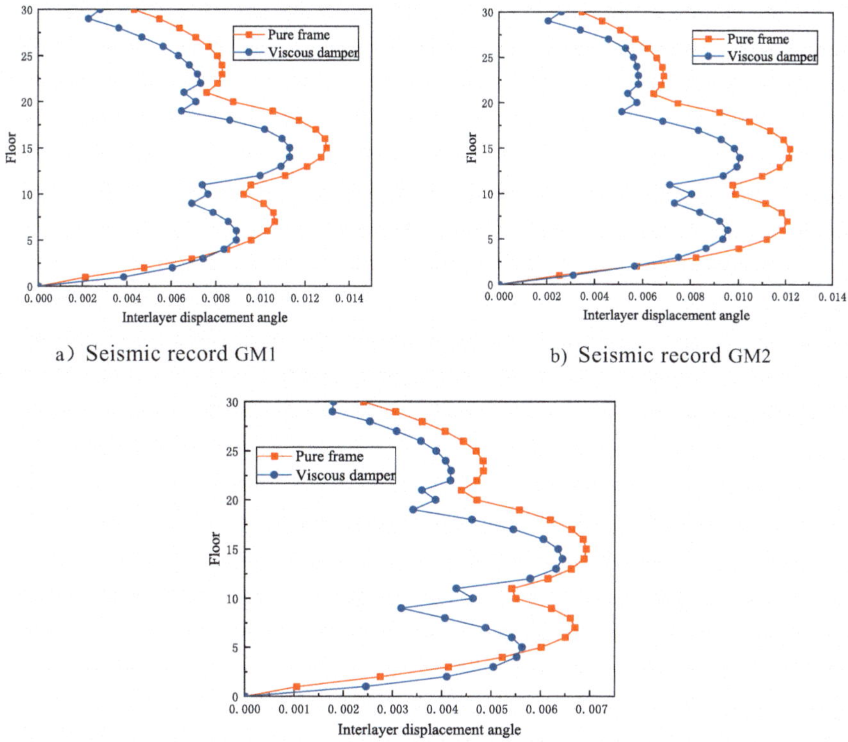

a) Seismic record GM1

b) Seismic record GM2

c) Seismic record GM3

Fig. 5.19 Comparison of interlayer displacement angles (PGA = 510 gal)

Table 5.12 Percentage of yield of secondary frame beams at PGA = 220gal

Structure	GM1 (%)	GM2 (%)	GM3 (%)	Average value (%)	Control effect (%)
Pure frame	80.02	76.14	72.22	76.13	–
Viscous damper	17.34	15.03	12.97	15.11	80.15

Table 5.13 Percentage of yield of secondary frame beams at PGA = 400 gal

Structure	GM1 (%)	GM2 (%)	GM3 (%)	Average value (%)	Control effect (%)
Pure frame	95.36	94.32	93.51	94.40	–
Viscous damper	59.04	55.25	50.93	55.07	41.66

Table 5.14 Percentage of yield of secondary frame beams at PGA = 510 gal

Structure	GM1 (%)	GM2 (%)	GM3 (%)	Average value (%)	Control effect (%)
Pure frame	96.75	95.58	94.44	95.60	–
Viscous damper	73.23	70.43	67.59	70.42	26.33

the secondary frame beam which is not connected to the main frame column is not as obvious as that of the secondary frame beam directly connected to it, and the yield state can not be well controlled, which means other energy dissipation measures need to be taken for protection.

5.3.3 Damping Performance of Primary and Secondary Frame Structures with Secondary Frame Isolation

Because the bottom sub-frame is separated from the whole frame after isolation, equivalent to an independent isolated structure, it does not affect each other with the whole frame. Therefore, in this section, it is only necessary to arrange the isolation bearing with appropriate parameters at the bottom of the base isolation sub-frame to meet the seismic response requirements. In order to facilitate the discussion, it is not elaborated in this article. There are three different schemes to change the layout of the upper isolation frame. The first scheme uses the top-level frame isolation, the second scheme uses the middle-level frame isolation, and the third scheme uses the top-level and middle-level frame double isolation. Only the horizontal ground motion in the X direction is considered, and in the time history analysis, the modal damping is 5%. The vertical view of the three structures is shown in Fig. 5.20.

The tuning ratio of the isolated secondary frame structure to the main structure is 0.98, the damping ratio of the isolation layer of the top isolation structure is 0.3, and the damping ratio of the isolation layer of the middle-level frame isolation structure is 0.1. The damping of the secondary frame structure is concentrated in the isolation layer.

The maximum inter-story displacement angles of the four kinds of structures under the action of large earthquakes are shown in Fig. 5.21. It can be seen from the figure

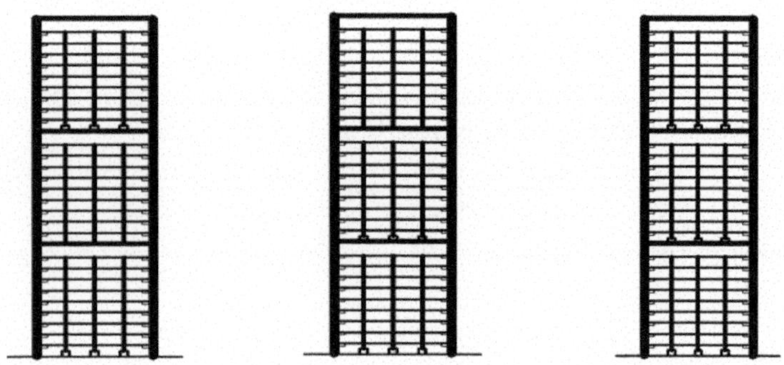

a) Top and bottom isolation b) Middle and bottom isolation c) Three-story isolation

Fig. 5.20 Schematic diagram of primary and secondary frame isolated structure that under the action of large earthquake, the maximum inter-story displacement angle of the sub-frame isolated structure is significantly lowered than that of the pure frame structure. Compared with the pure frame structure, the maximum inter-story displacement angle of the top–bottom sub-frame isolation structure, the middle-bottom sub-frame isolation structure and the three-level frame isolation structure decreases by 38.32%, 32.93% and 46.11%, respectively. The damping effects of the three kinds of structures are as follows: three-level frame isolation structure > top–bottom frame isolation structure > middle-bottom sub-frame isolation structure. Therefore, it can be seen that the secondary frame isolation can effectively reduce the maximum inter-story displacement angle of the primary and secondary frame structure.

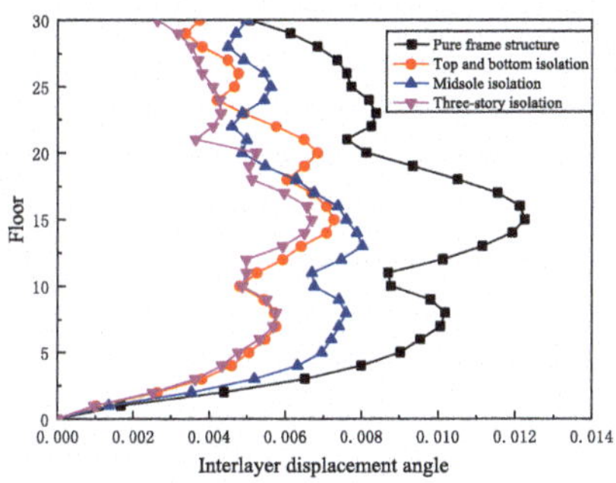

a) Seismic record GM1

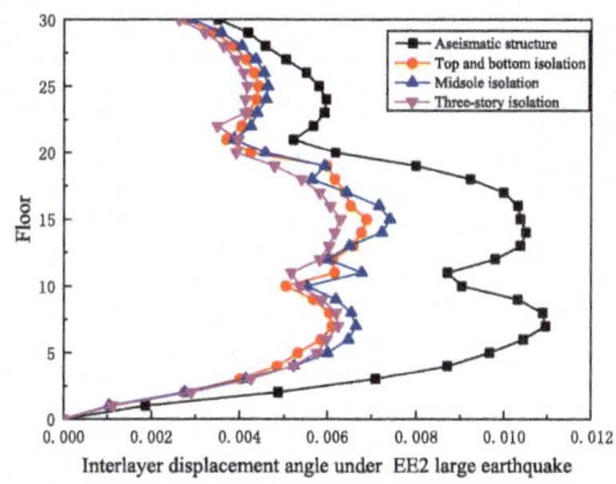

b) Seismic record GM2

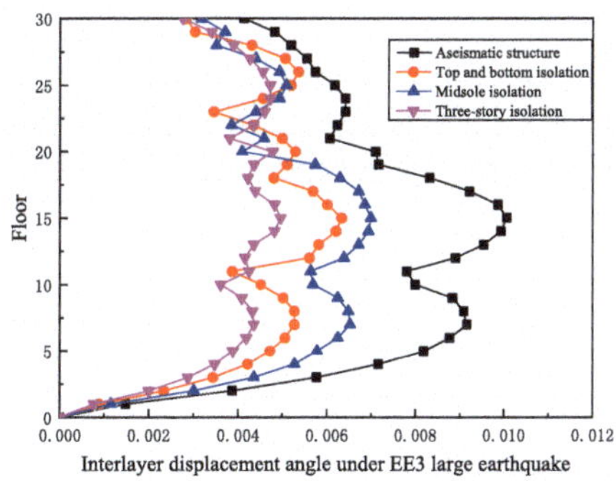

c) Seismic record GM3

Fig. 5.21 The interlayer displacement angles of 4 structures (PGA = 400 gal)

5.4 Conclusions

Based on the two-order stress system of the main frame and the secondary frame in the mega-frame structure, this chapter puts forward the primary and secondary frame structure system based on the energy dissipation hinge joint, the prefabricated structure system of the cast-in-place main frame, and precast sub-frame respectively. In this way, seismic demand can be met and damage of the sub-frame of 8-degree prefabricated frame structure can be limited.

(1) Assembling and reciprocating bending energy dissipation hinge joints are adopted to connect the main frame and the secondary frame. Plastic deformation of the structure is concentrated on the energy dissipation hinge joints to protect other members of the sub-frame at elastic or slight damage stage. Through the parametric analysis of energy dissipation hinge joints, the design idea of mega frame structure based on energy dissipation hinge joints is given. The study shows that under large and super-large earthquakes, the interstory deformation of energy dissipation hinged joints is limited, and the number of moderate damage sub-frame beams is reduced, which effectively reduce the seismic damage of sub-frames.

(2) The skateboard corbel and adjustable liquid damper are used to connect the main frame and the sub-frame beam to change the lateral deformation coordination relationship between the sub-frame and the main frame, which significantly reduces the sub-frame damage. At the same time, the prefabricated assembly is used to tune the secondary frame, and the bottom column of the secondary frame is connected with the floor of the main frame through the isolation bearing to form the tuned secondary frame structure. The study shows that under the strong earthquake, the sub-frame isolation can reduce the lateral deformation of the cast-in-place main frame-precast assembled sub-frame structure system and significantly improve the seismic performance of the structure system.

References

1. Chen L, Zhang Y (2003) Mega-structure system and its development trend. J Harbin Inst Technol 11:1307–1310 (in Chinese)
2. Englekirk RE (2003) Seismic design of reinforced and precast concrete buildings
3. Arditi D, Ergin U, Günhan S (2000) Factors affecting the use of precast concrete systems. J Architectural Eng 6(3):79–86
4. Choi HK, Choi YC, Choi CS (2013) Development and testing of precast concrete beam-to-column connections. Eng Struct 56(6):1820–1835
5. Zhu Z, Guo Z (2012) Seismic test and analysis of joint of new precast concrete shear wall structures. Chin Civil Eng J 1:69–76 (in Chinese)
6. Zhu Z, Guo Z (2015) Research on seismic performance of a spatial model of a new precast concrete shear wall structure. Eng Mech 32(4):153–159 (in Chinese)
7. Ma J, Pan J, Yin W et al (2017) Experimental study on seismic behavior of wholly precast RC frame-shear wall structure. J Build Struct 38(06):12–22 (in Chinese)

8. Korkmaz HH, Tankut T (2005) Performance of a precast concrete beam-to-beam connection subject to reversed cyclic loading. Eng Struct 27(9):1392–1407
9. Pampanin S, Nigel Priestley MJ, Sritharan S (2001) Analytical modelling of the seismic behaviour of precast concrete frames designed with ductile connections. J Earthquake Eng 5(3):329–367
10. Perez FJ, Pessiki S, Sause R et al (2002) Lateral load tests of unbonded post-tensioned precast concrete walls. Adv Build Technol 49(2):423–430
11. Shen X, Ou J (2009) Failure mode analysis of super-tall huge steel frame structure. J Southeast Univ (Nat Sci Edn) 39 (in Chinese)
12. Wang C (2016) Seismic performance of prefabricated beam-to-column connection with hinge and energy-dissipating plates and the weak beam strong column frames.Harbin Institute of Technology (in Chinese)

Chapter 6
Prefabricated Rocking Wall Structural System

Abstract Prefabricated rocking wall structural system has the features of low seismic damage and quick energy dissipation, harnessing the unique response pattern of rocking. In this chapter, a new type of damage-controllable rocking wall system is proposed. The system consists of multiple walls embedded inside a frame structure, with the prestressed tendon providing the self-resetting feature, replaceable damping blocks between walls providing energy dissipation, and dry contact surfaces lowering residual deformation. Firstly, both the monotonic and cyclic responses are investigated theoretically. Secondly, quasi-static experiments are conducted using large-scale specimens. Thirdly, finite element simulations are carried out. The results show that the proposed system experiences efficient dissipation, low structural damage, and low residual displacement when subjected to earthquakes.

Keywords Rocking wall structure · Post-tensioned precast concrete wall · Self-resetting · Replaceable energy-dissipating block · Quasi-static experiment · Low structural damage

6.1 Introduction

Earthquake disasters usually cause great damage to buildings, resulting in very large economic and social losses. However, the traditional architectural form adopts ductile design methods to dissipate seismic energy at the cost of component failure. Many earthquake damage investigations have found that buildings that are finely designed following the specifications are often too damaged to meet the requirements of repair and continuous use. Designing for the prevention of damage or for damage that can be repaired quickly has become an important direction of research for the sustainable development of earthquake engineering. Therefore, it is urgent to find a new structural system to meet the increasing seismic requirements of buildings. The self-resetting rocking structure is in line with such a design concept.

The discovery of the rocking structure can be traced back to the Chile earthquake in 1960. There were high flume structures that swayed due to the foundation lifting, thus avoiding structural damage. In 1963, Housner [1] proposed that rocking made the structure more stable, which was the prelude to research

G. Wu et al., *Novel Precast Concrete Structure Systems*, Springer Tracts in Civil Engineering, https://doi.org/10.1007/978-981-19-6821-1_6

on rocking structures. The self-resetting rocking wall is a typical rocking structure and a new seismic structure system.

By relaxing the constraints on the interface between the structure and foundation or the interface of structural components, the section of the self-resetting rocking wall can only be compressed and not tensioned. It will sway under seismic action, and the residual deformation of the structure will be small. The application of prestress (or vertical force) causes the wall to self-reset, thus reducing the residual deformation of the structure. The rocking wall does not consume energy, but the rocking of the wall causes some specific parts to be relatively displaced under seismic action, which provides the possibility for the subsequent installation of energy dissipation components and enhancement of the energy dissipation capacity of the structure. The description in this book does not strictly distinguish between self-resetting rocking walls with energy dissipation devices and self-resetting rocking walls without energy dissipation devices.

Research on the structure of rocking walls began in 1996. Kurama et al. [2] proposed the analysis model of a posttensioned unbonded prestressed concrete wall. Later, in 1999, Kurama et al. [3] began to systematically study the working performance of the wall, which relaxed the constraints between the wall and the foundation. The foundation and wall panels of the prefabricated reinforced concrete were connected into a whole by applying prestress, and spiral stirrups were added on both sides at the bottom of the wall to constrain the concrete in this part. Under the horizontal loading, the wall rotated around the central axis, and the self-reset was realized by the prestressed steel strand. This rocking wall had almost no damage under large lateral deformation, had a good self-resetting ability, and showed good seismic performance, as shown in Fig. 6.1. In 2000, Kurama et al. [4] proposed improvement measures based on previous studies to increase the energy consumption of viscous dampers. The research showed that the increase in energy dissipation damping effectively reduced the maximum interstory displacement and prevented excessive damage to the wall, as shown in Fig. 6.2. In 2001, Kurama [5] also tried a friction damper between wall slices.

In 2002, Perez et al. [6], based on the study of Kurama et al., designed a low cyclic experiment to compare with the analysis results. The experimental model is shown in Fig. 6.3, and the experimental results are shown in Fig. 6.4. Experiments showed that the damage to the wall was very small under low cyclic loading and that the wall could realize a self-reset.

In 2003, Holden et al. [7] carried out quasi-static experiments on self-resetting reinforced concrete rocking walls and ordinary reinforced concrete shear walls, setting energy-dissipating steel bars inside the rocking wall. The experiment showed that the residual displacement of the wall could be effectively controlled by the self-setting rocking wall so that the damage of the wall was much less than that of the ordinary shear wall. The addition of energy-dissipating steel bars increased the energy dissipation capacity of the self-resetting rocking wall. In 2007, Restrepo and Rahman [8] further designed three groups of control experiments on this basis and compared the difference between the energy-dissipating steel bars and different levels of prestress. The experimental results showed that the more energy-dissipating steel

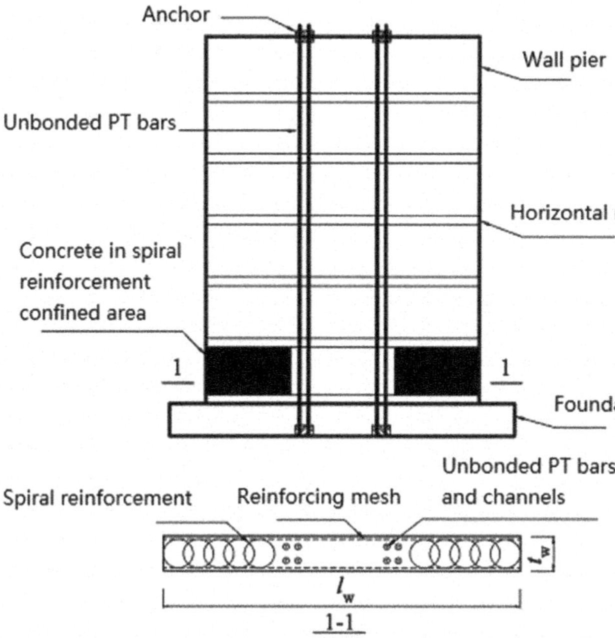

Fig. 6.1 Self-resetting rocking wall structure [3]

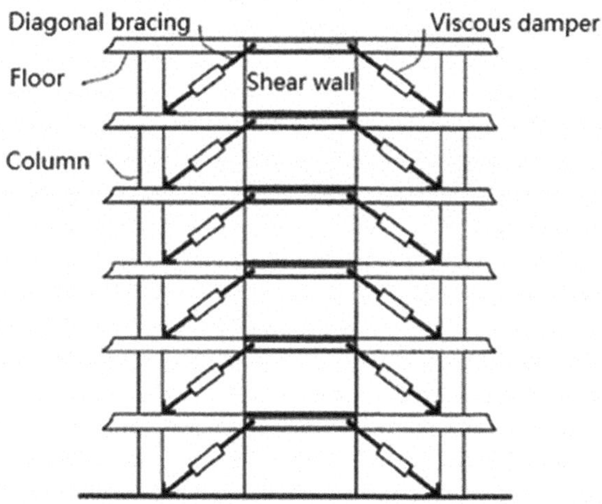

Fig. 6.2 Self-resetting rocking wall structure with viscous damper [4]

bars there were, the better the energy dissipation capacity and the fuller the hysteresis loop. The greater the initial prestress was, the more obvious the self-reset effect.

In 2004, Ajrab et al. [9] first proposed the concept of a rocking wall frame structure, which combined a rocking wall with a cable and set damping units at the bottom of the cable. They used the performance-based design method to design and analyze a

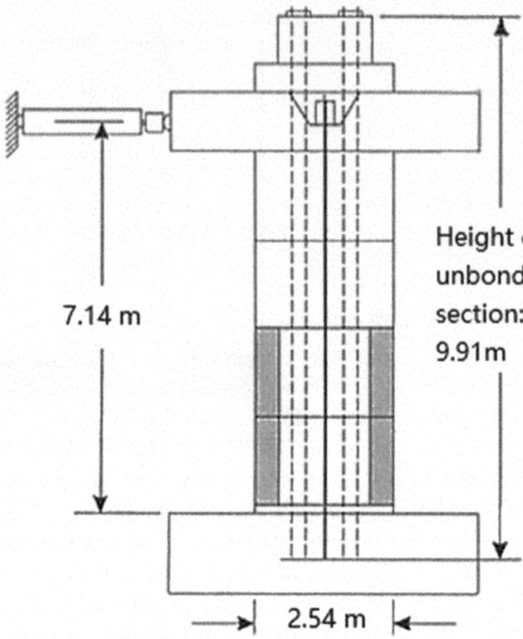

Fig. 6.3 Schematic diagram of the experimental model [6]

six-story rocking wall frame structure. As shown in Fig. 6.5, this form of rocking wall frame could reduce the response of the structure under seismic action and control the deformation mode of the frame.

In 2008, Marriott et al. [10] experimented with a self-resetting rocking wall on a shaking table and compared the four components with no damping, mild steel damping, viscous damping, and the combination of the two energy dissipation elements. Then, they designed the energy dissipation elements to be replaceable and pointed out that the wall was suitable for existing structures and new structures.

In 2009, Wiebe and Christopoulos [11] pointed out that multiple rocking points could be set on the rocking wall to release the bending moment and achieve better performance. The simplified calculation model is shown in Fig. 6.6. In 2015, Khanmohammadi et al. [12] also proposed an analysis model for rocking walls with multiple rocking points, as shown in Fig. 6.7. The model was more detailed and was in good agreement with the experimental results of Holden et al. [7]

In 2012, Nicknam et al. [13] combined buckling-restrained braces (BRB) with rocking walls and proposed an analysis model, as shown in Fig. 6.8. Subsequently, Nicknam et al. [14, 15] conducted shaking table experiments on this model. They conducted an in-depth analysis of the system and obtained some beneficial results. In 2016, Yooprasertchai et al. [16] conducted a quasi-static experimental analysis of the combination of BRB and rocking walls, which was different from the previous model with the unilateral addition of BRB, as shown in Fig. 6.9.

In 2012, Preti and Meda [17] conducted a low cyclic loading test of a full-scale reinforced concrete rocking wall. During the experiment, the concrete at the corner

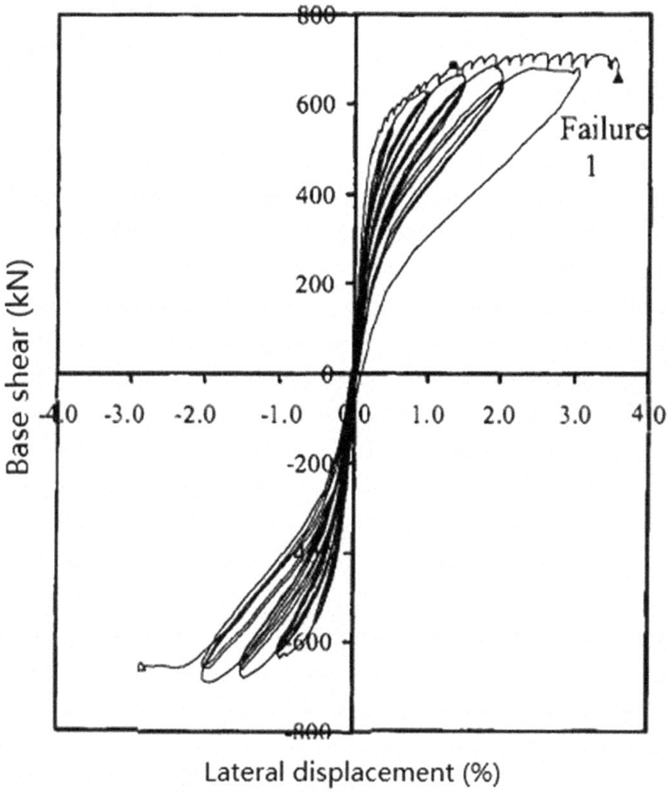

Fig. 6.4 Loading displacement curve [6]

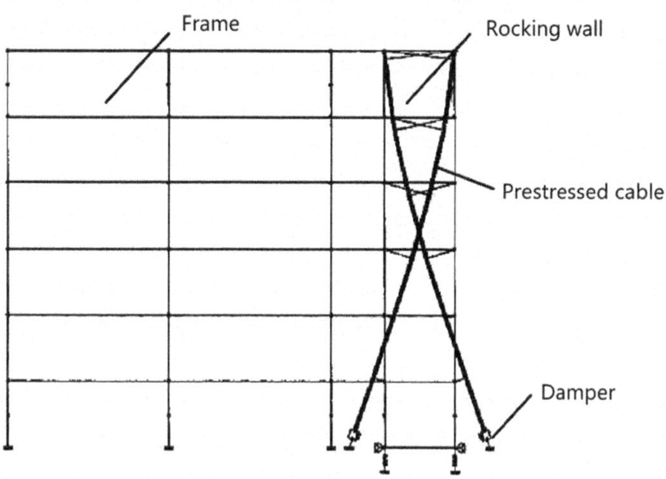

Fig. 6.5 Rocking wall frame structure [9]

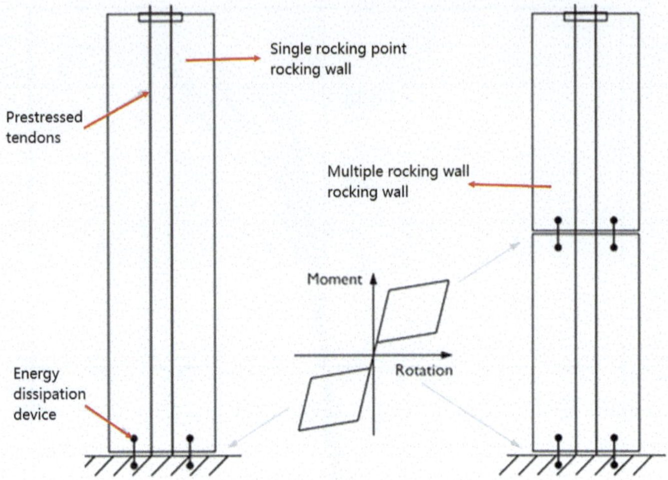

Fig. 6.6 Schematic diagram of the multi-rocking system [11]

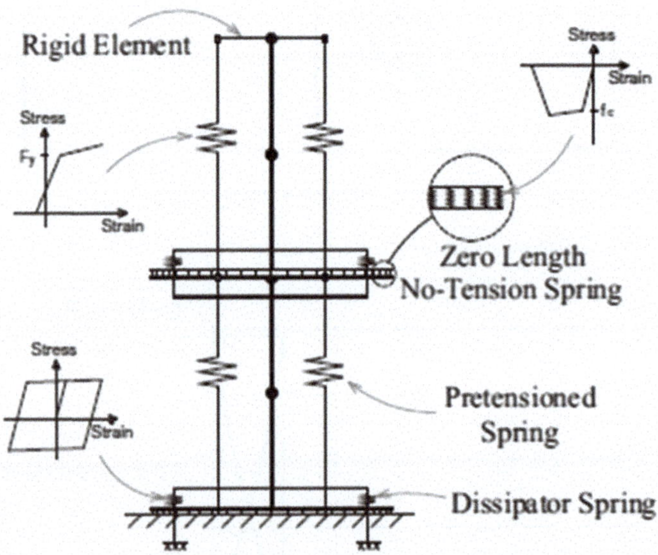

Fig. 6.7 Simplified model of the multi-rocking system proposed by Mohammad et al. [12]

was gradually damaged and peeled off. Then, high-performance fiber-reinforced concrete was used to repair the corner of the rocking wall, and the repaired wall was tested. The results showed that this kind of repair was effective, the performance of the repaired wall was stable, and there was less corner damage.

In 2014, Belleri et al. [18] arranged a prefabricated 7.01 m rocking wall longitudinally on both sides of a half-scale three-story four-span prefabricated concrete

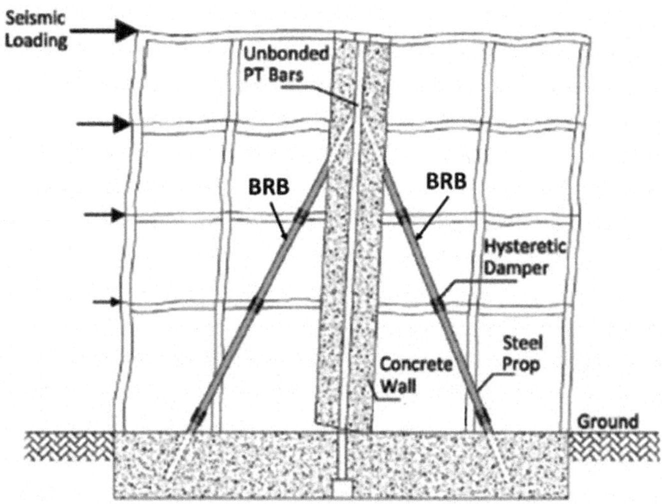

Fig. 6.8 Self-resetting rocking wall with BRB on both sides [13]

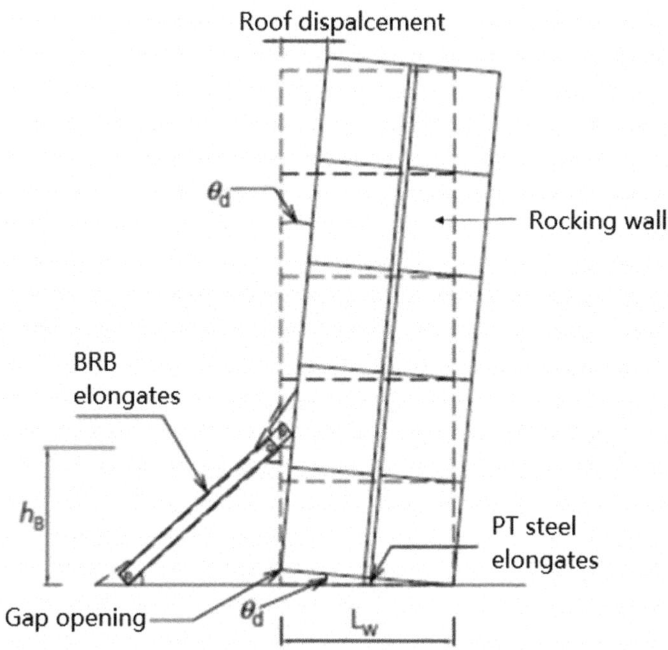

Fig. 6.9 Self-resetting rocking wall with BRB on one side [16]

frame structure. Then, they conducted a shaking table experiment analysis as shown in Fig. 6.10 and obtained some beneficial conclusions.

In 2014, Loo et al. [19] designed a rocking wooden wall (without prestressing) with sliding steel joints and conducted low cyclic loading tests, as shown in Fig. 6.11. In 2014, Moroder et al. [20, 21] proposed an improved design method to solve the problem of floor damage caused by uncoordinated displacements at the joints between the rocking wall and the floor during the rocking process. The method was to set connection points between the rocking wall and the beam and carry out experimental verification.

In 2016, Qureshi et al. [22] proposed a finite element model to predict the dynamic performance of a rocking wall under peak acceleration. In the same year, Buddika et al. [23] conducted a study on the seismic performance evaluation of self-resetting

Fig. 6.10 Experimental frame rocking wall on shaking table [18]

Fig. 6.11 The rotation node of a rocking wooden wall designed by Loo et al. [19]

frame rocking wall structures and frame-shear structures. The research showed that the structural damage of the frame rocking wall was more obvious under the action of ground acceleration. It was also pointed out that the vertical seismic component was not the key factor affecting the peak response of these two types of structures under seismic action.

In 2009, Weda and Qu et al. [24] adopted the joint reinforcement technology of rocking walls and steel dampers for the G3 building of the Tsuda Campus of Tokyo Institute of Technology. A schematic diagram before and after reinforcement is shown in Fig. 6.12. Qu et al. [25] made a useful summary and analysis of this, which showed that the average response of the structure under different seismic actions was effectively reduced after reinforcement. The paper pointed out that during the 2011 Tohoku-Pacific earthquake in Japan, the rocking wall of the G3 building, which was approximately 400 km away from the epicenter, performed well after experiencing a certain degree of ground motion.

The successful application of rocking walls in this project has produced and upsurge of research on rocking walls in China. From 2010 to 2011, Qu [26, 27] conducted a series of studies on rocking wall-frame structures and carried out theoretical analysis and numerical simulation analysis on the damage control mechanism and design method of rocking wall-frames. The results showed that compared with the pure frame structure, the rocking wall-frame structure could effectively control the lateral deformation mode of the whole structure and made the deformation distribution of each floor of the frame structure more uniform, which could prevent the occurrence of floor yielding, make full use of the seismic bearing capacity of each floor, and improve the seismic performance of the whole structure. Compared with the frame-shear wall structure, the rocking wall-frame structure had greater deformation capacity and could avoid the design difficulties caused by the simultaneous

(a) G3 building before reinforcement (b) G3 building after reinforcement

Fig. 6.12 3D renderings of the G3 building before and after reinforcement [24]

bearing of a large bending moment and shear force at the bottom of the wall. This can effectively avoid brittle shear failure at the bottom of the wall, as shown in Fig. 6.13. In 2012, Cao et al. [28] proposed a connection node between the rocking wall and frame and conducted relevant experimental studies to show its effectiveness. In addition, in 2015, Wu and Peng [29] proposed a rocking-filled wall-frame structure. In this structure, part of the filled wall was cast in place to form a rigid filled wall, which was arranged along the full height of the structure. The bottom of the frame column on both sides of the rigid-filled wall was disconnected from the foundation, and the two were touching but not connected. In 2016, Wu et al. [30] proposed the distributed parameter model of a frame-rocking wall structure, studied the application of rocking walls in seismic reinforcement and retrofitting in China, and used the general finite element software ABAQUS to carry out elastoplastic time history analysis.

In 2011, Lu et al. [31] proposed the concept of a recoverable functional structure, which referred to a structure that could be restored to use without repair or with little repair after an earthquake. In 2011, Zhou and Lu [32] conducted a more detailed review of research on rocking structures and self-reset structures, which

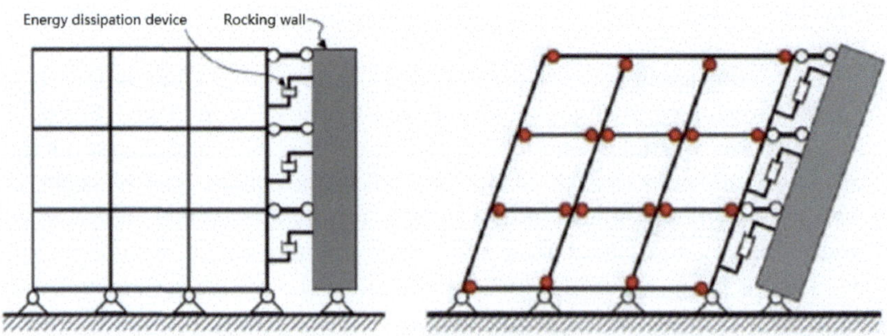

(a) Rocking wall frame structure (b) Overall destruction mechanism

Fig. 6.13 Rocking wall frame structure system and its failure mechanism [26]

pointed out directions for follow-up research. In 2013, Chen and Lu [33] conducted a finite element comparative analysis of the seismic performance of frame self-resetting walls, including a comparison of frame-shear wall structures, frame-self-resetting wall structures, and pure frame structures. The model is shown in Fig. 6.14. The results showed that the frame-self-resetting wall structure had a more uniform interstory displacement angle, but its energy dissipation capacity was weak and the structural deformation was large. Calculations showed that the introduction of dampers could effectively improve its energy dissipation capacity, thereby reducing the displacement of each floor of the frame-self-resetting wall. In the same year, Xu [34] further analyzed the seismic responses of frame-rocking wall and frame-shear wall structures using energy-based methods. The finite element results showed that the energy dissipation of the frame-rocking wall structure depended more on the kinetic energy, potential energy, and hysteresis energy of the damper of the overall rocking of the rocking wall, and the seismic performance of the frame-rocking wall structure was better than that of the frame-shear wall structure. In addition, from 2013 to 2014, Dang et al. [35–38] carried out a finite element analysis and test analysis for self-resetting shear walls. The test results showed that the prestressed self-resetting shear wall with horizontal joints at the bottom had the same lateral bearing capacity as the common shear wall and had a better self-resetting ability. A model of the self-resetting shear wall is shown in Fig. 6.15.

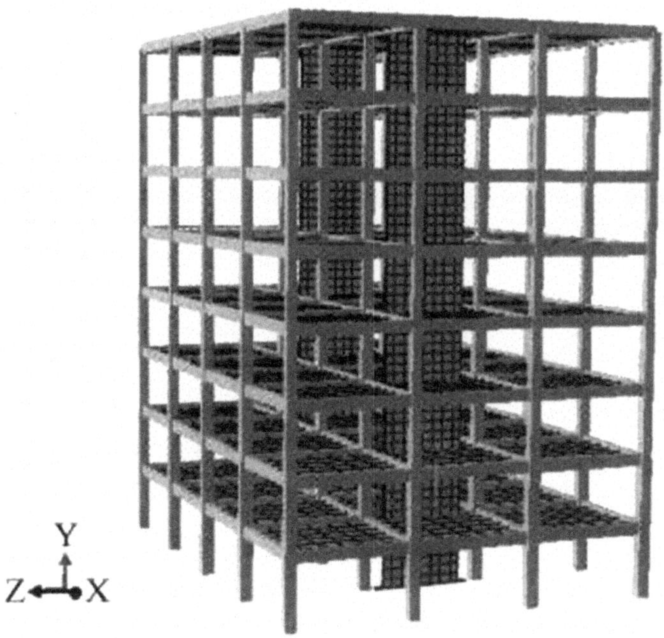

Fig. 6.14 Finite element model diagram of frame-self-resetting wall [33]

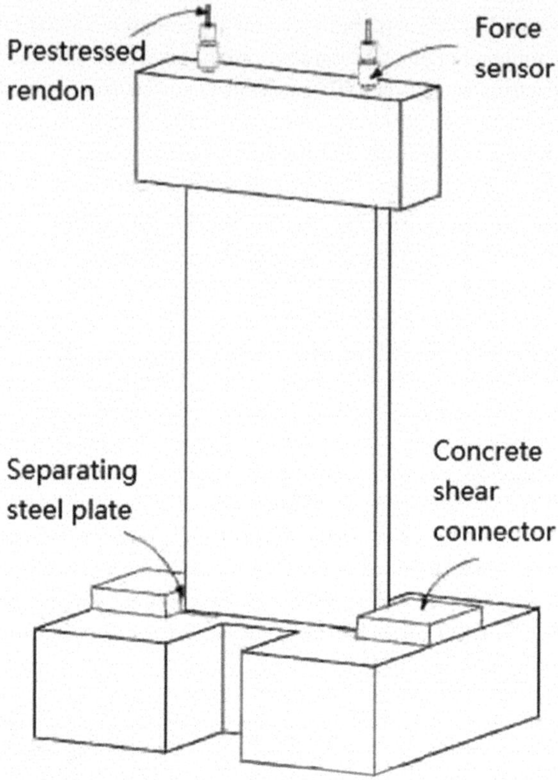

Fig. 6.15 Schematic diagram of a bottom slotted prestressed self-resetting shear wall [35]

In 2014, Yang et al. [39] proposed a simplified calculation method for the frame-rocking wall structure. They carried out corresponding research on the frame-rocking wall structure and then studied the influence of the connection method of the rocking wall [40], the stiffness of the rocking wall [41], etc., on the seismic performance of the structure. In addition, Yang et al. [41] also conducted a shaking table test study on an embedded frame-rocking wall structure. In the same year, Jia et al. [42] conducted a numerical simulation analysis on the seismic performance of frame rocking wall structures with different numbers of stories. In 2015, Zhao [43] conducted a finite element simulation of the seismic performance of a double-limb rocking wall structure. Then, in 2016, Yang et al. [44] introduced a profiled steel damper into a double-limb rocking wall structure and performed a corresponding finite element analysis.

In 2015, Feng et al. [45] proposed the structural form of the continuous rocking wall-buckling restrained bracing frame and used OpenSees to conduct numerical simulations, comparing the continuous rocking wall-buckling restraint bracing frame (CRW-BRBF), rocking wall-buckling restraint bracing frame (RW-BRBF) and buckling restraint bracing frame (BRBF). The results showed that the continuous rocking wall could reduce the nonuniformity factor of the interstory displacement angle

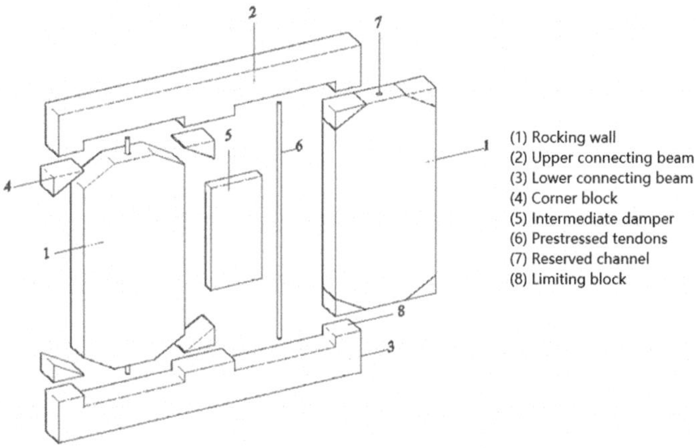

Fig. 6.16 Schematic diagram of the decomposition of the new damage-controllable rocking wall

of the buckling-restrained braced structure and that the use of continuous rocking walls could reduce the bending moment and shear requirements of the rocking wall compared to the rocking wall-buckling restrained braced frame.

This book proposes a new type of damage-controllable rocking wall, which belongs to the category of embedded self-resetting rocking structure. It is set in the floor and is a vertical rocking member inside the structure. By loosening the connection at the end of the wall, the deformation of the wall during an earthquake is concentrated on the rocking interface. The self-reset of the structure after an earthquake is realized by the application of the self-weight and prestress of the structure, and the residual displacement is reduced. The replaceable energy dissipation component is attached to the middle of the wall to protect the main body of the structure by dissipating energy during the earthquake, and the energy dissipation device can be quickly replaced after the earthquake. The corner of the wall is made of an elastic material with high ductility so that damage to the wall can be controlled. Therefore, this new type of rocking wall has the characteristics of low damage under seismic action, high energy dissipation capacity, controllable structural damage, and quick repair after an earthquake, as shown in Figs. 6.16 and 6.17.

6.2 Research on a New Type of Prefabricated Damage-Controllable Rocking Wall

6.2.1 Monotonic Loading Curve

The geometric dimensions of the new damage-controllable rocking wall proposed in this book are shown in Fig. 6.18. It is assumed that shear-type energy dissipation

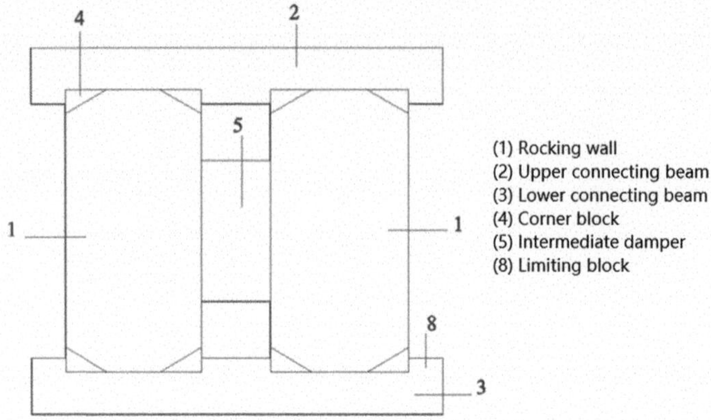

Fig. 6.17 Schematic diagram of the composition of the new damage-controllable rocking wall

damping is ideal elastic–plastic and that there is no relative sliding between the two rocking walls and the upper and lower connecting beams during the rocking process. The influence of the shear damping axial pressure is ignored. When the rocking wall rotates by an angle of θ, the geometric relationships are shown in Fig. 6.19, where:

$$s = L + t + 2a \tag{6.1}$$

$$l = H \sin\theta + L(1 - \cos\theta) \tag{6.2}$$

$$h = L \sin\theta - H(1 - \cos\theta) \tag{6.3}$$

$$t' = s \cos\theta - L - 2a \tag{6.4}$$

$$\Delta H_t = s \sin\theta \tag{6.5}$$

The force on the rocking wall is shown in Fig. 6.20, which can be solved as follows:

$$F = F_1 = F_2 \tag{6.6}$$

$$F = \frac{1}{H+h}\left[(N_1 + G_1 + G_2 + N_A)(L - l) + 2N_M a + 2PL\cos\tfrac{\theta}{2} + Ts\cos\theta\right] \tag{6.7}$$

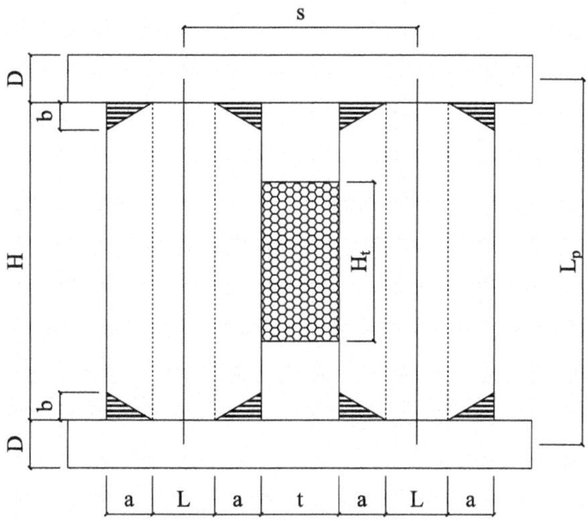

Fig. 6.18 Geometry of the new damage-controlled rocking wall

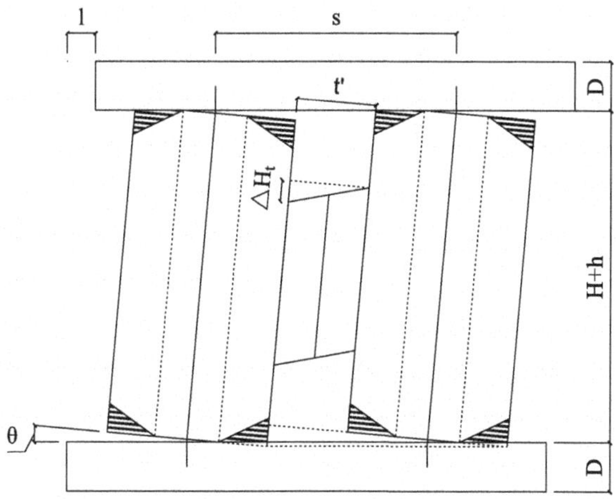

Fig. 6.19 Swing state of the new damage-controllable rocking wall

In formula (6.7):

$$P = P_0 + ES_P \frac{2L \sin \frac{\theta}{2}}{L_P} \tag{6.8}$$

$$N_A = K_A h \tag{6.9}$$

$$N_M = K_M \theta \tag{6.10}$$

$$T = \begin{cases} K_T \Delta H_t & \theta < \arcsin(T_y/(sK_T)) \\ T_y & \theta \geq \arcsin(T_y/(sK_T)) \end{cases} \tag{6.11}$$

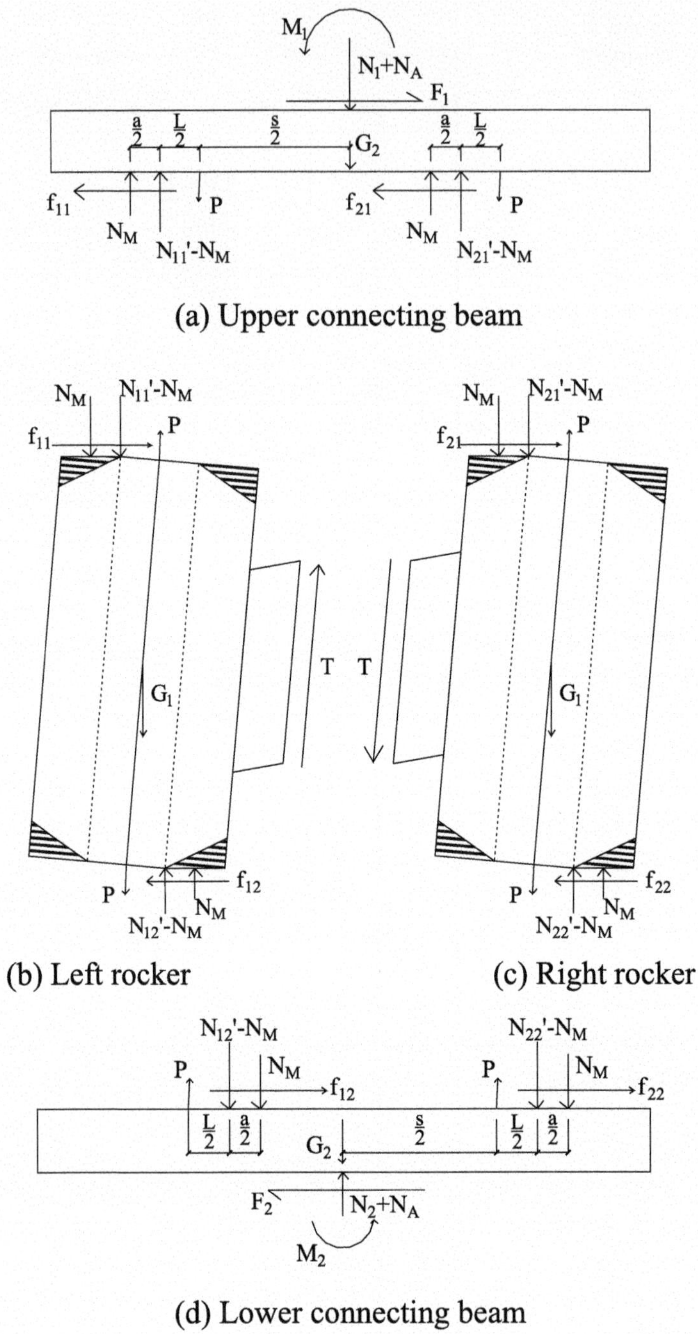

Fig. 6.20 Force analysis of the new damage-controllable rocking wall

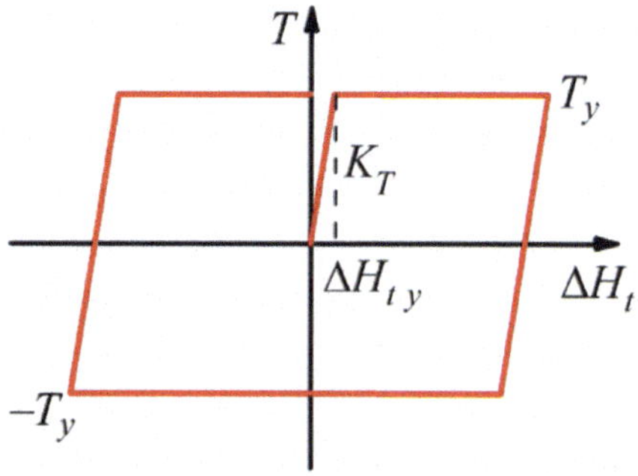

Fig. 6.21 Damped hysteresis curve

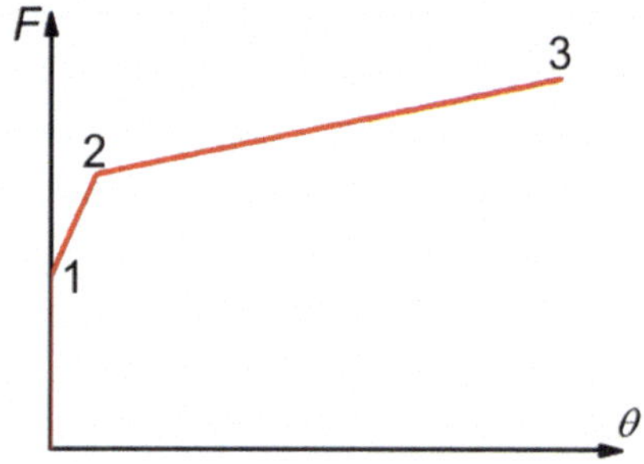

Fig. 6.22 $F - \theta$ curve of the rocking wall

where K_A is the increment of the restraining force of the structure when the rocking body is raised per unit height. K_M is the normal stress resultant of the contact surface of the corner when the rocking body rotates by a unit angle. K_T is the shear force when the shear damping produces a unit relative displacement. T_y is the damping yield force, as shown in Fig. 6.21. The $F - \theta$ relation of the damage-controllable rocking wall is shown in Fig. 6.22. Point 1 is the starting point of rocking, point 2 is the damping yield point, and point 3 is the ending point of loading (the starting point of unloading).

6.2.2 *Hysteresis Curve*

The loading path of the damage-controllable rocking wall can be calculated by formula (6.7), and the unloading path is related to the ending position θ_r of loading. The formula for horizontal thrust during unloading is basically the same as the loading formula, but the expression of damping force in the formula (6.7) becomes:

$$T = \begin{cases} T_y - K_T s \sin(\theta_r - \theta) & \theta_r - 2\arcsin(T_y/(sK_T)) < \theta \le \theta_r \\ -T_y & \theta \le \theta_r - 2\arcsin(T_y/(sK_T)) \end{cases} \tag{6.12}$$

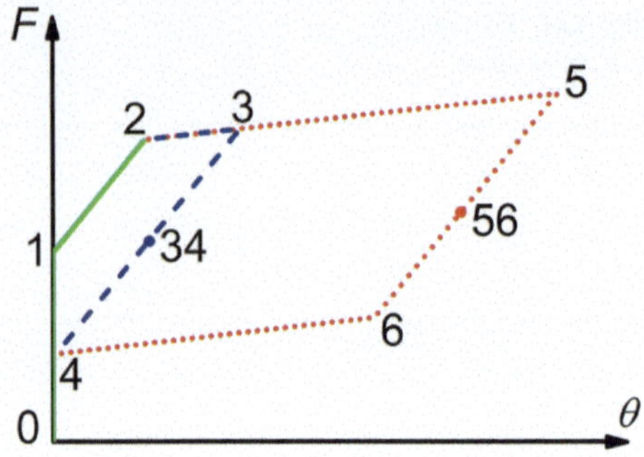

Fig. 6.23 Hysteresis curve of the new damage-controllable rocking wall

As shown in Fig. 6.23, when $\theta_r = \arcsin(T_y/(sK_T))$ at point 2 of the loading end position, the loading and unloading path is 0–1–2–1–0. When $\theta_r = 2\arcsin(T_y/(sK_T))$ at point 3 of the loading end position, the loading and unloading path is 0–1–2–3–34–4–0, where the damping force T yields at point 2, the direction of damping force reverses at point 34, and conversely yields at point 4. When $\theta_r > 2\arcsin(T_y/(sK_T))$ at point 5 of the loading end position, the loading and unloading path is 0–1–2–5–56–6–4–0, where the damping force T yields at point 2, the direction of damping force reverses at point 56, and conversely yields at point 6. The area of the envelope of the hysteresis curve 0–1–2–5–6–4–0 is:

$$S_{0125640} = \int_0^{\theta_5} F_{0125} d\theta - \int_0^{\theta_5} F_{0465} d\theta \tag{6.13}$$

6.2.3 Condition of Implementing the Self-Reset

According to the above analysis, there is no residual deformation after theself-reset of the damage-controllable rocking wall; that is, to realize the flag-shaped hysteresis curve, the point 4 of the hysteresis curve in Fig. 6.23 must be above the origin point:

$$F = \frac{1}{H}\left[(N_1 + G_1 + G_2 + 2P_0)L - T_y s\right] \geq 0 \tag{6.14}$$

which is:

$$N_1 + G_1 + G_2 + 2P_0 \geq \frac{sT_y}{L} \tag{6.15}$$

In practical engineering, considering factors such as cracking yield of reinforced concrete and wall slippage, a safety factor K_s greater than 1 should be considered, and formula (6.15) becomes:

$$N_1 + G_1 + G_2 + 2P_0 \geq K_s \frac{sT_y}{L} \tag{6.16}$$

6.3 Experimental Analysis of the New Damage-Controllable Rocking Wall

6.3.1 Experimental Design

To verify the feasibility of the proposed new damage-controllable rocking wall, this book studies its seismic performance through low cyclic loading tests. The test consists of three sets of new damage-controllable rocking walls. The total height of the designed specimen is 2.8 m, and the width × height × thickness of the single-piece small wall of the rocking wall is 800 mm × 2000 mm × 150 mm. The test performs low cyclic loading tests on three specimens: a new rocking wall with energy dissipation damping in the middle but no out-of-plane limit block (RW-D-NL), a new rocking wall with energy dissipation damping in the middle and an out-of-plane limit block (RW-D-L), and a new rocking wall with no energy dissipation damping in the middle but with limit blocks outside the plane (RW-ND-L). The basic schematic diagram of the test specimen is shown in Fig. 6.24. Among them, the new damage-controllable rocking wall is compared by setting different parameters, and the parameter settings are shown in Table 6.1. During the rest, only the energy dissipation damping is dismantled and replaced, and the prestress is retensioned after each test.

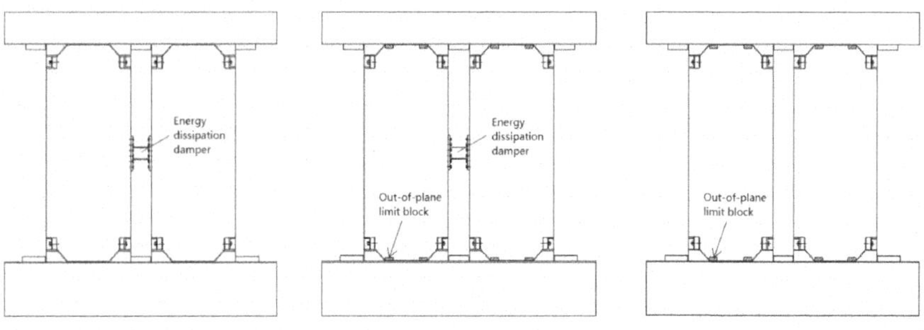

(a) RW-D-NL (b) RW-D-L (c) RW-ND-L

Fig. 6.24 Basic schematic diagram of the test specimen

Table 6.1 Model specimen parameters

Specimen number	Monolithic wall initial prestress (kN)	Damping	Out-of-plane limit
RW-D-NL	300	yes	no
RW-D-L	350	yes	yes
RW-ND-L	350	no	yes

6.3.2 Experimental Device

A schematic diagram of the loading device for the rocking wall is shown in Fig. 6.25. The multichannel electrohydraulic servo structure test system is used for loading in this test, and the loading plan is formulated with reference to the *Building Seismic Test Regulations* (JGJ/T 101-2015).

Displacement loading is used in this test, and a displacement of 0.5 mm is preloaded before the test to eliminate the internal nonuniformity of the test piece and check whether the test equipment and measuring instruments are operating normally. For the new damage-controllable rocking wall, no destructive test is carried out, and 1/50 of the height of the rocking wall, that is, 40 mm, is the maximum displacement loading point. Loading is carried out in 10 stages, which are 0.5 mm, 1 mm, 1.5 mm, 2 mm, 4 mm, 8 mm, 16 mm, 24 mm, 32 mm, and 40 mm. Cycling is performed 3 times for each stage, as shown in Fig. 6.26.

The specimen before loading in the final test is shown in Fig. 6.27. It should be noted that the erection of the scaffold is only to prevent accidents during the test and does not participate in the force.

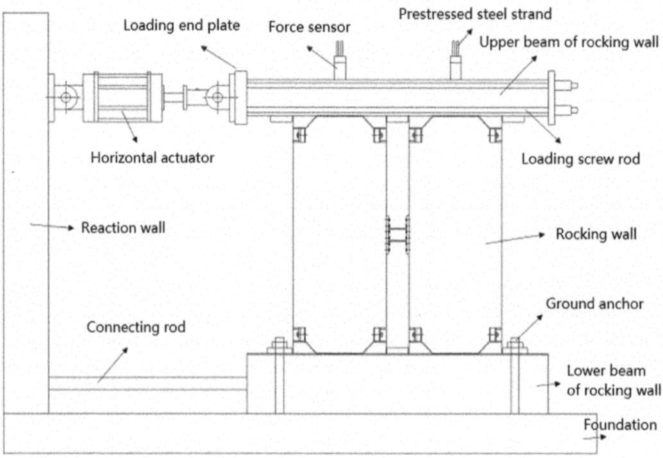

Fig. 6.25 Schematic diagram of loading device of rocking wall

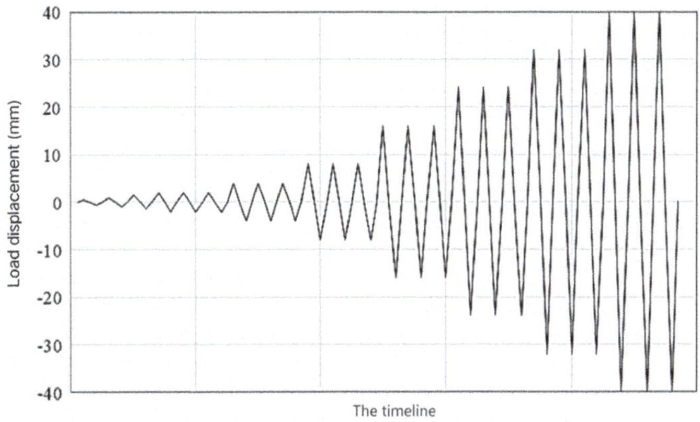

Fig. 6.26 Loading system of the rocking wall

(a) RW-D-NL

(b) RW-D-L

(c) RW-ND-L

Fig. 6.27 Actual specimen before loading

6.3.3 Experimental Analysis

The hysteresis curve obtained from the test is shown in Fig. 6.28.

In Fig. 6.28a, RW-ND-L is a basic reference group without damping. It can be seen that the new damage-controllable rocking wall has good self-reset performance, and the curve does not fully return to the loading origin because there is slippage

Fig. 6.28 Hysteresis curve

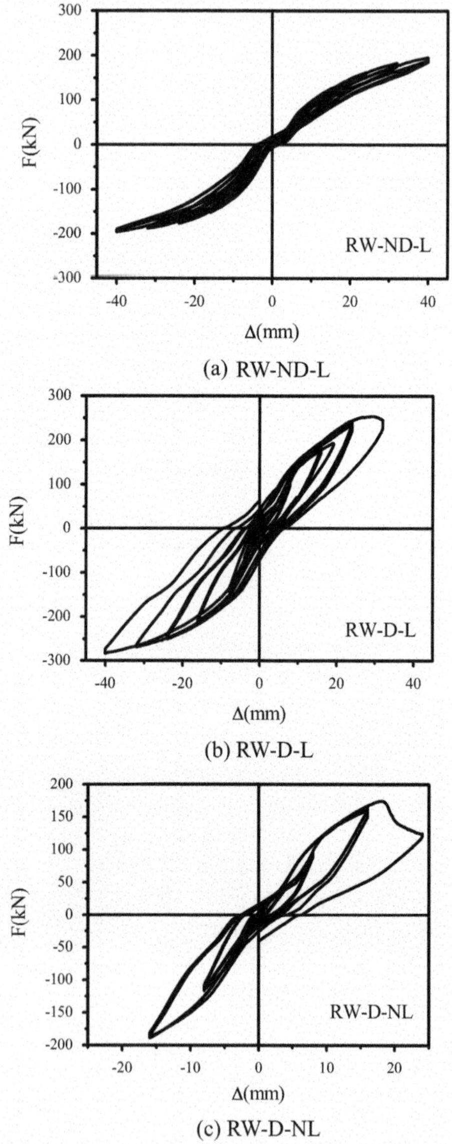

during the loading process. The main reasons for the slippage are the manufacturing and assembly errors of the various parts of the rocking wall and the nonuniformity of the actuator. In addition, the rocking wall consumes almost no energy, which means the wall is not damaged from the side.

In Fig. 6.28b, RW-D-L is under forward loading, that is, in the first quadrant area. The displacement loading of +32 mm presents an overall out-of-plane rollover and a decline in horizontal lateral bearing capacity, while the reverse loading is relatively stable and shows a good energy dissipation capacity. When the load is approximately 2 mm, the curve begins to show an inflection point. When loading repeatedly at 4 mm, the damping begins to show obvious energy dissipation. When a displacement loading of 8 mm is carried out, the curve of the third quadrant has a second inflection point, the curve of the first quadrant has an inverted S-shaped hysteresis loop, and the stiffness shows a relatively obvious increase. It is more intuitive to see this point in Fig. 6.28c, and it can be judged that the structure in this stage eliminates the slip gap. Starting from a displacement loading of 24 mm, there is an obvious displacement of the structure. On the one hand, the damping force is relatively large, and the prestress is not enough to pull it back completely. Meanwhile, according to the data recorded by the force sensor, the prestress will be lost when each level of displacement is loaded back to the starting point. On the other hand, the deformation after damping yield will affect the clearance of the rocking wall specimen itself.

Compared with the traditional shear wall, the rocking wall has a better self-resetting ability, and the damage degree of the rocking wall itself is much smaller than that of the traditional shear wall. The energy dissipation of the undamped rocking wall is much smaller than that of the shear wall, which can be reasonably improved by adding damping devices.

By comparing Fig. 6.28a, b, as shown in Fig. 6.29, it can be seen that with the increase in damping, the energy dissipation of the rocking wall increases significantly, and the bearing and poststage stiffness also increase accordingly.

Fig. 6.29 Comparison of hysteresis between RW-ND-L and RW-D-L

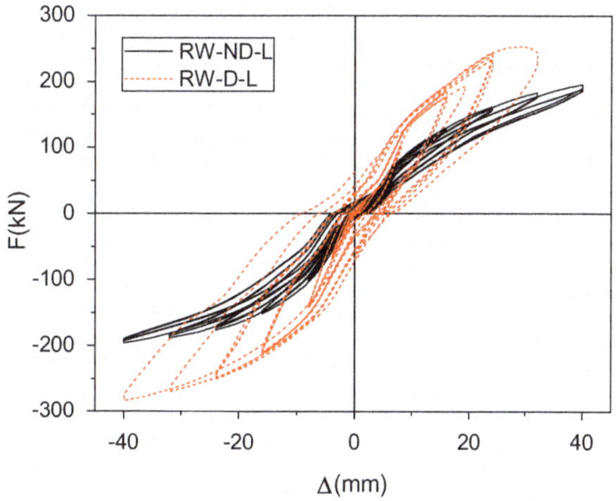

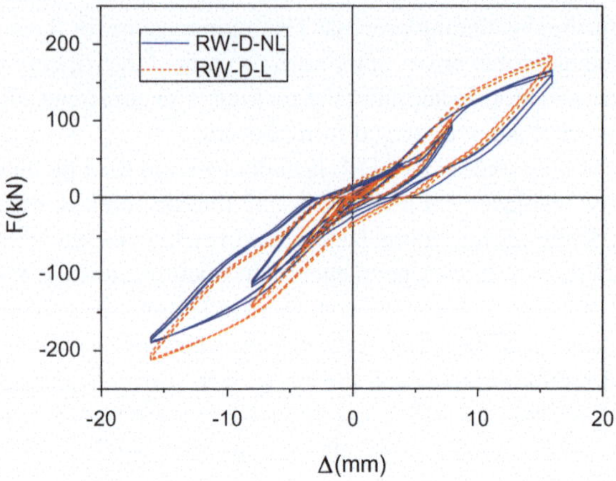

Fig. 6.30 Comparison of hysteresis between RW-D-NL and RW-D-L

By comparing Fig. 6.28b, c, as shown in Fig. 6.30, only the loading displacement of 16 mm is compared. The initial prestress value of model RW-D-L is 350 kN, and the initial prestress value of model RW-D-NL is 300 kN. It can be seen that with the increase in the initial prestress, the lateral bearing capacity of the new damage-controllable rocking wall increases, and the horizontal force required at the starting point of rocking increases, but the change in energy dissipation capacity is not obvious from the perspective of the hysteresis of the outermost ring.

6.4 Finite Element Simulation of Damage-Controllable Rocking Wall

6.4.1 Establishment of Finite Element Model

The ABAQUS finite element model of the new damage-controllable rocking wall is shown in Figs. 6.31 and 6.32. The model mainly includes the rocking wall and its reinforcement, upper and lower connecting beams, corner elastic bodies, intermediate shear energy dissipation components, prestressed tendons, and their anchors. Among them, concrete, corner elastic body, intermediate shear energy dissipation components, and anchors are all built with 3D solid modules, and the eight-node hexahedral linear reduced-integration element C3D8R is selected for mesh division. The ordinary steel bar and prestressed tendon are established by the truss module in the three-dimensional solid line, and the T3D2 element is selected for division. C40 is selected for concrete, HRB400 for steel bar, and a nominal diameter of d = 15.2 mm for prestressed steel strand.

When ABAQUS conducts solid element calculations, surface–surface contact is generally used to simulate the contact between different parts. The interaction

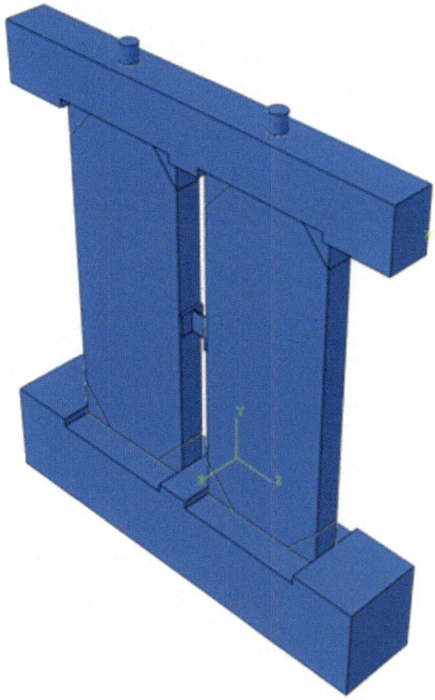

Fig. 6.31 Model of the new damage-controllable rocking wall

between the contact surfaces consists of two parts, which are the normal interaction between contact surfaces and the tangential action between contact surfaces. Tangential action also includes relative motion and possible friction between the contact surfaces. In this chapter, there is no bond between the rocking wall and the upper and lower beams but only through the prestress in the middle, so the rocking wall and the upper and lower beams are in face-to-face contact. The normal contact behavior adopts "hard" contact. When the contact gap between the two surfaces is zero, the contact constraint is imposed. When the contact pressure between the contact surfaces becomes zero or negative, the two contact surfaces are separated, and the contact constraint is removed. For the tangential behavior, according to the theoretical analysis in Chap. 2, the friction of the contact interface of the rocking wall is the key to ensuring that it does not slide, so the selection of the friction coefficient should not be too small. In this paper, the "penalty" friction contact is adopted, and the friction coefficient is taken as 0.4.

The prestress simulation method in this chapter adopts the cooling method, that is, the temperature load (cooling) is applied to the prestressed tendons so that the prestress tendons shrink, generating the prestress. The model uses the posttensioned unbonded prestress, which is arranged in the middle of each wall. The prestressed tendons are prestressed steel strands with a yield strength of 1860 MPa and an elastic modulus of 1.95×10^5 MPa. In the model, the prestressed tendons and the anchors are directly connected by MPC, that is, the two end points of the prestressed tendons

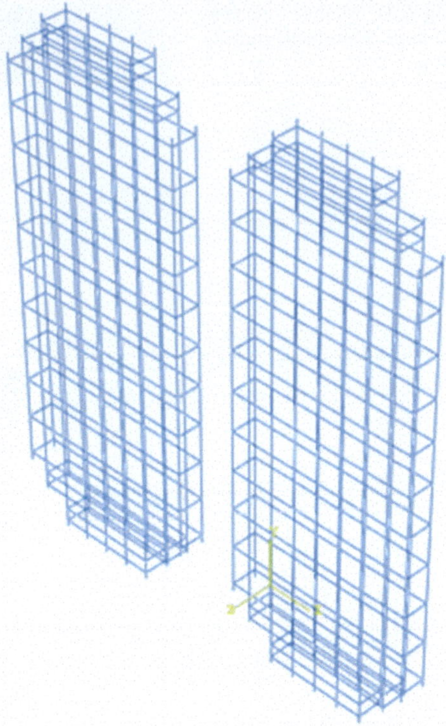

Fig. 6.32 Rebar model of the rocking wall

are anchored into the corresponding anchors, and the contact surfaces of the anchors with the upper and lower beams are connected by a "tie". In this way, the prestress applied by cooling can be transmitted to the structure through the anchor.

6.4.2 Comparison of Experimental and Theory and Finite Element Results

Theoretical calculations and ABAQUS simulations were carried out based on the parameters of the RW-ND-L and RW-D-L specimen models. The main parameters are set as follows: 0.3 for the height of the upper connecting beam, 0.5 for the height of the lower connecting beam, 2 for the height of the rocking body, 0.4 for the contact width, 0.2 for the width of the corner, 3.33 for the length of the prestressed bars, 556 for the area of the prestressed bars, 195 for the elastic modulus of the prestressed bars, 350 for the initial prestressed force, 0.2 for the damping width, 7091 for the corner stiffness, 6.0 for the weight of the rocking body, and 4.8 for the weight of the connecting beam. The RW-D-L model has a shear damping stiffness of 120 and a damping yield force of 185. RW-ND-L does not set damping and has a damping yield force of 0. It should be noted that in the test, force sensors and some reaming

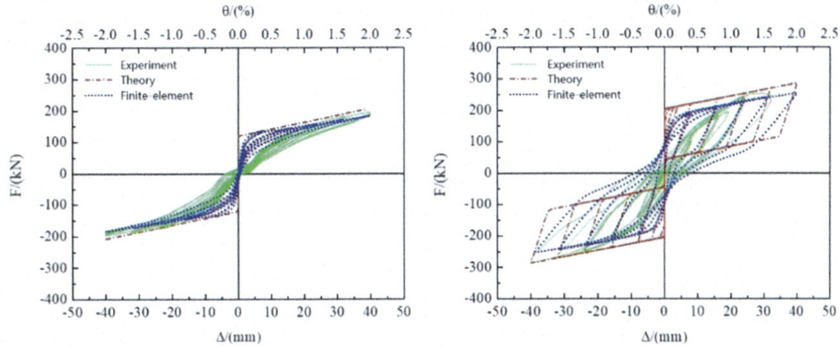

(a) RW-ND-L (b) RW-D-L

Fig. 6.33 Comparison of theory, experiment, and the finite element model

devices are added under the anchor cup, which causes the actual length of prestressed tendons to reach 3.33 m.

Figure 6.33 shows that the hysteresis curves of the three are relatively consistent. The theoretical bearing capacity is slightly larger than that of the test specimen, while the simulated initial stiffness is greater than the experimental value. The reasons for these phenomena are summarized from two aspects. First, some simplification assumptions make the structure more rigid and stronger. These assumptions mainly include the rigidity assumption of the wall and the ideal elastic–plastic assumption of damping. Meanwhile, the elastic deformation section in the early stage of the rocking wall is not considered in the theoretical calculation. Second, there are errors in the fabrication and assembly of the test specimens. In addition, there are prestress losses after each stage of displacement loading, which greatly reduces the bearing capacity of the new damage-controllable rocking wall.

6.5 Conclusions

With research on the sustainable development of earthquake engineering, this chapter proposes a new type of damage-controllable rocking wall by reviewing and summarizing previous literature. Through the arrangement of structural forms and the design of vulnerable parts, it will have the characteristics of low damage under seismic action, high energy dissipation capacity, controllable structural damage, and quick repair after an earthquake.

The main design parameters of the new damage-controlled rocking wall include the height-width ratio (geometric dimensions), damping parameters, material properties of the corner, area of prestressed tendons, and initial prestress. In theoretical analysis, the key points of hysteresis performance are the starting point of rocking, the damping yield point, the ending point of loading, the yielding point of the reverse position of the damping during unloading, and the intersection between the origin and the longitudinal axis when unloading.

The test results show that damage control of the new damage-controllable rocking wall is achieved and that the wall is still not damaged after three sets of experiments. First, the design of rubber blocks in the corner improves the stress concentration at the foot of the concrete rocking wall. Second, the setting of the embedded parts in the corner of the rocking wall and the embedded steel plate on the side of the wall effectively control damage to the wall. Rapid replacement of the intermediate energy dissipation damping is carried out in the experiment. The corner rubber block could also be quickly replaced if necessary.

References

1. Housner GW (1963) The behavior of inverted pendulum structures during earthquakes. Bull Seismol Soc Am 53(2):403–417
2. Kurama Y, Pessiki S, Sause R et al (1996) Analytical modeling and lateral load behavior of unbonded post-tensioned precast concrete walls. Research Report No. EQ-96–02, Department of Civil and Environmental Engineering, Lehigh University, Bethlehem, PA
3. Kurama Y, Sause R, Pessiki S et al (1999) Lateral load behavior and seismic design of unbonded post-tensioned precast concrete walls. ACI Struct J 96(4):622–632
4. Kurama YC (2000) Seismic design of unbonded post-tensioned precast concrete walls with supplemental viscous damping. ACI Struct J 97(4):648–658
5. Kurama YC (2001) Simplified seismic design approach for friction-damped unbonded post-tensioned precast concrete walls. ACI Struct J 98(5):705–716
6. Perez FJ, Sause R, Pessiki S et al (2002) Lateral load behavior of unbonded post-tensioned precast concrete walls. In: Advances in building technology. Elsevier, pp 423–430
7. Holden T, Restrepo J, Mander JB (2003) Seismic performance of precast reinforced and prestressed concrete walls. J Struct Eng 129(3):286–296
8. Restrepo JI, Rahman A (2007) Seismic performance of self-centering structural walls incorporating energy dissipators. J Struct Eng 133(11):1560–1570
9. Ajrab JJ, Pekcan G, Mander JB (2004) Rocking wall–frame structures with supplemental tendon systems. J Struct Eng 130(6):895–903
10. Marriott D, Pampanin S, Bull D et al (2008) Dynamic testing of precast, post-tensioned rocking wall systems with alternative dissipating solutions. Bull New Zealand Soc Earthq Eng 41(2)
11. Wiebe L, Christopoulos C (2009) Mitigation of higher mode effects in base-rocking systems by using multiple rocking sections. J Earthquake Eng 13(S1):83–108
12. Khanmohammadi M, Heydari S (2015) Seismic behavior improvement of reinforced concrete shear wall buildings using multiple rocking systems. Eng Struct 100:577–589
13. Nicknam A, Filiatrault A (2012) Seismic design and testing of propped rocking wall systems. In: Proceedings of the 15th world conference on earthquake engineering, Lisbon, Portugal
14. Nicknam A (2015) Seismic analysis and design of buildings equipped with propped rocking wall systems. State University of New York at Buffalo
15. Nicknam A, Filiatrault A (2015) Seismic fragility analysis of buildings equipped with propped rocking wall systems. In: SECED 2015 conference, Cambridge UK
16. Yooprasertchai E, Hadiwijaya IJ, Warnitchai P (2015) Seismic performance of precast concrete rocking walls with buckling restrained braces. Mag Concr Res 68(9):462–476
17. Preti M, Meda A (2015) RC structural wall with unbonded tendons strengthened with high-performance fiber-reinforced concrete. Mater Struct 48(1–2):249–260

18. Belleri A, Schoettler MJ, Restrepo JI et al (2014) Dynamic behavior of rocking and hybrid cantilever walls in a precast concrete building. ACI Struct J 111(3):661–671
19. Loo WY, Kun C, Quenneville P et al (2014) Experimental testing of a rocking timber shear wall with slip-friction connectors. Earthq Eng Struct Dynam 43(11):1621–1639
20. Moroder D, Sarti F, Palermo A et al (2014) Seismic design of floor diaphragms in post-tensioned timber buildings. In: World conference on timber engineering, Canada
21. Moroder D, Sarti F, Palermo A et al (2014) Experimental investigation of wall-to-floor connections in post-tensioned timber buildings. In: NZSEE conference, Auckland, pp 21–23
22. Qureshi IM, Warnitchai P (2016) Computer modeling of dynamic behavior of rocking wall structures including the impact-related effects. Adv Struct Eng 19(8):1245–1261
23. Buddika HADS, Wijeyewickrema AC (2016) Seismic performance evaluation of post-tensioned hybrid precast wall-frame buildings and comparison with shear wall-frame buildings. J Struct Eng 142(6):04016021
24. Wada A, Qu Z, Motoyui S et al (2011) Seismic retrofit of existing SRC frames using rocking walls and steel dampers. Front Archit Civil Eng China 5(3):259
25. Qu Z, He T, Ye L (2011) Seismic retrofit of frame structures using rocking wall system. J Build Struct 09:11–19 (in Chinese)
26. Qu Z (2010) Study on seismic damage mechanism control and design of rocking wall-frame structures: [Doctoral Dissertation]. Department of Civil Engineering, Tsinghua University, Beijing (in Chinese)
27. Qu Z, Ye L (2011) Seismic damage mechanism control of rocking wall-frame system. Earthq Eng Eng Dynam 04:40–50 (in Chinese)
28. Cao H, Pan P, Wu S et al (2012) Experimental study of connections of frame-rocking wall system. J Build Struct 12:38–46 (in Chinese)
29. Wu S, Pan P (2015) Seismic performance evaluation of rocking infilled wall-frame structure. J Build Struct 10:81–87 (in Chinese)
30. Wu S, Pan P, Zhang X (2016) Characteristics of frame rocking wall structure and its application in aseismic retrofit. Eng Mech (06):54–60 (in Chinese)
31. Lu X, Chen Y, Mao Y (2011) New concept of structural seismic design: earthquake resilient structures. J Tongji University (Nat Sci) (07):941–948 (in Chinese)
32. Zhou Y, Lu X (2011) State-of-the-art on rocking and self-centering structures. J Build Struct 32(9):1–10 (in Chinese)
33. Chen K, Lu X (2013) Analysis of seismic performance of the self-centering wall frame structure. Struct Eng 04:118–124 (in Chinese)
34. Xu J, Lu X (2013) Energy based seismic response of frame-rocking-wall structure and frame-shear-wall structure. Build Struct S2:418–422 (in Chinese)
35. Dang X, Lu X, Zhou Y (2013) Study on seismic performance of a rocking wall with bottom horizontal slits. Earthq Eng Eng Dynam 05:182–189 (in Chinese)
36. Dang X, Lu X, Qian J et al (2014) Finite element analysis with solid and plane element of seismic performance of self-centering pre-stressed shear walls. J Build Struct 05:17–24 (in Chinese)
37. Dang X, Lu X, Zhou Y (2014) Experimental design and measured behavior analysis of self-centering shear walls with horizontal bottom slit. Earthq Eng Eng Dynam 06:103–112 (in Chinese)
38. Dang X, Lu X, Zhou Y (2014) Experimental study and numerical simulation of self-centering shear walls with horizontal bottom slit. In: National conference on earthquake engineering (in Chinese)
39. Yang S, Yu D, Jia J et al (2014) Simplified calculation method of frame rocking-wall structure system. Earthq Resist Eng Retrofitting 02:94–99 (in Chinese)
40. Yang S, Wei Z, Xie B (2014) Influence of different connections of swaying wall on seismic performance for frame structure. Indust Constr 11:99–103 (in Chinese)
41. Yang S, Yan L, Jia J et al (2014) Influence of rocking wall stiffness on seismic behavior of frame rocking wall structure. World Earthq Eng 04:27–33 (in Chinese)

42. Jia J, Yan L, Yang S et al (2014) Study on seismic performance of rocking-wall frame structure with different stories. Earthq Eng Eng Dynam 02:97–103 (in Chinese)
43. Zhao Y (2015) Analysis of rocking coupled wall structure: [Master's Dissertation]. College of Civil Engineering, Hubei University of Engineering, Handan (in Chinese)
44. Yang S, Zhang X, Zhu X et al (2016) Application research of profile steel dampers in rocking coupled wall structure. Build Sci (09):108–113 (in Chinese)
45. Feng Y, Wu J, Meng S (2015) Seismic performance analysis of continuously rocking wall-buckling restrained braced frames. In: Proceedings of the 24th national conference on structural engineering (vol ii). Xiamen, Fujian, China, p 5 (in Chinese)

Chapter 7
Prefabricated Concrete Cassette Structure

Abstract This chapter proposed the components, behavior, and practical applications of a novel large-span structure called cassette structure. Composed of open-web sandwich slabs and grid frame walls, the cassette structure can reach over 30 m spans with only a floor system height of 1–1.5 m, much lower than current frame structures, and shows a good seismic performance. The test results of the open-web sandwich slab and grid frame wall were discussed, and the entire structural performances of the cassette structure under earthquakes were analyzed and assessed based on the test results.

Keywords Cassette structure · Large-span structure · Open-web sandwich slabs · Grid frame wall · Experiment study · Seismic assessment

7.1 Introduction

The cassette structure is a new type of structure, which was named for its cassette-like shape. As shown in Figs. 7.1 and 7.2, a cassette structure is composed of two main parts: the floor system called an open-web sandwich slab and the vertical system called a grid frame wall (GFW). The open-web sandwich slab is a crossing beam system where the space between the beams is approximately 2–2.5 m. Unlike a traditional frame structure or prestressed frame structure, the beam web of the open-web sandwich slab, which is inefficient in bearing a bending movement, is hollowed, and the upper and bottom slabs are connected by shear keys rigidity. Therefore, its self-weight can be reduced by approximately 40% without obviously affecting the performance of the slab. Because of its light and high-strength characteristics, the open-web sandwich slab can span over 40 m with a relatively small height, usually approximately 1/25 of its span. Furthermore, a high-stiffness new vertical system consisting of interlayer beams and dense columns is applied and placed around the structure, which means there are no extra columns inside the building and the huge span space can be achieved. In addition, caused by its hollowed web and extra interlayer beams, the cassette structure can reduce concrete and steel usage by approximately 25% when achieving the same seismic performance and a much larger span compared to traditional structures. As of now, this new structure has been

G. Wu et al., *Novel Precast Concrete Structure Systems*, Springer Tracts in Civil Engineering, https://doi.org/10.1007/978-981-19-6821-1_7

applied in large span public buildings and mid-rise office buildings for more than one million square metres. As shown in Figs. 7.3 and 7.4, the Sports Hall of Shandong Weifang Sports School is a multilayer, large span concrete cassette structure, whose planned view size is 64 m × 56 m. The span of the sports hall is 32 m and its floor height is 1250 mm, which is approximately 1/25 of its total span. Moreover, by using the cassette structure, over 1200 m^3 of concrete and 600 t of steel are saved compared to traditional frame structures or prestressed frame structures.

Fig. 7.1 Cassette structure

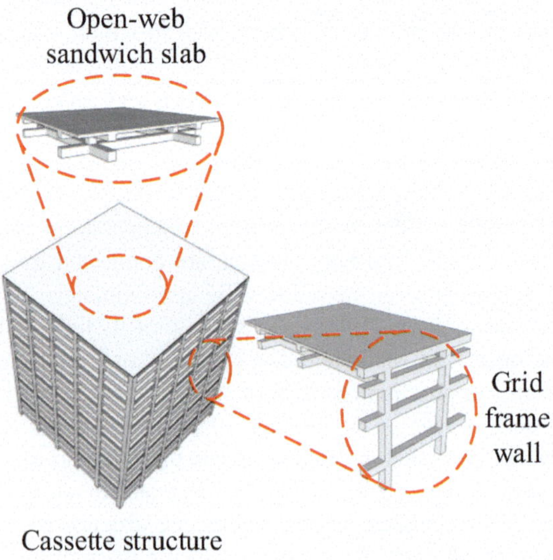

Fig. 7.2 The component of cassette structure

Fig. 7.3 The sports hall in Weifang, Shandong

Fig. 7.4 The open-web sandwich slab in the sports hall

Currently, mid-rise and high-rise buildings (height between 24 and 100 m in Chinese Code JGJ 3–2010) are widely seen in large cities, especially for use as apartments, hotels and offices. Frame structure is widely used among these buildings, but the large element cross-section and fixed space division limit the architectural design and functional usage of it. To address these concerns, the cassette structure is applied in mid-rise and high-rise buildings. According to the present study, a cassette structure can reduce the self-weight of the entire construction by approximately 20% and allow users to design the inner space freely. Furthermore, building industrialization is one of the priorities developed area in China. Because the element cross section in cassette structure is comparatively small, and easy to be manufactured, transported and assembled, it is a very suitable structure system for the development of building industrialization. Meanwhile, the number of nodes and elements in cassette structure is much more than traditional structures, and the traditional cast-in-place method is complicated for cassette structure. Therefore, using building industrialization to modify the manufacture processing of cassette can largely upgrade its economic efficiency and practicability.

Current studies of cassette structure mainly focus on large span structures, and the dynamic characteristics, failure mode and appropriate height in tall buildings are rarely considered. This limits the application and development of cassette structure in high-rise buildings. Therefore, the hysteresis performance, seismic performance and economic performance of the cassette structure in high-rise buildings are studied, and some engineering practice are discussed to perform the performance of cassette structure in mid-rise and high-rise buildings.

7.2 Development and Component of Cassette Structure

Cassette structure developed from an open-web grid, which is a high (usually 1/8 of the span), truss-like space structure and was proposed in 1986 [1]. After years of research and practice, the mechanical and economic characteristics of open-web grids were well-performed. However, the large structural height and comparatively low stiffness limit application in large multilayer structures. In 1995, to solve these problems, open-web sandwich slab was proposed [1]. By increasing the cross-section of the top and bottom truss and adding a stiff solid element, the height of open-web sandwich slab can be decreased to 1/30 of the span, and the stiffness is strong enough to be applied in large span structures. The cassette structure was further developed from an open-web sandwich slab in 2009 [2] The normal frame columns were changed to slab-like elements in cassette structure, and the dynamic characteristics were largely enhanced. The details of open-web sandwich slab and cassette structure are as follows.

The configuration of cassette structure is shown in Fig. 7.5a, b, which mainly consists of OWSS and GFW.

The construction of an OWSS is shown in Fig. 7.5c, d. It has a similar configuration to rib floor system, but the entire beam web is hollowed. Its mechanical property is a sandwich slab considering shear deformation. There are three main components of the slab: top rib, bottom rib and shear key. The two ribs can be regarded as the upper and bottom fibers of a solid beam. To make them work together, shear keys are added. A shear key is a block element whose height-to-width ratio should be less than one. The bending and shear stiffness of shear keys are much larger than that of the two ribs, and both ribs can be connected rigidly by the shear keys (Zhang et al. 2006). When bearing loads, through the connection of shear keys, the two ribs can work together like the upper and bottom fiber of a solid beam. A pair of tension and pressure force is produced in the two ribs and can resist bending force, and both ribs can resist shear force together. To maintain the stiffness and overall performance of the slab, the distance between two shear keys (i.e., the grid size of the open web sandwich slab) should be below 4 m, and the range from 1.5 m to 3 m is recommended. According to current research, experiments and application, a properly designed OWSS can reduce 50% self-weight of the floor system without noticeably affecting structural

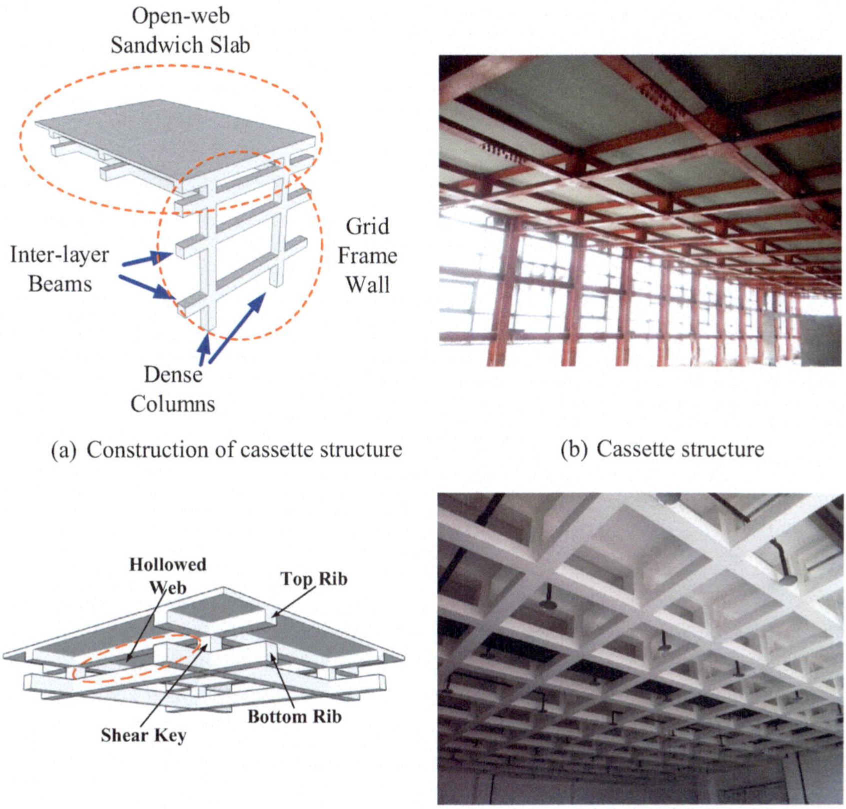

(a) Construction of cassette structure (b) Cassette structure

(c) Construction of OWSS (d) OWSS

Fig. 7.5 Components of cassette structure

performance (Ma et al. 2009). In large span structures, 50% or more vertical load comes from self-weight of beams and large beam cross-section restrain the structural span ability. However, as the beam web of OWSS is hollowed, it can largely decrease the load of the total floor system and span up to 40 m with a comparatively small height (usually 1/25–1/30 of the total span). Furthermore, as the entire beam web is hollowed, all the pipelines and equipment can be placed in it, and this can increase the net height and structural function of the story.

The GFW, which is shown in Fig. 7.5a, is a dense frame with interlayer beams. Compared with traditional frame structure, the column space of GFW is 4 m or less, and one or two interlayer beams are added (shown in Fig. 7.5b). As extra columns and interlayer beams are used, the structure stiffness increases rapidly (seen in next section). Another character of GFW is that all the columns are placed around the structure and no column is placed inside the structure. By combining OWSS and GFW together rigidly, cassette structure has the characteristic of high-stiffness and large span capacity.

7.3 The Test of Open-Web Sandwich Slab

In order to investigate the behavior of OWSS, several element tests are conducted. Because testing the entire open-web sandwich slab is difficult, the single open-web beam (OWB) is selected as the test specimen.

7.3.1 *Specimen Design*

To examine the overall performance of the OWSS, four experiment studies were conducted at Southeast University [3]. The fundamental elements, OWB, were tested to represent the structural behaviour of the entire OWSS. The design of the prototype model is based on an existing 15 m span public building that is located in Guiyang, China. The length and width of all the specimens are the same, which is 15 m in length and 0.4 m in width. Restricted by the experimental condition, all the specimens were scaled down with a ratio of 1:2. The similitude laws were established based on the scaling theory and are shown in Table 7.1.

The four specimens are numbered from *S*-1 to *S*-4, and *S*-1 is the reference specimen. The dimension of the shear keys in *S*-2 is decreased from 250 to 150 mm to investigate the influence of the shear key dimension. The height-to-width ratio of *S*-3 is increased from 1 to 1.4 to investigate the linear stiffness of shear keys. Finally, the interval between the shear keys is increased from 1500 to 3000 mm in *S*-4, and this specimen is used to study the influence of the linear stiffness of the chords. All the detailed dimensions of the specimens are shown in Figs. 7.6 and 7.2, including the total specimen length (L), specimen width (b), specimen height (H), shear key length and width (S_H and S_b), and interval between the shear keys (l). In Table 7.3, the reinforcement layout is shown, and the subscripts u, b, s indicate longitudinal rebar in upper chord, bottom chord and shear key, the additional subscripts su, sb, ss denote the stirrup in the three parts, and the symbol D represents the diameter of the used rebar (Table 7.2).

Table 7.1 Main similitude parameters

Item	Dimension	Similitude factor
Length	L	0.5
Force	F	0.25
Stress	FL^{-2}	1
Elasticity modulus	FL^{-2}	1
Bending moment	FL	0.125
Deflection	L	0.5

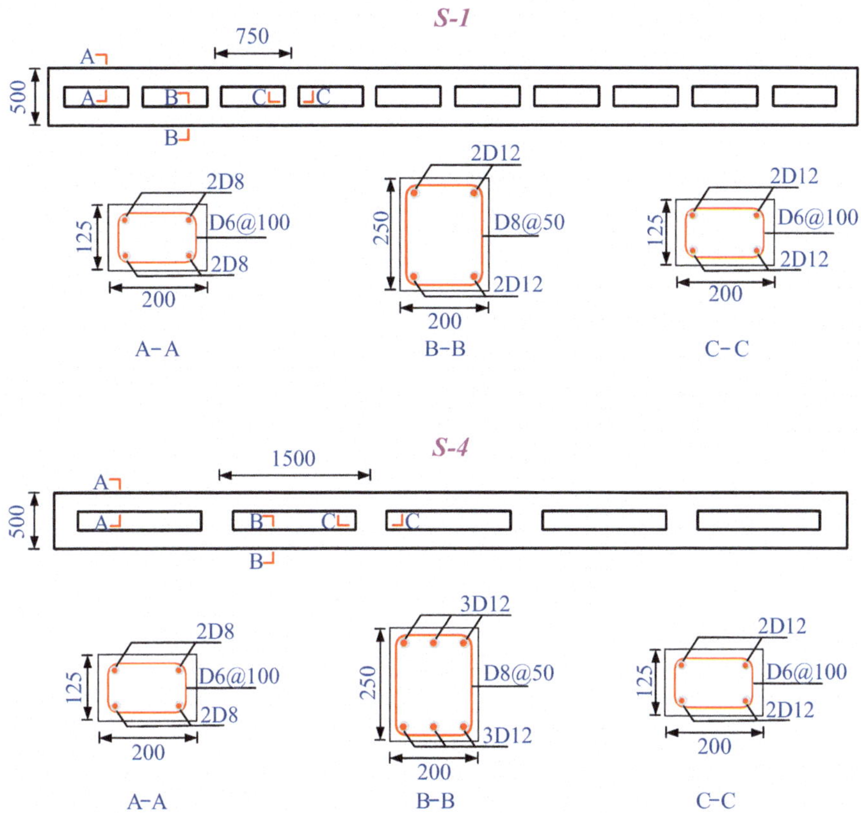

Fig. 7.6 Geometric dimensions of the typical specimens

Table 7.2 Geometric dimensions of the specimens

Specimen No	L (mm)	H (mm)	b (mm)	S_H (mm)	S_b (mm)	l (mm)
S-1	7500	500	200	250	200	750
S-2	7500	500	200	150	200	750
S-3	7500	600	200	250	200	750
S-4	7500	500	200	250	200	750

Table 7.3 Reinforcement layout of the specimens

Specimen No	ρ_u (mm^2)	ρ_{su} (mm^2)	ρ_s (mm^2)	ρ_{ss} (mm^2)	ρ_b (mm^2)	ρ_{sb} (mm^2)
S-1	4D8	D6@100	4D12	D8@50	4D12	D6@100
S-2	4D8	D6@100	4D12	D8@50	4D12	D6@100
S-3	4D8	D6@100	4D12	D8@50	4D12	D6@100
S-4	4D8	D6@100	6D12	D8@50	4D12	D6@100

Table 7.4 Material properties used in the test

Material	Diameter (mm)	Elastic modulus (MPa)	Yield strength (MPa)	Ultimate strength (MPa)
Rebar	6	213,583	294	526
	8	209,634	291	533
	12	198,794	367	652
C40 concrete	–	28,687	–	23.6

7.3.2 *Material Properties*

In this experiment, HPB300 rebar was used for D6 and D8 rebar, and HRB400 rebar was used in D12 rebar. The concrete used in this test was C40 as defined in Chinese code GB50010-2010. The elastic modulus, ultimate strength and yield strength of the rebar were obtained through material tests, and the elastic modulus and 28-day compressive strength of the C40 concrete were tested through 150 mm cubes based on the codes. All the results of the material tests are shown in Table 7.4.

7.3.3 *Loading and Measuring Scheme*

As shown in Fig. 7.7, the simply supported specimens were loaded into two symmetric points. The load-controlled scheme was used in the elastic stage, in which the load incensement in every step was 10 kN. After every load step, crack development and gauge reading were recorded. After yielding, the displacement-controlled scheme was used, and the loading rate was 3 mm/min. The layout of the linear-variable-differential-transformers (LVDT) and strain gauges is shown in Fig. 7.8.

7.3.4 *Test Results*

7.3.4.1 Failure Modes

The three specimens *S*-1, *S*-2 and *S*-3 showed the same failure mode, in which an expected ductile failure was observed. The Fig. 7.9 shows the cracking development and failure mode of the first specimen, and the characteristics of the first three specimens are similar to the solid beam. In the tests, the cracking development showed the upper chord beard pressure stress and the bottom chord beard tensile stress, and the shear keys achieved an object that combined the two chords rigidly. Although some local cracks could be observed in the shear keys and chords, the entire specimen failed due to the integral moment. After yielding, the deformation of the bottom steel

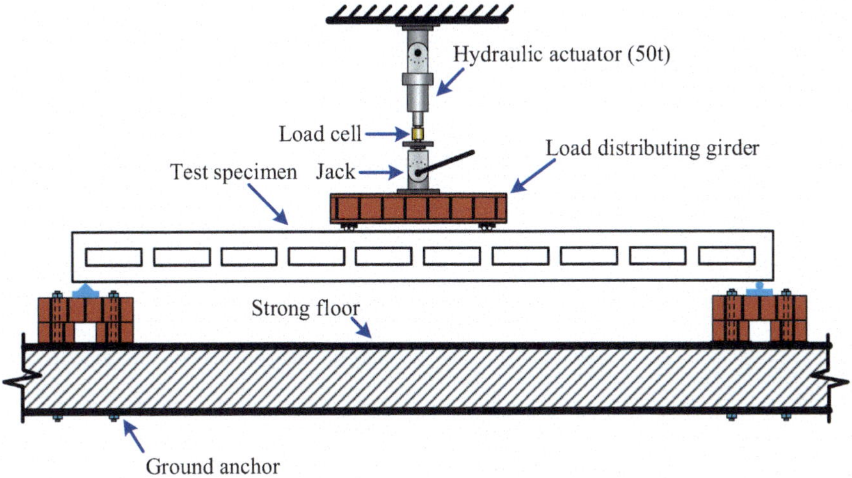

Fig. 7.7 Loading condition of the specimen

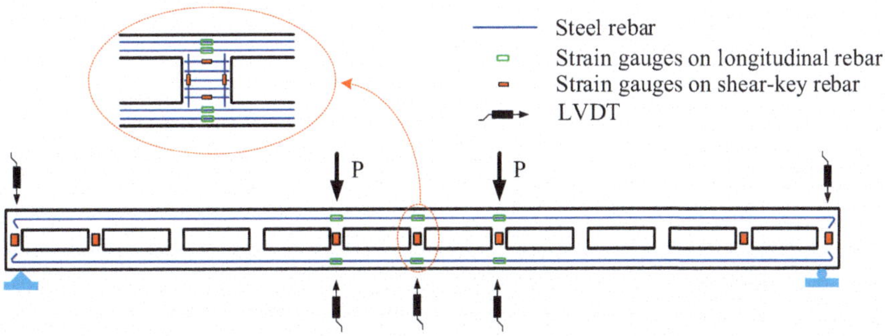

Fig. 7.8 Gauge layout of the specimen

rebar increased rapidly, causing a crushing of the concrete in the upper chord. All three specimens showed good ductile performance but were affected by the different parameters. The peak loads and stiffnesses were different, and this will be discussed later.

In contrast, a different cracking development and failure mode was observed in the fourth specimen (Fig. 7.10). The specimen failed due to local cracking, which was caused by the weakness of the chords and some initial defects. Because the horizontal displacement was restrained by the left support, the entire specimen was moved slightly to the left under vertical load, which made the loading point skew to the right. However, as the linear stiffness of the chords was much lower compared to the shear keys, when skewing, the chords could not provide enough horizontal

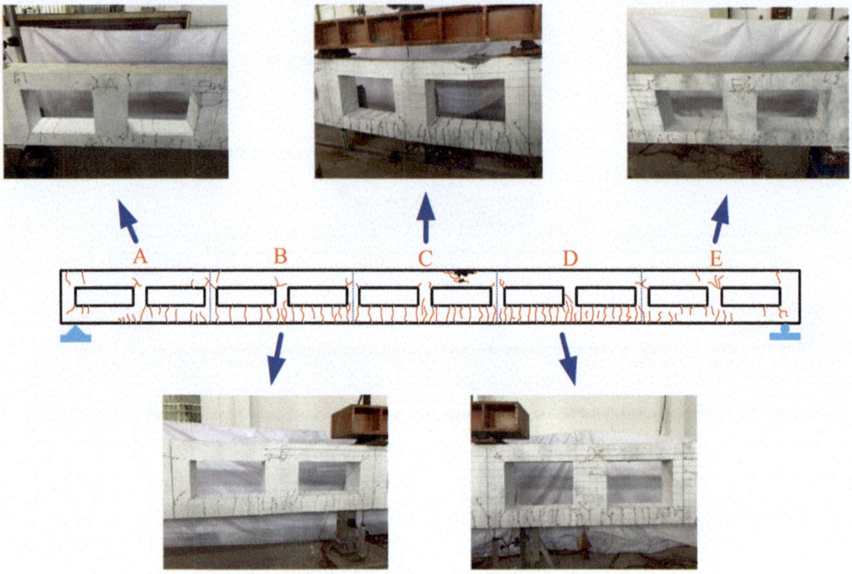

Fig. 7.9 Cracking development and failure mode of the first specimen

tension to make the shear keys move together. Therefore, as shown in Fig. 7.10, a large tension stress occurred at the connection section and eventually caused a local failure of the chord. This largely affected the integrality of the entire OWB and caused a large decrease in structural strength.

The load–deflection curves are shown in Fig. 7.11, and *S*-1 shows the largest stiffness in the first three specimens, and this well meets the test assumption that rigid shear keys can connect the chords rigidly and make them work as an entire element. Therefore, the stiffness and peak load of the OWB can be elevated. This can be seen in the comparison between *S*-1 and *S*-3, in which the peak load of *S*-3 is larger than *S*-1 due to the higher cross-section, but its stiffness is not increased as expected. Caused by the lower linear stiffness of the shear keys, the integrality of *S*-3 is weaker, and a 20% height increase does not bring the same stiffness increase as expected. In addition, as shown in *S*-4, the stiffness of chords is also important. If the chords are too weak to make the shear keys move together, the shear keys will then act as a rigid constraint, and the entire OWB is transformed to a multi-span beam, in which an unexpected local failure will occur.

7.3.4.2 Strain Analysis

The strain of the mid-span of the four specimens is shown in Fig. 7.12, and because the strain development is nearly the same in the first three specimens, *S*-1 is chosen as the representative. As discussed above, the strain distribution of a well-designed

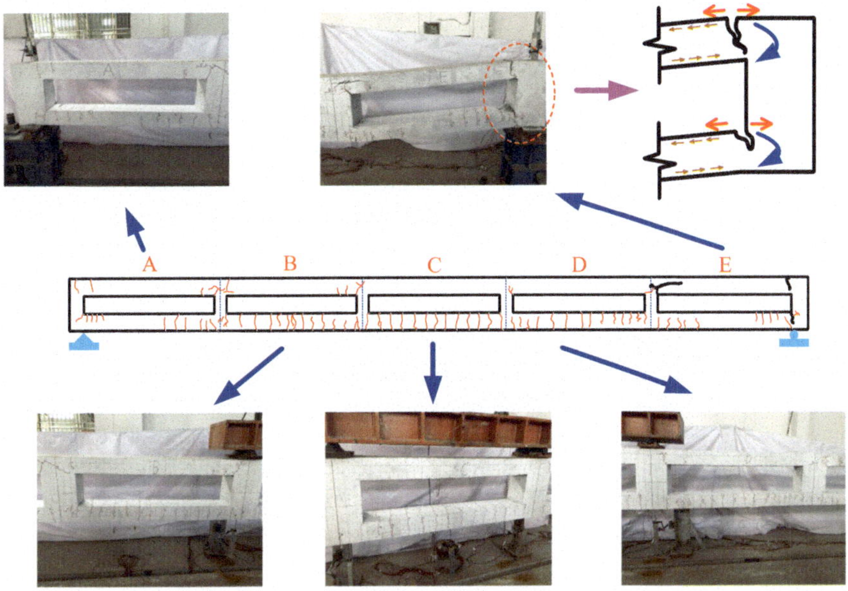

Fig. 7.10 Cracking development and failure mode of the fourth specimen

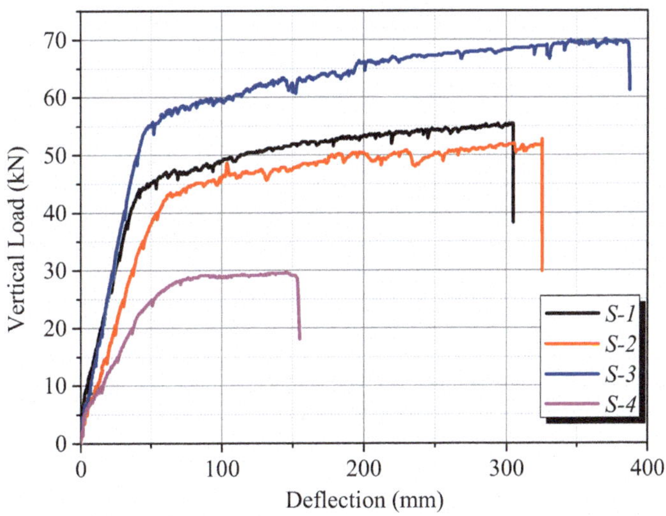

Fig. 7.11 Load–deflection curves of the four specimens

OWB is similar to a solid beam. Connected by the stiff shear keys, both chords work together, and the strain distribution satisfies the plane cross-section assumption. As shown in Fig. 7.12b, the neutral surface of *S*-1 is in the upper chord, which can match the failure mode of the first three specimens. However, the distribution of *S*-4 is much different. As shown in Fig. 7.12c, there is no pressure in the chords, and the strain distribution of the two chords is nearly the same. This shows that the integrality of the entire OWB nearly disappears as the linear stiffness of the chords decreases. When bearing loads, the flexible chords work independently, and the shear keys, which cannot rotate with the chords, act as the stiff boundaries. Therefore, the entire OWB works as a continuous beam, and as shown in Fig. 7.12d, serious stress concentration can be seen at the connecting parts of chords and shear keys, which eventually causes undesirable local failure. This can also be seen in Fig. 7.13, which shows the load-strain curve of the longitudinal rebar in shear keys. Among all four specimens, the strain of *S*-4 is the largest, which shows an obvious stress concentration at the shear key-chord connection point. Furthermore, the stress concentration in *S*-1 is acceptable, which shows a good integrality of the entire element, and the strain in *S*-3 is the smallest, indicating that an increase in element height may promote element strength while the shear keys are reliable.

7.4 The Hysteresis Test of the Grid Frame Wall

From the previous discussion, cassette structure is composed by open-web sandwich slab and grid frame wall. The performance and design method of open-web sandwich slab was studied and two local codes were published [4]. Therefore, the discussion is emphasis on the grid frame wall.

7.4.1 The Mechanical Principle of the Grid Frame Wall

The existing mechanical model of a GFW is shown in Fig. 7.14 and Eq. (7.1), where K_1, K_2, K_3, and K_4 are the stiffness of the four structures respectively. Through the equations, the stiffness of a structure can increase rapidly by adding extra constraints (i.e., extra beams and columns). When changing a bent structure (Fig. 7.14a) to a frame structure (Fig. 7.14b), the stiffness can increase by a factor of 4. Similarly, when adding n-1 beams (Fig. 7.14d), the stiffness can theoretically increase to $4n^2$ times, respectively (n is the total number of the beams). In this way, by changing a large beam to several small beams, the structural stiffness can be increased rapidly without using any extra material. Furthermore, although changing a large column to several small columns cannot increase the structure stiffness directly if the material usage is the same (shown in Eq. 7.1(c), where m is the total number of the columns), but this kind of design is meaningful. When changing a large beam to several small beams, the constraint between the beam and column will decrease and eventually the

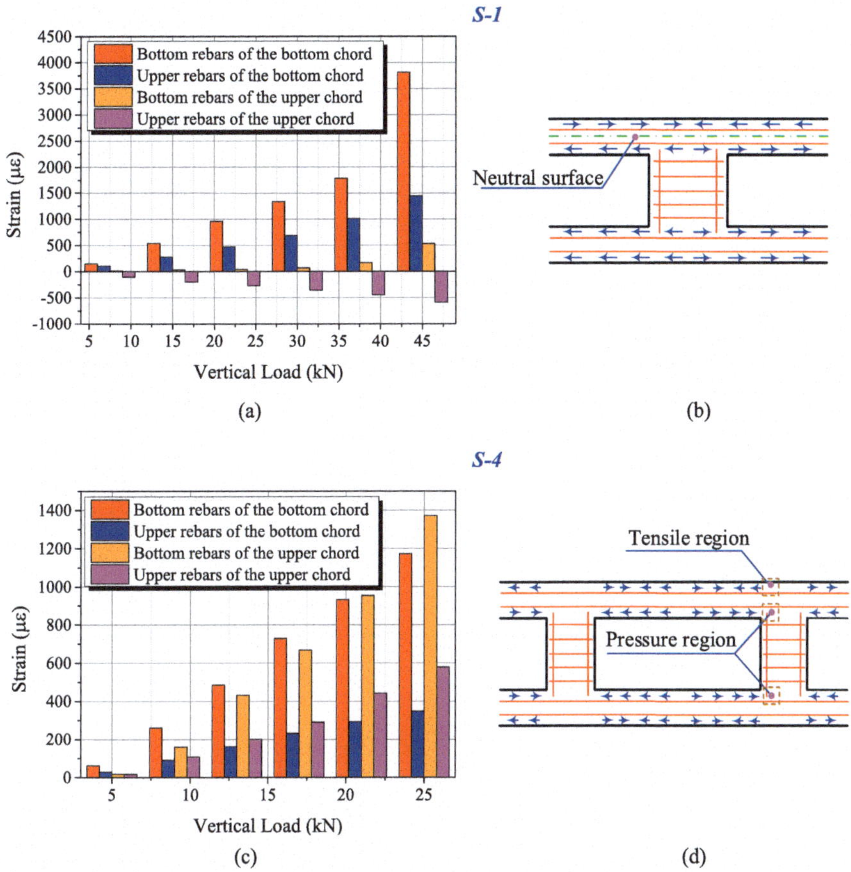

Fig. 7.12 The load-strain analysis in the mid-span of S-1 and S-4 specimens: **a** load-strain columns of S-1, **b** strain distribution of S-1, **c** load-strain columns of S-4, **d** strain distribution of S-4

stiffness of the whole structure will decrease. Therefore, dividing a large column to small columns to maintain an appropriate linear-stiffness ratio between the columns and beams are necessary.

$$\begin{cases} K_1 = \frac{F}{\Delta 1} = \frac{3E_c I_c}{h^3} \quad (a) \\ K_2 = \frac{F}{\Delta 2} = \frac{12E_c I_c}{h^3} = 4K_1 \quad (b) \\ K_3 = \frac{F}{\Delta 3} = \frac{12mE_c \frac{I_c}{m}}{h^3} = 4K_1 = K_2 \quad (c) \\ K_4 = \frac{F}{\Delta 4} = \frac{24n^2 E_c \frac{I_c}{2}}{h^3} = 4n^2 K_1 = n^2 K_2 \quad (d) \end{cases} \tag{7.1}$$

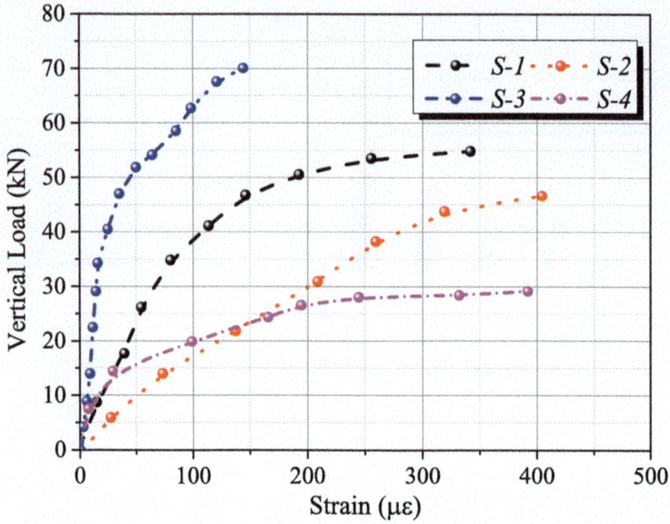

Fig. 7.13 The load-strain curves of shear key longitudinal rebar the four specimens

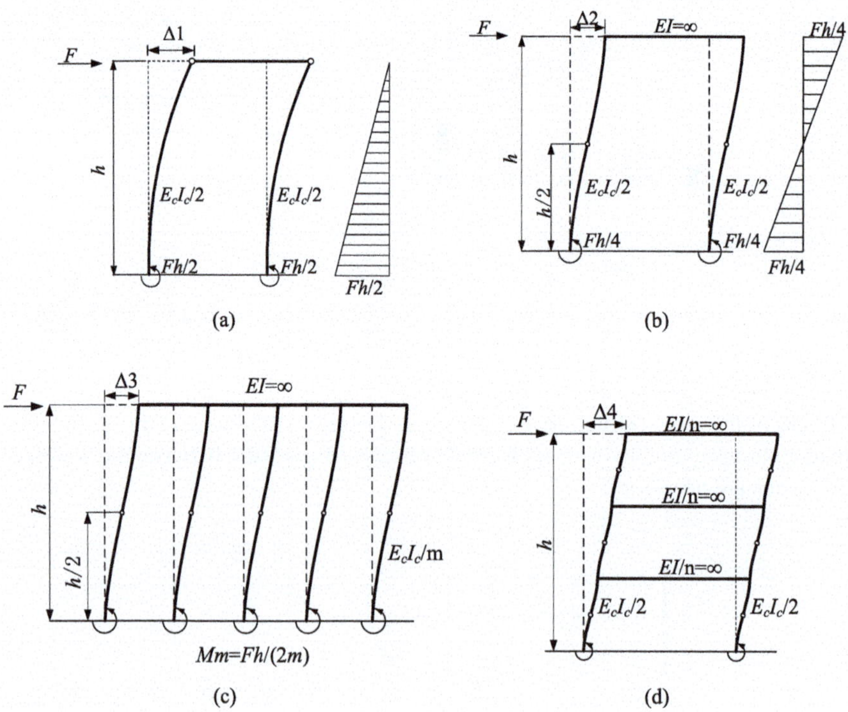

Fig. 7.14 Mechanical characteristics of the cassette structure: **a** bent structure; **b** frame structure; **c** adding inner columns in one floor; **d** adding interlayer beams in one floor

7.4.2 The Hysteresis Performance of the Grid Frame Wall

Based on the theoretical calculation, a grid frame wall and a traditional frame structure were tested through repeated load [5], and the failure mode and fracture distribution are shown in Fig. 7.15.

From the beginning to the end, both specimens undergone four phases, including elastic phase, cracking phase, yielding phase and fail.

Grid frame wall: The crack first appeared at the two middle columns when the lateral load reached 30 kN and the displacement was 1.6 mm in negative direction. The

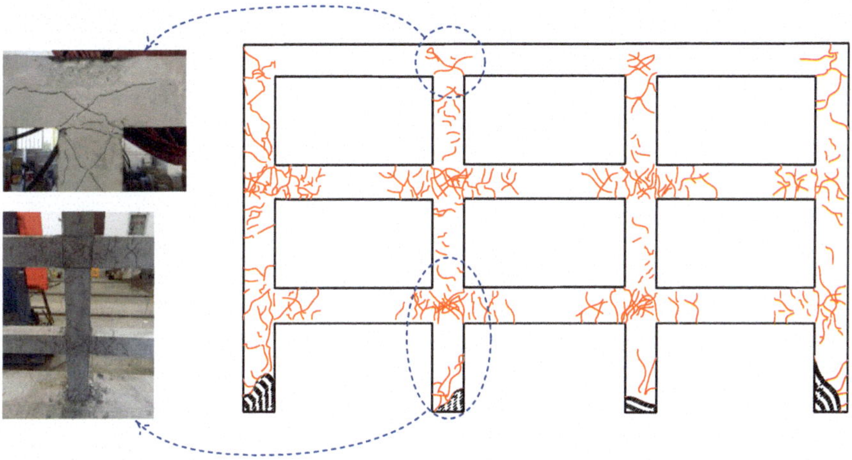

(a) Grid frame wall

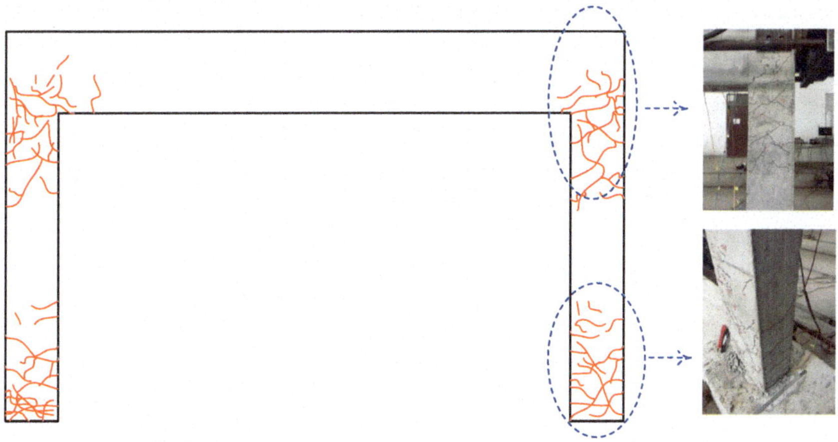

(b) Frame structure

Fig. 7.15 The failure mode and fracture distribution of the tests displacement and the corresponding load are defined as the cracking displacement and strength. When the displacement was loaded to 7.5 mm and the corresponding load was 76.81 kN, the strain of the reinforced bars at the column bottom was 2014 $\mu\varepsilon$. The displacement and corresponding load are defined as the yield displacement Δ_y and yield load P_y. At the first cycle of $2\Delta_y$, numerous shear diagonal cracks appeared at the top of the columns below beam 3. Here, the characteristics of shear deformation were very obvious in the beams and columns, which can be attributed to the short columns and short beams in the frames. The span-depth ratio of the beams was 5.3, and the span-depth ratio of the column was 3. At the first cycle of $3\Delta_y$, the grid frame wall specimen reached the maximum load of 116 kN and a plastic hinge first appeared at the column bottom. At the third cycle of $3\Delta_y$, local small-scale concrete crushing occurred at the column bottoms, and cracks appeared at the interface between the foundation and the column bottoms. At the third cycle of $7\Delta_y$, severe shear diagonal cracks at the column bottoms were developed, and the concrete cover was nearly falling off. At the first cycle of $8\Delta_y$, the strength decreased with increasing displacement before the specimen was loaded to the target displacement. The strength at this moment dropped below 85% of the maximum load, indicating that the specimen had failed.

Traditional frame structure: When the displacement reached 1.4 mm, the first shear diagonal crack appeared at the right beam-column joint. The corresponding load was 21.36 kN, which was defined as the cracking load. When the displacement reached 1.6 mm, horizontal flexural cracks appeared at the left column bottom. The flexural cracks also appeared at the top of the columns under the beam-column joint when the displacement reached 6 mm. When the displacement reached 7 mm, the corresponding load was 60.94 kN and the strain of the reinforced bar was 2135 $\mu\varepsilon$. Therefore, the displacement and corresponding load were defined as the yield displacement and load. At the first cycle of $3\Delta_y$, the specimen RC-MRF achieved the maximum load of 77.52 kN, which was approximately 70% that of the specimens RC-GMRF and PC-GMRF. At the second cycle of $4\Delta_y$, the local concrete at beam-column joints crushed, and a plastic hinge first appeared at the top of the column. At the second cycle of $5\Delta_y$, cracks appeared at the interface between the foundation beam and the column bottom. At the third cycle of $6\Delta_y$, obvious plastic hinges were observed at both the top and bottom of the columns. At the first cycle of $7\Delta_y$, the strength decreased to less than 85% of the maximum load.

In the traditional frame specimen, the plastic hinges at the top of the column were obvious, however, the crack in the beam end was not fully developed. This indicated that the ideal failure mode of "strong column and weak beam" had not been realized. There were two reasons: first, the specimen RC-MRF was designed using the same amount of construction materials as were the specimens RC-GMRF and PC-GMRF, including concrete and reinforcements in the beams and columns, respectively. Second, there was a steel beam upon the beam and anchor plates at both sides of the joints to apply reverse cyclic loading on the specimen. Therefore, the beam ends and beam-column joints were constrained in three directions.

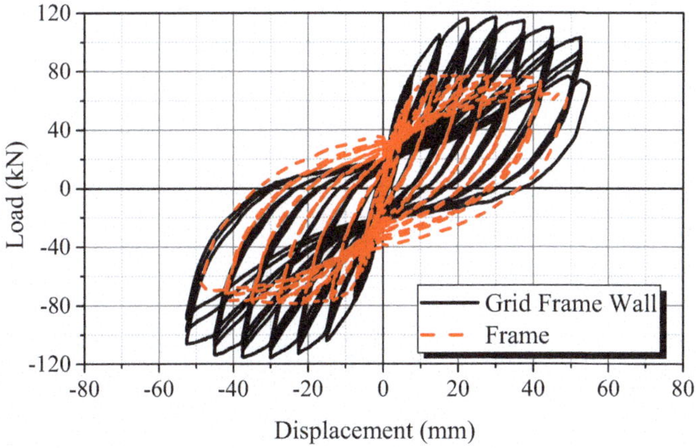

Fig. 7.16 The lateral load–displacement curves of the two specimens

7.4.3 Lateral Load–displacement Relationships

The lateral load–displacement curves of the three specimens are shown in Fig. 7.16. The peak load of the grid frame wall is much higher than the frame structure. After yielding of the grid frame wall, the obvious pinching of the hysteretic loops was mainly attributed to reinforced bars bond-slip degradation, diagonal shear cracks, reinforcement yielding, or a combination of those reasons in different loading stages. Furthermore, the hysteretic loop of the frame specimen showed slighter pinching throughout the loading process than the grid frame wall. The load–displacement envelope curves of the three specimens are plotted in Fig. 7.17. The initial stiffness and load-carrying capacity were greatly improved compared with the frame specimen. The peak-strength of the grid frame wall was 115.79 kN while the peak-strength of the frame structure was only 78.42 kN, about 70% of the grid frame wall. In the grid frame wall, the loading-carrying capacities gradually decreased after the peak-strength. Such post-peak softening behavior was caused by the loss of the bond strength and concrete crush at the plastic hinges at the top and bottom of the column.

7.4.4 Stiffness Degradation

The stiffness degradation could reflect the cumulative damage of a unit in the structure under seismic load and be assessed by the secant stiffness variation. The secant stiffness of three specimens is shown in Fig. 7.18. In the two specimens, the secant stiffness decreased continuously with increasing displacement due to the cumulative damage of the specimens under reverse cyclic loading. The initial stiffness of the frame structure was 11% less than that of grid frame wall, which confirmed the

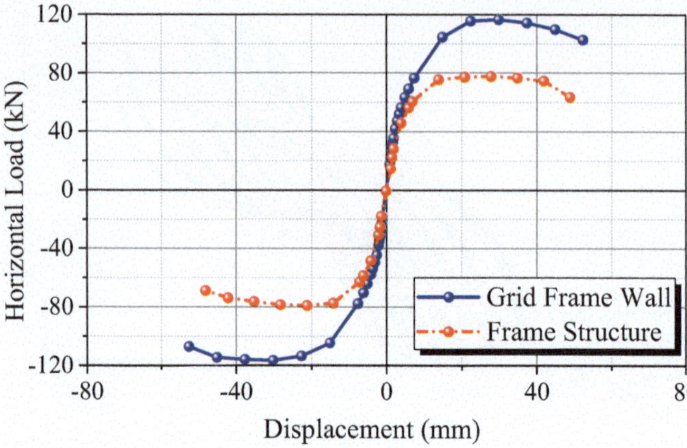

Fig. 7.17 Load–displacement envelope curves of the two specimens

theoretical analysis in the previous discussion. The secant stiffness of the two specimens deteriorated quickly before $2\Delta_y$, which was attributed to concrete cracking and reinforcement yielding. Ductility reflects the ability of a structure to undergo plastic deformations without significant loss of strength, which is a key element in earthquake-resistant design of structures. The maximum displacement of the grid frame wall was 52.5 mm, higher than the 42 mm displacement of the frame structure.

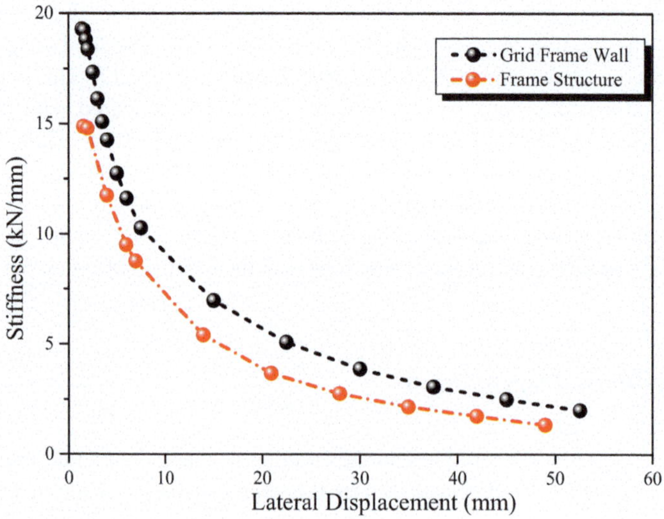

Fig. 7.18 The stiffness degeneration of the two specimens

7.4.5 Energy Dissipation Capacity

The energy dissipation during a certain cycle is represented by the enclosed area of the hysteretic loops. Moreover, the cumulative energy dissipation at a particular displacement is calculated by adding the energy dissipated per loop to the corresponding cycle. Figure 7.19a, b show the energy dissipation per load cycle and the cumulative energy dissipation, respectively, of the three test specimens. According to Ang and Park [6], the energy dissipation of RC members is mainly due to the reinforcing bars experiencing large plastic strains. As shown in Fig. 7.15, the majority of the plastic deformation and energy dissipation developed at the beam-column joint, as well as at the bottom and top of the columns. In Fig. 7.19a, it can be seen that the energy dissipation per load cycle of the frame structure was largely attributed to the large plastic deformation of the longitudinal bars and concrete cracks concentrated in the beam-column joint. The grid frame wall had larger load-carrying capacity than that of the frame structure but dissipated less energy per load cycle. This was attributed to the pinching of hysteretic loops and the characteristic of shear deformation under reverse cyclic loading. However, as shown in Fig. 7.19b, the cumulative dissipated energy of the grid frame wall eventually exceeded that of the frame structure due to the larger ultimate lateral displacement.

7.5 The Seismic Analysis of Cassette Structure

From the tests, the grid frame wall presents a well stiffness and ductility. However, in the real application, the height of the structure and the randomness of the earthquake will largely affect the performance of the cassette structure, and further studies are needed. Therefore, eight finite models in different height were established, and pushover analysis and IDA were used.

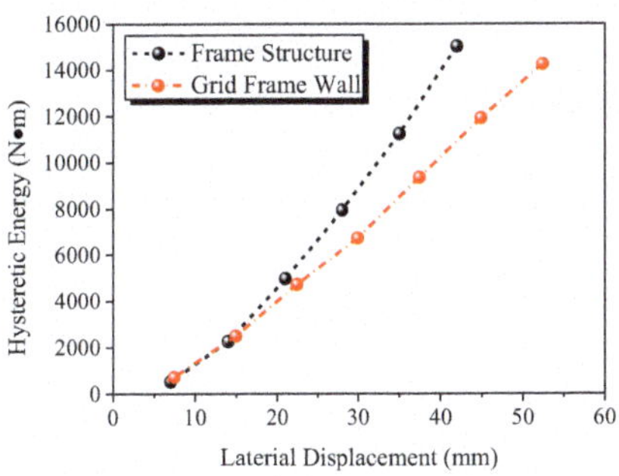

(a) Energy dissipation per load cycle

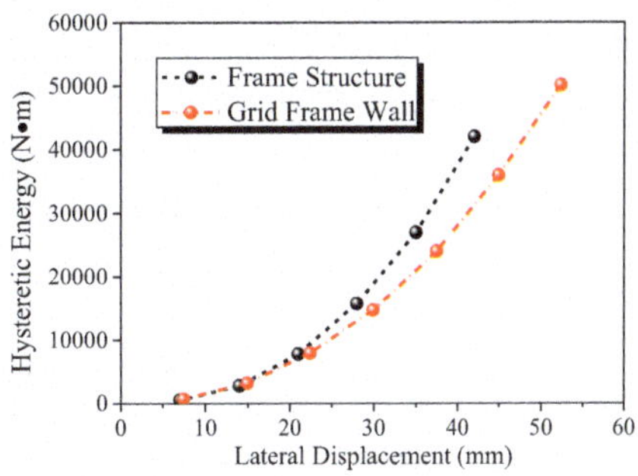

(b) Cumulative energy dissipation

Fig. 7.19 Energy dissipation of specimens

7.5.1 *Design of Prototype Structures*

To further investigate the seismic performance of this structure, three types of prototype buildings were designed and static pushover analysis and incremental dynamic analysis were performed. The wind pressure for all the models was 0.45. For the seismic load, the seismic intensity zone for the structures was 7°, class II and the first group. This means the equivalent shear-wave velocity is between 250 m/s and 500 m/s for 30 m soil (V_{s30}). The characteristic period of the site is 0.35 s and the corresponding peak ground acceleration (PGA) value of the fortification level earthquake (i.e., 10% probability of exceedance in 50 years) is 100 cm/s^2 [7].

7.5.1.1 50-M Models

According to technical specifications for concrete structures of tall buildings (CMC 2010b), the maximum height of the frame structure should be less than 50 m. So, a 50-m cassette structure and a 50-m frame structure were selected. The dimensions of the main structure members are shown in Tables 7.1 and 7.2. The plane dimension of both structures is 18 m × 18 m. For cassette structure, the height is 49.75 m and the column grid is 3 m with no columns inside the construction (shown in Fig. 7.21a, c). The height of the open-web sandwich slab is 600 mm, which is 1/30 of the structure span (18 m), and the top and bottom ribs are 215 mm × 300 mm. The dead load of the floor is 3.95 kN/m^2, and the live loads are 4.1 kN/m^2 for 1–5 floors and 2.6 kN/m^2 for others. Since simulating open-web sandwich slab directly is complicated, according to existing research, open-web sandwich slab can be transformed to an equivalent solid beam. The transformation principle is to make the height and in-plane inertia moment of the open-web sandwich slab the same as the solid beam (Fig. 7.20). This transformation does not change the mechanical characteristics of the slab. In this case, the open-web sandwich slab that is 600 m in height is replaced by a 600 mm × 476 mm solid beam.

The frame structure height is 49.1 m and the column grid is 6 m (shown in Fig. 7.21b, d). As shown in Table 7.5, the beam dimension and floor height of the frame structure are different than those in the cassette structure, but considering the 150-mm fire-control suspended ceiling, the net height is the same. The dead load is 4.55 kN/m^2 and the live loads are 3.5 kN/m^2 for 1–5 floors and 2 kN/m^2 for more

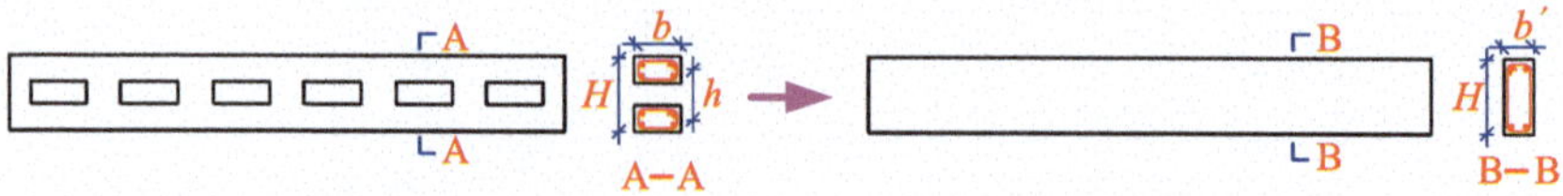

$$2\cdot\frac{(H-h)^3\cdot b}{12}+2b\cdot(H-h)^2\cdot\left(\frac{h}{2}\right)^2=\frac{H^3\cdot b'}{12}$$

Fig. 7.20 The transformation principle of open-web sandwich slab

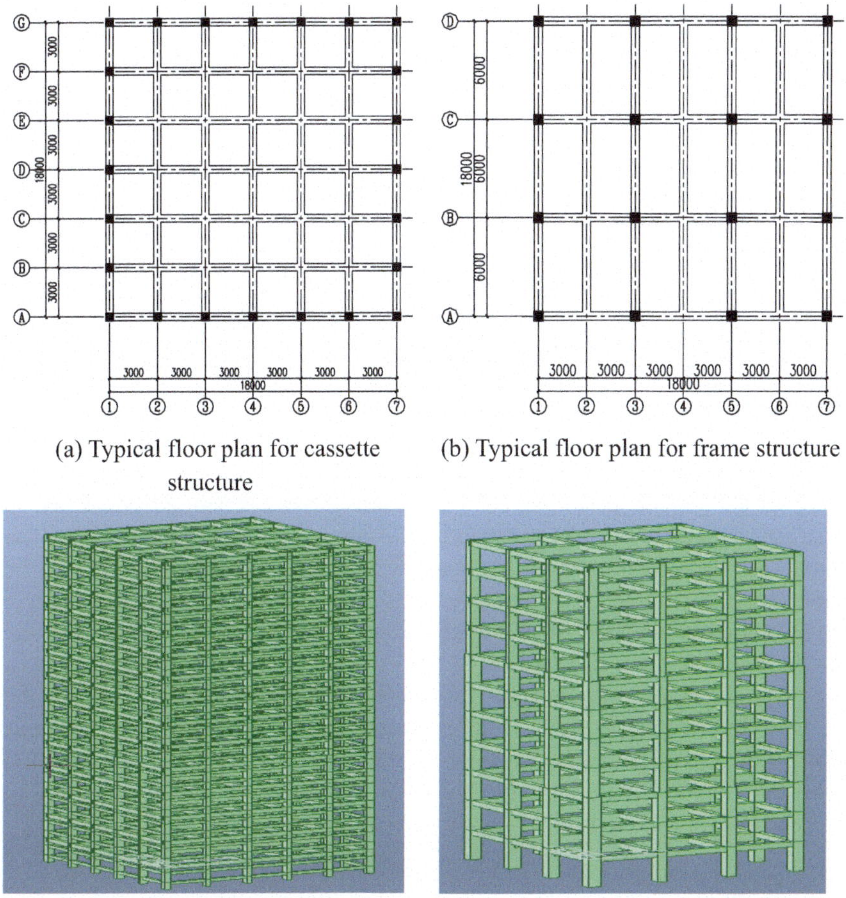

(a) Typical floor plan for cassette structure
(b) Typical floor plan for frame structure
(c) Cassette structure model
(d) Frame structure model

Fig. 7.21 The finite model of the 50-m structures

floors. The difference between the cassette structure and frame structure in load cases is due to considering the free-layout partition walls. The live load should increase to 0.6 kN/m^2, while the dead load decreases to 0.6 kN/m^2 in cassette structure (Table 7.6).

Table 7.5 50-m cassette structure (unit: mm)

Floor	Floor height	Column	Open-web sandwich slab	Interlayer beam
1 ~ 5	3650	600 × 600/600 × 450	600 × 300	400 × 300
6 ~ 14	3650	550 × 550/550 × 400	600 × 300	400 × 300

Table 7.6 50-m frame structure (unit: mm)

Floor	Floor Height	Column	Main Beam	Secondary Beam
1 ~ 5	3900	800 × 800/800 × 600	700 × 450	500 × 350
6 ~ 13	3700	750 × 750/750 × 550	700 × 450	500 × 350

7.5.1.2 90-M Models

90-m-tall buildings can be widely seen, and due to the large lateral loads, the structure requires a core tube. Therefore, three tube structures are selected for analysis.

Typical floor plans are shown in Fig. 7.22a–c. The plane dimensions are 36 m × 36 m and the tubes are 10.8 m × 10.8 m. The column grid is 3.6 m for cassette-core-tube structure and 7.2 m for frame-core-tube structure, and there are no columns inside the cassette-core-tube structure. A new type of structure called an open-web sandwich slab frame-core-tube structure (OSSFS) is selected. This structure is adopted to show the performance of open-web sandwich slab independently, so the structure consists of open-web sandwich slab and normal frame column. The column spacing is 7.2 m, which is the same as in the frame-core-tube structure. The total height of the frame-core-tube structure is 91 m and 91.45 m for cassette-core-tube structure and OSSFS. Since the net height of all structures is the same, the floor height of the frame-core-tube is higher because of the large beam dimensions. As a result, there are 25 floors in the frame-core-tube structure and 27 floors in the core-tube structure and OSSFS. Other parameters, such as loads, are the same as in the 50-m models. The dimensions of the main structure members are shown in Tables 7.7, 7.8 and 7.9. Furthermore, the open-web sandwich slab is replaced by a 550 mm × 420 mm solid beam using the same principle mentioned above.

7.5.1.3 145-M Models

As the height-width ratio increases, the mechanical performance of buildings significantly change, so three 145-m tall structures are selected. The structure types are the same as in the 90-m models and the plan layouts are shown Fig. 7.23a–c.

The plane dimensions are 36 m × 36 m and the core-tubes are 14.4 m × 14.4 m. In addition to the change in main beams used in the frame-core-tube structure, other parameters including loads, element dimensions, and story heights are the same as in the 90-m structures. The total height of frame-core-tube is 145.3 m, with 39 stories, and 145.05 m, with 43 stories, for the other two buildings. Other dimensions of the main structure members are shown in Tables 7.10, 7.11 and 7.12.

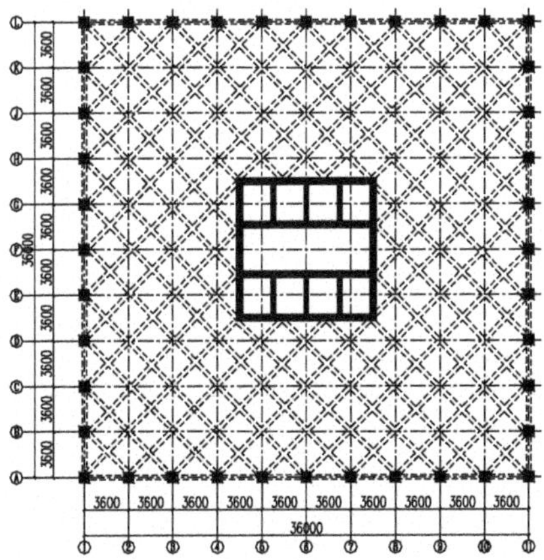

(a) Typical floor plan for cassette-core-tube structure

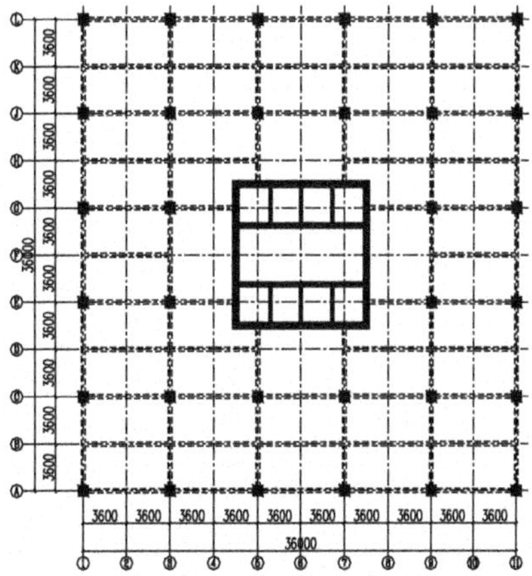

(b) Typical floor plans for frame-core-tube structure

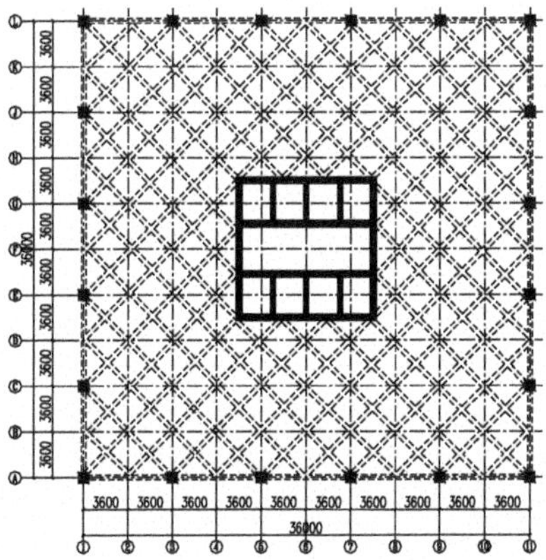

(c) Typical floor plan for open-web sandwich slab frame-core-tube structure

Fig. 7.22 Typical floor plan for 90-m structures

Table 7.7 90-m cassette structure (unit: mm)

Floor	Floor height	Column	Open-web sandwich slab	Interlayer beam	Shear wall
1 ~ 5	3550	650 × 650/650 × 500	550 × 300	300 × 200	300
6 ~ 15	3350	550 × 550/600 × 450	550 × 300	300 × 200	250
16 ~ 27	3350	500 × 500/550 × 400	550 × 300	300 × 200	200

Table 7.8 90-m OSSFS (unit: mm)

Floor	Floor height	Column	Open-web sandwich slab	Shear wall
1 ~ 5	3550	800 × 800/800 × 650	550 × 300	300
6 ~ 15	3350	750 × 750/750 × 600	550 × 300	250
16 ~ 27	3350	700 × 700/700 × 550	550 × 300	200

Table 7.9 90-m frame-core-tube structure (unit: mm)

Floor	Floor height	Column	Main beam	Secondary beam	Shear wall
1 ~ 5	3800	800 × 800/800 × 650	700 × 550	500 × 300	300
6 ~ 15	3600	750 × 750/750 × 600	700 × 550	500 × 300	250
16 ~ 25	3600	700 × 700/700 × 550	700 × 550	500 × 300	200

7.5.2 Results of Pushover Analysis

The results of pushover analysis are shown in Fig. 7.24. As shown in Fig. 7.24a, the drift ratio of cassette structure is much smaller than the frame structure in 50 m analysis. When using the same amount of concrete and steel, the maximum drift of cassette structure is only half of it in frame structure. However, when the structure height increases, the displacement curve of both structure change from shear type to bending type, in which the maximum drift occurs at the upper section of the structures, and the tension and compression deformation of the columns has become the majority instead of bending deformation. In 145 m structures (Fig. 7.24b), the drift of the three structures nearly is the same, which indicates that when the main displacement comes from tension and compression, the large bending stiffness of grid from wall cannot obviously increase the stiffness of the whole structure. Therefore, using an open-web sandwich slab independently is the best choice. Open-web sandwich slab can save 20% of concrete and steel and achieve a large span structure (i.e., no column in the entire structure, shown in Figs. 7.22, 7.23 and 7.25. Since the equipment and pipelines can be placed in the open-web, OSSFS can build four additional stories at the same height of the frame-core-tube structure, so the advantage is obvious.

In total, when the structure is comparatively low (low than 100 m), using cassette structure can largely increase the stiffness of the structure when the same amount of steel and concrete are used. As building height increases, the deformation curve changes to bending type, and the advantage of cassette begins to decrease. In this case, using open-web sandwich slab independently can decrease the seismic behavior and maintain the large span space effectively.

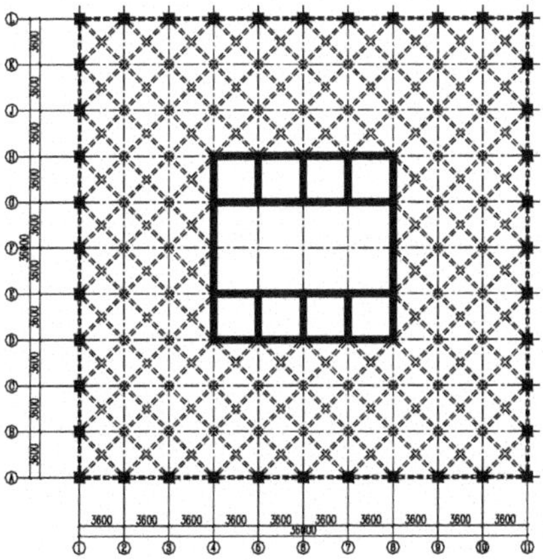

(a) Typical floor plan for cassette-core-tube structure

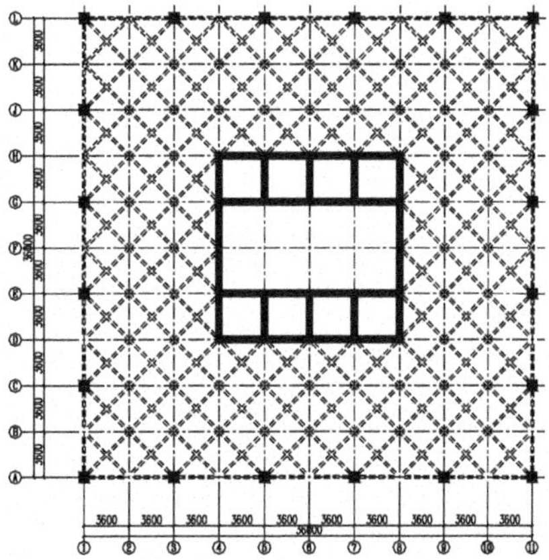

(b) Typical floor plan for OSSFS

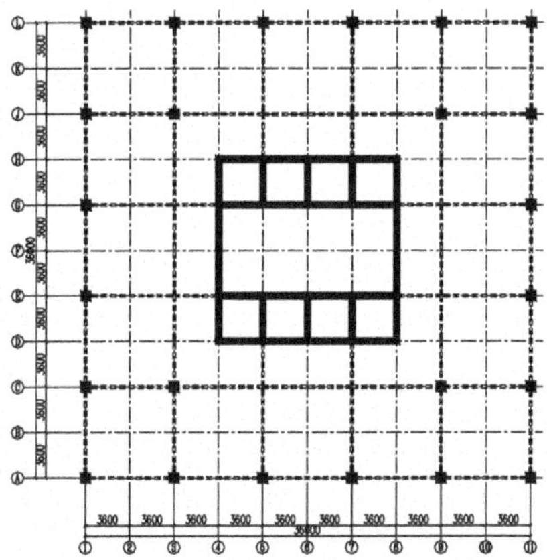

(c) Typical floor plan for frame-core-tube structure

Fig. 7.23 Typical floor plan for 145-m structures

Table 7.10 145-m cassette structure (unit: mm)

Floor	Floor height	Column	Open-web Sandwich slab	Interlayer beam	Shear wall
1 ~ 15	3550	650 × 650/650 × 500	550 × 300	300 × 200	500
16 ~ 30	3350	550 × 550/600 × 450	550 × 300	300 × 200	400
31 ~ 43	3350	500 × 500/550 × 400	550 × 300	300 × 200	300

Table 7.11 145-m OSSFS (unit: mm)

Floor	Floor height	Column	Open-web sandwich slab	shear Wall
1 ~ 15	3550	800 × 800/800 × 650	550 × 300	600
16 ~ 30	3350	750 × 750/750 × 600	550 × 300	500
31 ~ 43	3350	700 × 700/700 × 550	550 × 300	400

Table 7.12 145-m frame-core-tube structure (unit: mm)

Floor	Floor height	Column	Main beam	Secondary beam	Shear wall
1 ~ 15	3900	800 × 800/800 × 650	750 × 550	550 × 300	600
16 ~ 30	3700	750 × 750/750 × 600	750 × 550	550 × 300	500
31 ~ 39	3700	700 × 700/700 × 550	750 × 550	550 × 300	400

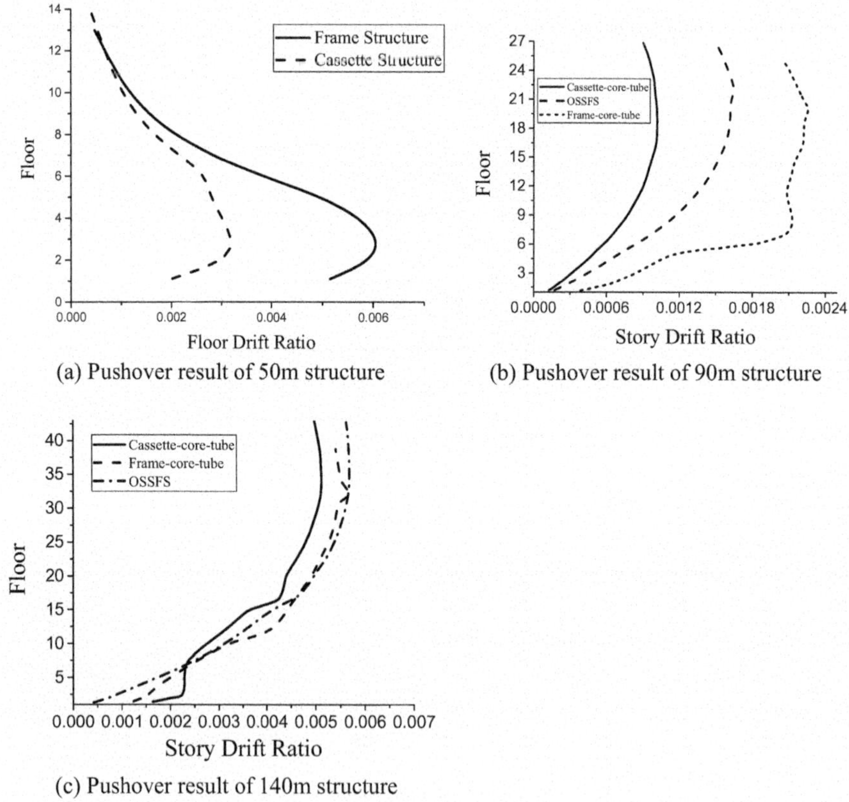

(a) Pushover result of 50m structure

(b) Pushover result of 90m structure

(c) Pushover result of 140m structure

Fig. 7.24 Pushover results

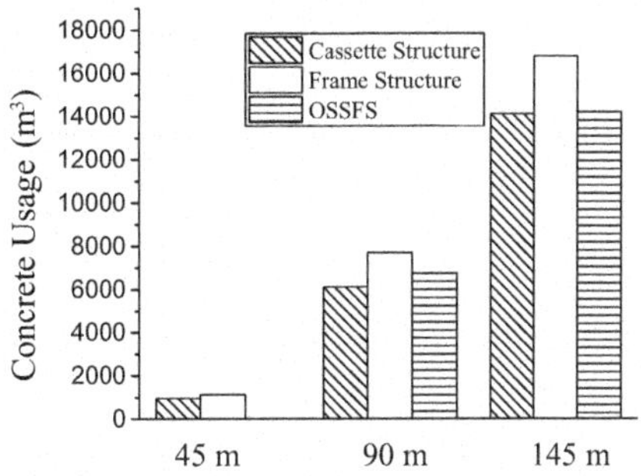

Fig. 7.25 Concrete usage of three structures at different heights

7.5.3 IDA Analysis

Since the stiffness of 50-m structures is relatively large, the seismic performance of far-field ground records is inconspicuous. As a result, 18 near-field ground motion records recommended by FEMA are selected. The selected records are shown in Table 7.13.

Table 7.13 Time-history records

Number	Name	Date	Station	PGA (g)
1	Imperial valley 06	1979	EI Centro Array #6	0.44
2	Irpinia, Italy 01	1980	Sturno	0.31
3	Superstition hills-02	1987	Parachute Test Site	0.42
4	Loma Prieta	1989	Saratoga-Aloha	0.38
5	Cape Mendocino	1992	Erzican,Turkey	0.63
6	Landers	1992	Luceme	0.79
7	Northridge	1994	Rinaldi Receiving Sta	0.87
8	Kocaeli, Turkey	1999	Lzmit	0.22
9	Chi-Chi, Taiwan	1999	TCU065	0.82
10	Duzce, Turkey	1999	Duzce	0.52
11	Gazli USSR	1979	Karakyr	0.71
12	Imperial Valley 06	1979	Bonds corner	0.76
13	Imperial Valley 06	1989	Chihuahua	0.28
14	Loma Prieta	1989	BRAN	0.64
15	Loma Prieta	1989	Corralitos	0.51
16	Cape Mendocino	1992	Cape Mendocino	1.43
17	Northridge-01	1994	LA-Sepulveda VA	0.73
18	Chi-Chi, Taiwan	1999	TCU067	0.56

Suitable IM and DM are other basic parameters of an IDA study. The IM selected for IDA is an important factor that reflects the real behavior of a structure under time-history record. In this case, the peak ground acceleration (PGA) and peak ground velocity (PGV) are used for IM and perform well. Although selecting first-mode spectral acceleration (SA) can reduce the dispersion of the result in some cases, it does not fit for these two structures. In addition, the maximum story drift ratio and roof displacement are selected as DM according to Chinese code in this study.

After the entire IDA and curves are extracted, further analysis is necessary. The IDA curves display a wide range of behavior with large record-to-record variability, so it is essential to summarize the data and quantify the randomness introduced by the records. Appropriate summarization techniques must be employed to reduce the data to the distribution of DM given IM. Based on mean value independently is not reasonable due to the uncertainty of DM at high levels of IM. Thus, statistical 16%, 50% and 84% fractiles are used to represent the DM values. The 50% fractile (mean value) is the most famous and common form of fractile, which means half of all quantities are greater than the value. Similarly, 16% of all curves are greater than the 16% fractile while 84% of all curve are greater than the 84% fractile.

The capacity points in IDA curves can be calculated through varieties of methods, one is called '20% slope' criteria (Vamvatsikos 2002). The first point of the IDA curve in which the tangent slope is equal to or less than 20% of the initial elastic slope is defined as capacity point in this method.

7.5.4 IDA Results

The results of IDA are shown in Figs. 7.26, 7.27 and 7.28. From the results, the basic trend of IDA and pushover analysis is the same, and the advantage of cassette structure under various seismic loads is obvious. From the results, when the two structures are in the elastic stage (story drift ratio is less than 0.001), the advantage of cassette structure is considerable and the story drift ratio in the 50% fractile curve of cassette structure is approximately 70% that of the frame structure. When the structures become plastic (story drift ratio is approximately 0.02), the advantage begins to decrease, and the story drift ratio of the cassette structure is approximately 85% that of the frame structure. When the structures become completely plastic and nonlinear (story drift ratio is greater than 0.05), the advantage begins to increase rapidly and eventually exceeds 100%, which means the story drift ratio of the frame structure is twice as large as that of the cassette structure (shown in Fig. 7.28). This is because when the structures are elastic, the stiffness of the cassette structure is very large, so the drift is relatively small. However, when the structures become plastic, the elements of the cassette structure are comparatively small, so they become plastic easier and faster, which induces the rapid decrease in cassette structure stiffness. After the two structures become completely plastic, the stiffness of the cassette structure becomes relatively large again. When the two structures become plastic, the large stiffness and vast nodes can largely increase the energy-discipline capacity of the cassette structure so the performance of cassette structure under seismic load is comparatively small.

Through the PGA and PGV curves, the advantage of cassette structure is obvious. Comparing the 16% and 84% fractile curves, if the PGA is used for DM, the dispersion of the cassette structure is larger. This case is especially obvious in the plastic phase. This is due to the uncertainty of the vast elements. As shown in Fig. 7.28, the 16% fractile curve of the cassette structure is almost the same as the 50% fractile curve of the frame structure, and performs better in the plastic phase, so the dispersion will not weaken the advantage of the cassette structure. Furthermore, when using PGV as DM, similar dispersion is observed in both structures, and the cassette structure is even better when story drift ratio is used for IM. This infers that although both indices have coincident results, PGV is a much better parameter for cassette structure.

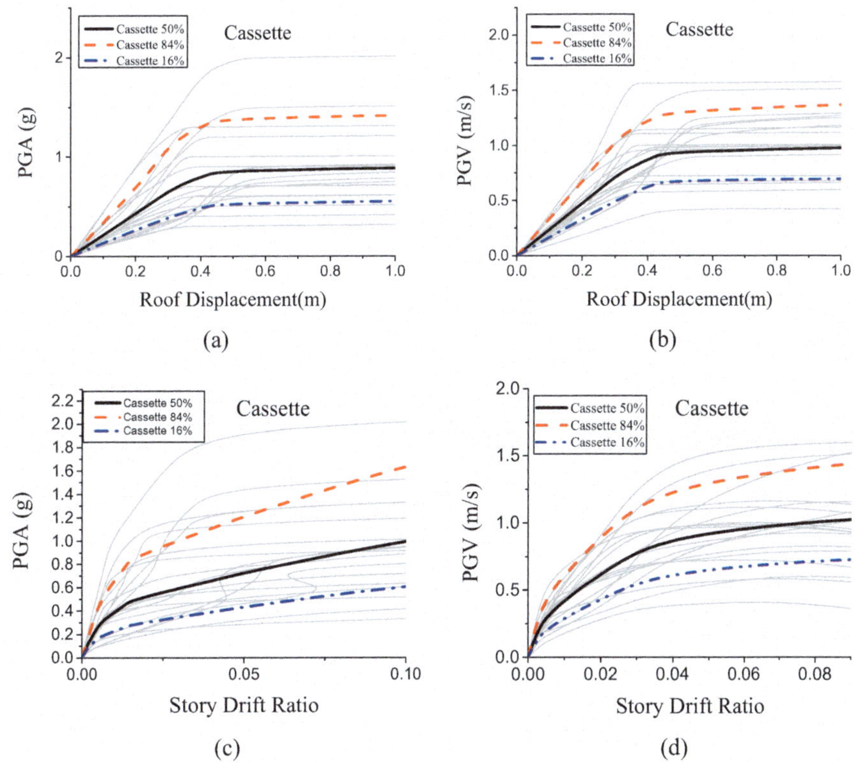

Fig. 7.26 IDA results of cassette structure

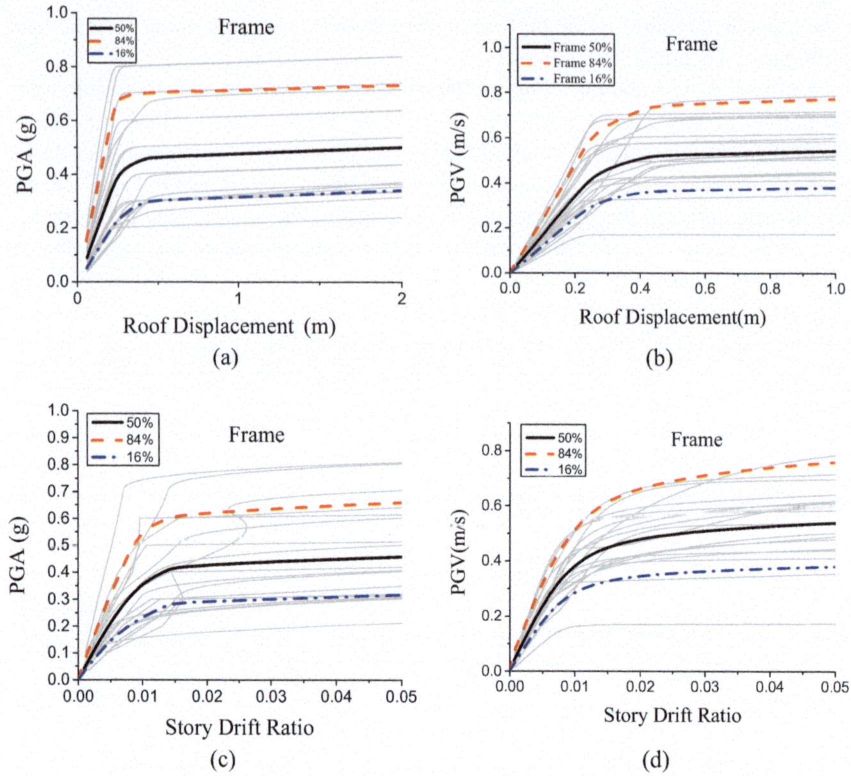

Fig. 7.27 IDA results of frame structure

7.5.5 Seismic Fragility Analysis

IDA data of each structure is selected to calculate the fragility curve. The number of exceedances of a defined limit for each performance level were counted. These performance levels include immediate occupancy (IO), life safety (LS) and collapse prevention (CP). The corresponding maximum story drift limits for each performance level, based on FEMA440 and JGJ 3–2010, are summarized in Table 7.14. Because the IO limit in FEMA440 is much greater than in Chinese code JGJ 3–2010, 1/550 for the IO limit from Chinese code is selected. Other limits are based on FEMA440.

According to the study mentioned above, story drift ratio and PGV are the best indices for IM and DM, so they are selected to calculate the fragility curve. The curve is shown in Fig. 7.29.

From the fragility curve, several observations are made: in the IO limit, cassette structure performs better than frame in the elastic phase. For the same reason mentioned earlier, the performance of cassette structure decreases slightly, but it still satisfies the code limit. In the LS and CP limits, a major advantage of the cassette

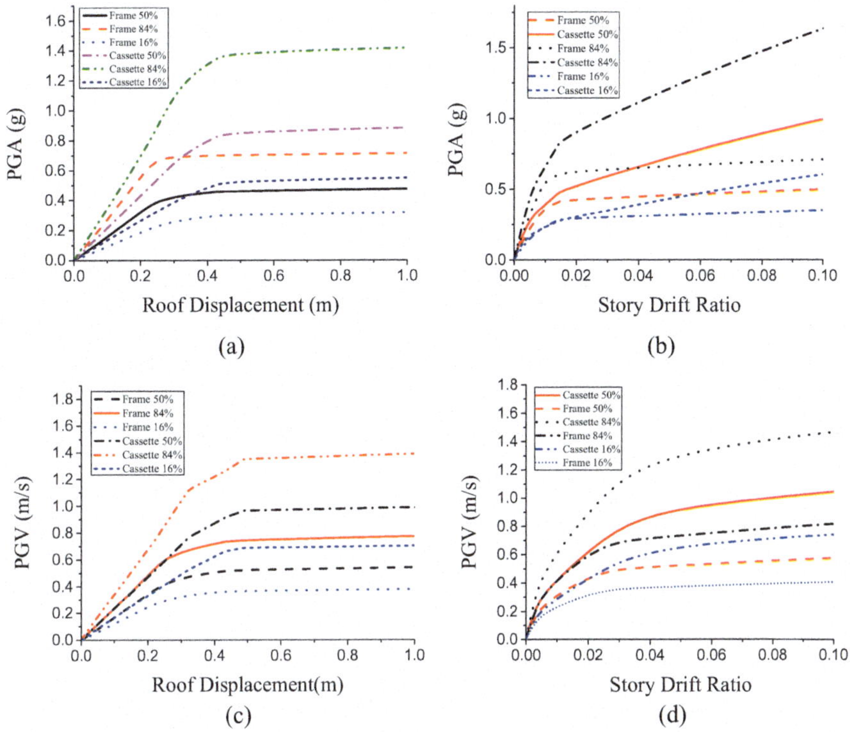

Fig. 7.28 IDA results of the two structures

Table 7.14 Limits of each performance level

Performance level	Drift ratio limit
IO	1/550
LS	0.01
CP	0.04

structure is observed. After both structures become plastic, the large stiffness and good energy-discipline capacity of cassette structure allow it to perform much better than the frame structure. Both structures satisfy the code limits, and the cassette structure performs better in the elastic and plastic phase.

7.6 The Comparison Study of Cassette Structure and Traditional Frame Structure

According to the discussion above, the advantage of cassette structure is obvious. In order to investigate the structural and economic performance of cassette structure,

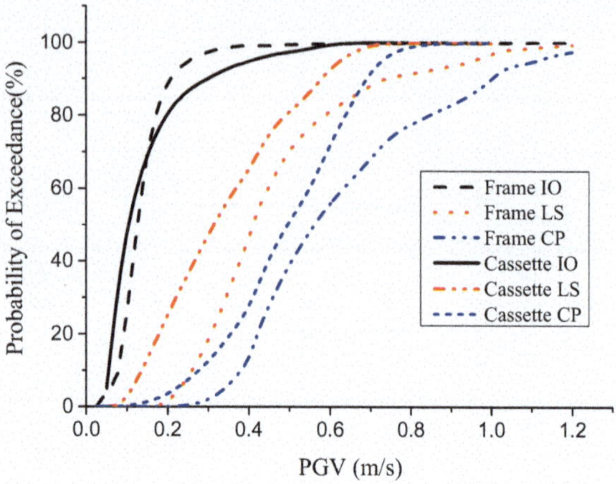

Fig. 7.29 Fragility curve of both structures

an existing frame structure is redesigned to a cassette structure in this section, and their overall performance are compared and discussed.

7.6.1 Design of the Frame Structure

The 47 m frame building is used as an office building and the typical floor plan is shown in Fig. 7.30. The entire building plane is 58.5 m × 22 m, and the column grid is generally 8400 mm along the x-axis and 7400 mm or 7300 mm along the y-axis. The dimensions of major element are listed in Table 7.15. The dead load, which includes self-weight of the slab, waterproof layer and other coating, is 5 kN/m^2 while the live load is 2 kN/m^2. Considering the partition walls, windows and the self-weight of beams, for the first three floors, a 12.5 kN/m linear load is added on the inner beams and a 22 kN/m load is added on the surrounding beams. For other floors, a 9 kN/m and 15 kN/m liner load are used for inner beams and surrounding beams, respectively.

7.6.2 Design of the Cassette Structure

As shown in the previous design, there is one major design weakness in the frame structure. The inefficient bending resistance mode and large self-weight limit its span ability and columns are needed inside the structure. All of these limits the room usage, restrain the span ability and lower its economic performance. To remedy this, the cassette structure is used.

The plan of the redesigned cassette structure is shown in Fig. 7.31 and there are several major differences between the cassette structure and the older one. First, all

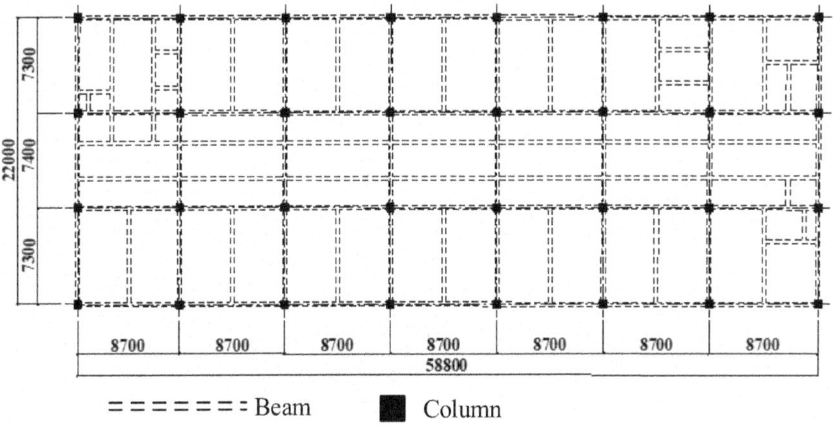

Fig. 7.30 Type floor plan of the frame structure

Table 7.15 Dimensions of element of the frame structure (unite: mm)

Floor	Floor height	Column	Surrounded beam	Main beam	Secondary beam
1–3	5400	1000 × 800	950 × 300	900 × 300	650 × 300
4	3800	800 × 800	950 × 300	650 × 350	550 × 250
5–6	3800	700 × 700	950 × 300	650 × 350	550 × 250
7–11	3800	600 × 600	950 × 300	650 × 350	550 × 250

the columns are placed at the edges of the cassette structure and there is no column inside it, which means the total span of the structure reaches 22 m. Second, the partition wall can be placed unconstrained, which means the interior space need not be previously designed and can be arranged as per user's wishes. Third, as listed in Table 7.16, although the dimensions of the cassette structure are comparatively small, but the mechanical properties are better than those of frame structures.

The column grid of the cassette structure is 3725 in x-ray, and 3667 in y-ray, and a 450 mm × 400 mm interlayer beam is used. The top and bottom ribs of the open-web sandwich slab used in the first three stories are 250 mm × 300 mm and 200 mm × 300 mm, respectively. The dimension of the shear key is 1000 mm × 1000 mm and the height-to-width ratio is 1. Though the height of the open-web sandwich slab is 1500 mm, the element section is small and most portions of slab are empty. Furthermore, considering the 150 mm fire sprinkler and 450 mm equipment pipeline can be placed in the open-web of the slab, the net height of the cassette structure and the frame structure are same. For other floors, the top and bottom ribs are both 250 mm × 300 mm, and the shear key is 450 mm × 450 mm with a height of 400 mm. The entire open-web sandwich slab is 900 mm in height, which presents 200 mm more net height than the original frame structure. As with the load cases, the dead load of the cassette structure is 5 kN/m^2, and a 3.8 kN/m liner load considering

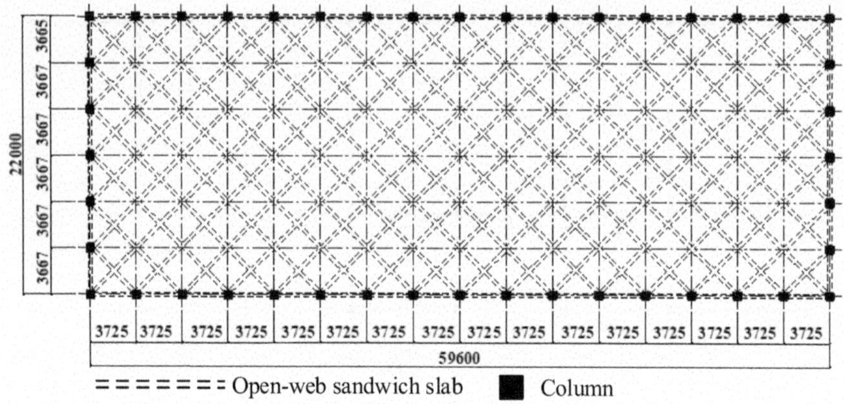

Fig. 7.31 Type floor plane of the cassette structure

Table 7.16 Dimensions of element of the cassette structure (unit: mm)

Floor	Floor height	Column	Open-web sandwich slab	Interlayer beam
1–3	5400	800 × 700	1500	300 × 200
4–6	3800	700 × 500	900	300 × 200
7–11	3800	650 × 450	900	300 × 200

the self-weight is added. The live load is 3.6 kN/m^2, which includes the free placed partition wall. Other dimensions of the elements are listed in Table 7.16.

7.6.3 *Performance of the Two Structures Under Earthquake*

To compare the structural performance under ground motions, 18 records suggested by Federal Emergency Management Agency (FEMA) mentioned above are selected as seismic excitation. The elastic acceleration spectra are shown in Fig. 7.32. In accordance with the Chinese code, all the time-history records are scaled into three levels. These three levels in Chinese code respond to levels of immediate occupancy (IO), life safety (LS) and collapse prevention (CP), respectively. Because these two buildings are designed on 7° seismic intensity zone as mentioned above, the scaled PGA limits are 35 cm/s^2,100 cm/s^2 and 220 cm/s^2, which represent a probability of exceedance in 50 years of 63%, 10% and 2%, respectively. When scaling, the seismic response of LS and CP level is relatively small, so the seismic records are rescaled to the limit of each phase (100 cm/s^2 and 220 cm/s^2 PGA). In addition to these three levels, a 0.4 g PGA analysis is added to study the structural performance during heavy earthquake.

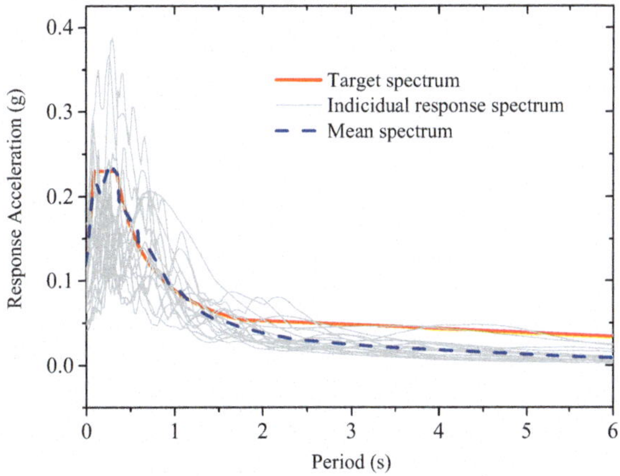

Fig. 7.32 Spectrum of the selected time-history records

The seismic analysis results of the two structures are shown in Fig. 7.33. Both structures have similar stiffness in both the x and y axes, which means the structure is well designed. Comparing these results, the advantage of the cassette structure is obvious. The average maximum drift of the cassette structure in IO level is 0.0012(Fig. 7.33a), which is only 70% of that in the frame structure. This advantage is observed in LS and CP level. In 0.4 g PGA level, which is a high intensity level and both the structures are in completely plastic stage, the drift of the cassette structure is 0.034 while that of the frame structure is 0.051. Caused by numerous nodes in cassette structure, the advantage of cassette structure is even larger in plastic stage. Numerous nodes and elements can dissipate vast energy and make cassette structure stable during earthquake. The drift limit of IO, LS and CP level is 1/550, 0.02 and 0.04. Response of cassette structure and frame structure both satisfy the drift limit but the drift of cassette structure is much smaller compared to frame structure. In addition, even in 0.4 g PGA level, while the drift of frame structure largely exceeds the drift limit, the cassette structure is still stable and the drift of it still satisfies the limit.

The development of plastic hinges of the two structures in three different levels is shown in Fig. 7.34. Figure 7.34 shows the overall development of plastic hinges in two structures. To make the figures look clear, two inter layer beams in cassette structure are combined and drawn as one fine line due to their similar performance, and the open-web sandwich slab and beams are drawn as thick lines. In these figures, large circles mean that the elements are totally plastic and dangerous while the small circles represent that the elements reach 60% of their plastic limit. It is obvious that plastic hinges develop more seriously in frame structure. In LS phase, only a small part of elements in cassette structure reach 60% of their plastic limit while in frame structure, more than 50% floors have elements in plastic phase, and some of beams in bottom floors even reach their plastic limit. As load increases, plastic hinges develop rapidly in frame structure, but in the cassette structure, only some elements in bottom floors reach plastic limit while others either only reach 60% of their plastic limit or

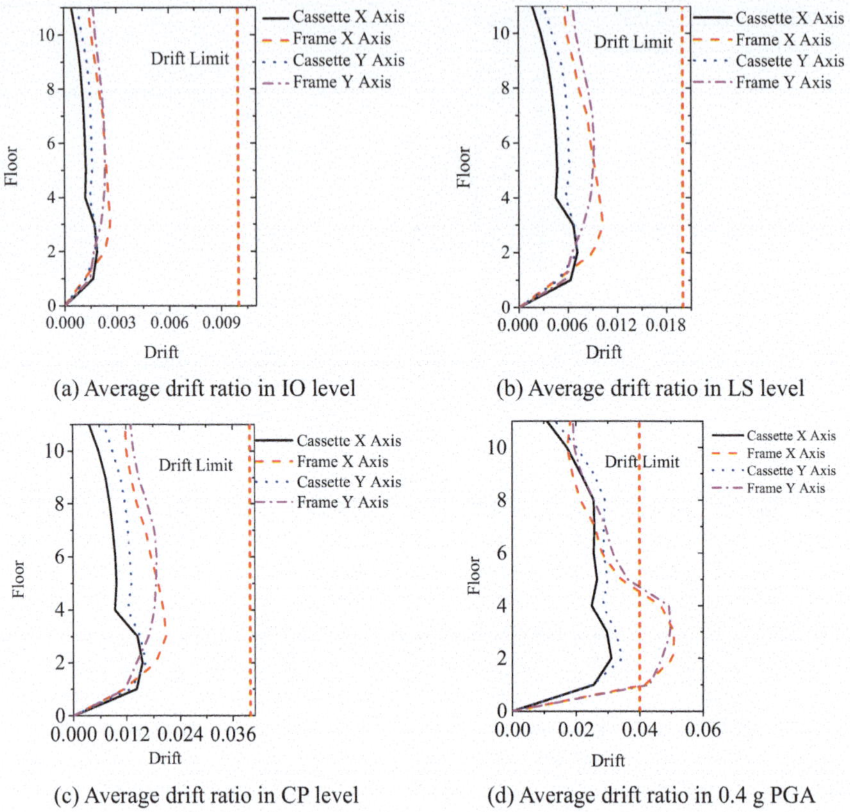

(a) Average drift ratio in IO level

(b) Average drift ratio in LS level

(c) Average drift ratio in CP level

(d) Average drift ratio in 0.4 g PGA

Fig. 7.33 Results of time-history analysis

just stay elastic. Furthermore, while the entire frame structure is nearly plastic and failed at the 0.4 g PGA phase, the plastic hinges in cassette structure only develop in bottom floors and the entire structure's function is maintained.

The plastic development can also be seen in the energy scale maps, which can provide the energy dissipation information directly. In these maps, dissipated energy, which can be the response of a structure to an earthquake, is divided into two parts. For elastic structures, it is usually assumed that energy is dissipated by viscous damping (it is a modeling approximation, except for structures with actual viscous dampers). For inelastic structures, it is still assumed that there is viscous damping, and additional energy is dissipated by inelastic effects such as yielding, friction, etc. As shown Fig. 7.35, in IO phase, most dissipated energy in cassette structure is strain and damping energy, which indicates that the whole structure is still in elastic phase. However, some dissipated inelastic energy appears in frame structure, which shows some small damages occurred. In LS phase, which is shown in Fig. 7.35c, d, the inelastic energy in cassette structure is about 5% of the total dissipated energy while in frame structure, about 30% dissipated energy is inelastic energy. Similarly, in CP

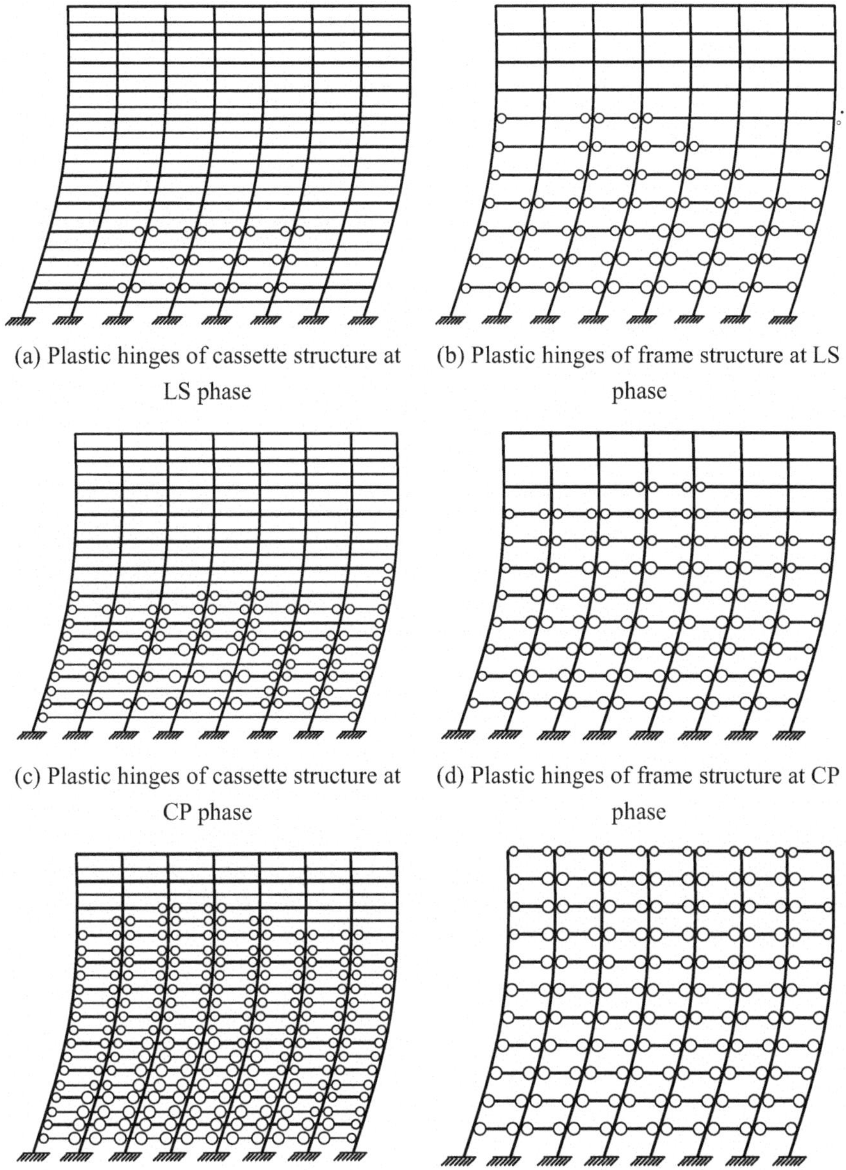

(a) Plastic hinges of cassette structure at LS phase

(b) Plastic hinges of frame structure at LS phase

(c) Plastic hinges of cassette structure at CP phase

(d) Plastic hinges of frame structure at CP phase

(e) Plastic hinges of cassette structure at 0.4 g PGA phase

(f) Plastic hinges of frame structure at 0.4 g PGA phase

Fig. 7.34 Development of the plastic hinges in two structures

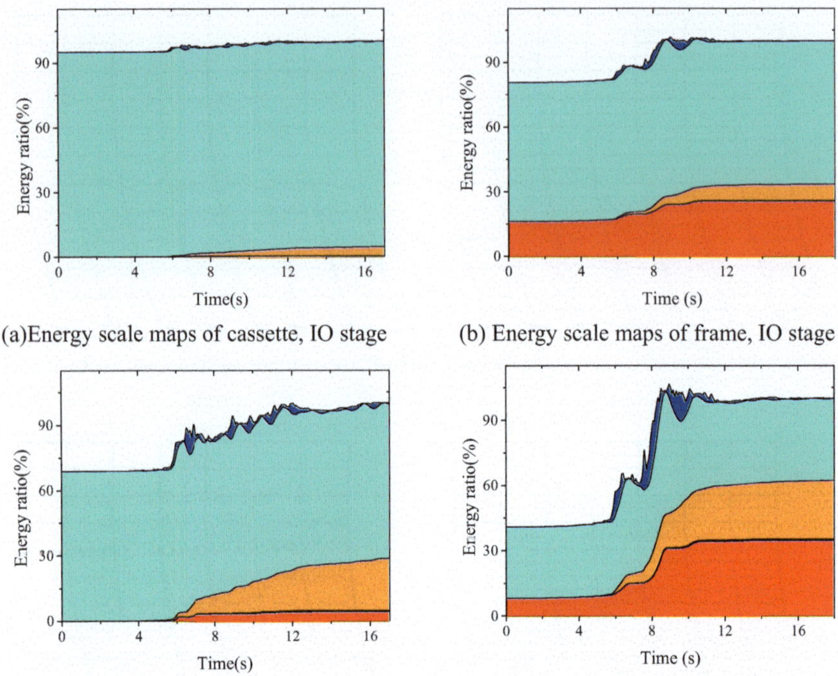

(a)Energy scale maps of cassette, IO stage (b) Energy scale maps of frame, IO stage

(c) Energy scale maps of cassette, LS stage (d) Energy scale maps of frame, LS stage

Fig. 7.35 Energy scale maps of the structures

phase, inelastic energy is much larger in frame structure than the cassette structure, and in 0.4 g phase, about 60% dissipated energy in frame structure is inelastic energy, which indicates the structure is largely damaged. However, only 40% dissipated energy in cassette structure is inelastic energy and shows the structure is undergoing moderate damage, which can be concluded that cassette structure has a better seismic performance than frame structure.

7.6.4 Park-Ang Damage Analysis

Park–Ang damage index was proposed by Park and Ang in 1985 and has been widely accepted in seismic studies. The index is shown as

$$\mathrm{D} = \frac{X_m}{X_u} + \beta \frac{\varepsilon}{F_y X_u} \tag{7.3}$$

where X_m is the maximum displacement of an element during earthquake, X_u is the maximum displacement of an element under monotonic load, β is an index, which

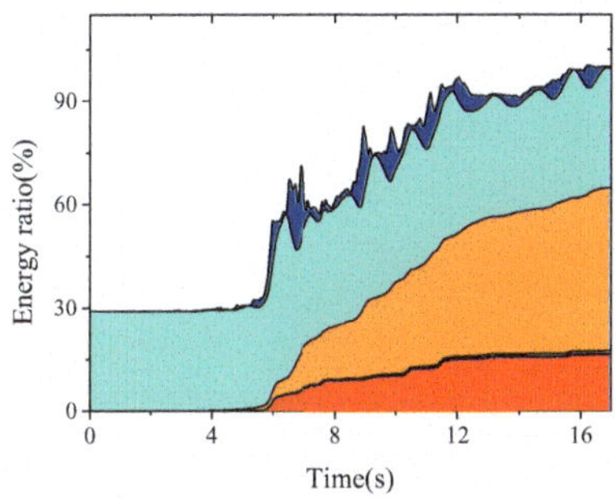

(e) Energy scale maps of cassette, CP stage

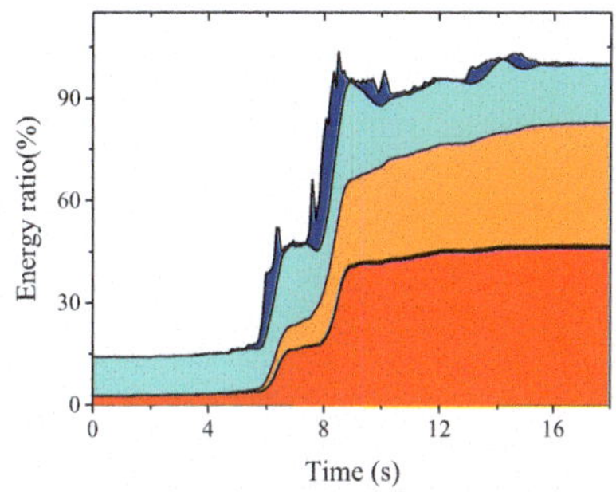

(f) Energy scale maps of frame, CP stage

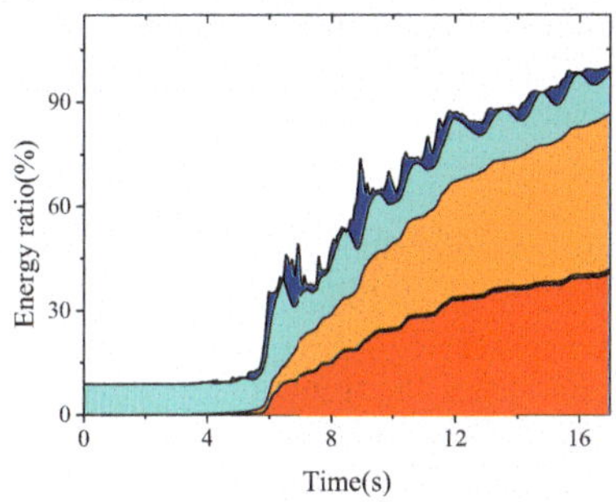

(g) Energy scale maps of cassette, 0.4 g PGA stage

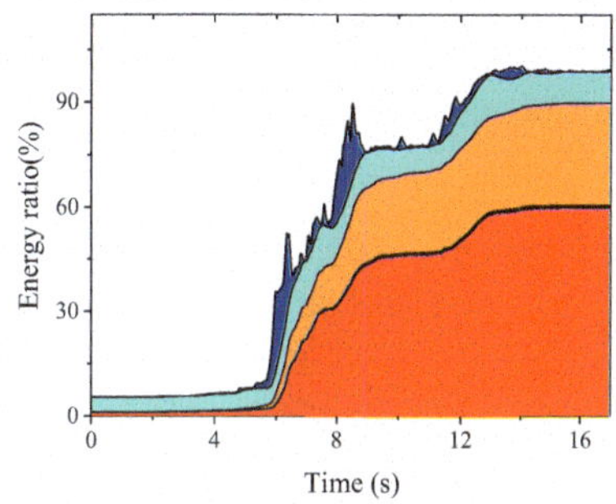

(h) Energy scale maps of frame, 0.4 g PGA stage

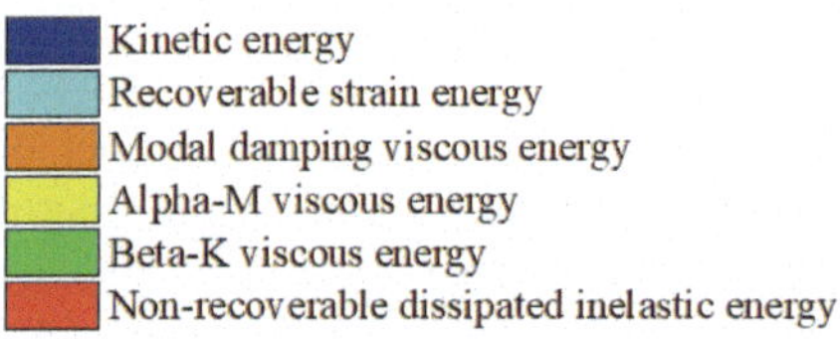

Fig. 7.35 (continued)

is 0.15 for frame structure, ε is the energy dissipation and F_y is the yield shear force of element.

The Park–Ang damage indexes are shown in Fig. 7.36. The damage indexes have the same tendency as the drift ratio, and the advantage of the cassette structure is obvious. When in IO level, both indexes are under 0.1, which means structures only have minor damage. With the increase of PGA, a much larger damage is observed in the frame structure. When in LS and CP level, the indexes of the frame structure are 0.18 and 0.5 respectively, which are much larger than 0.1 and 0.36 in the cassette structure. Furthermore, the index of the frame structure is approximately 1 in 0.4 PGA level, which means the entire structure is in failure stage and unstable. However, the index of the cassette structure is below 0.9 in this stage, which indicates that the structure is largely damaged, but the main element is still stable and there is no risk of collapse. The Park–Ang damage index time-history records are shown in

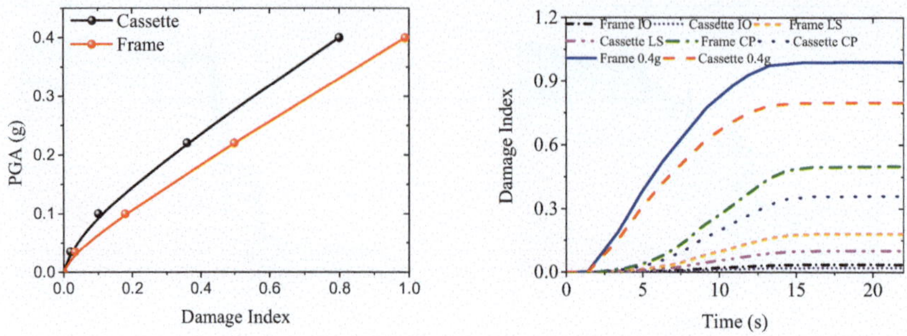

(a) Park–Ang damage indexes for the two structures

(b) Park–Ang damage indexes development for the two structures

Fig. 7.36 Park-Ang damage analysis for the two structures

Fig. 7.36b. All the curves show that the development of damage is much faster and larger in the frame structure, especially in 0.4 g PGA level. In all, based on both the final damage index and the time-history development, cassette structure presents a much better performance in the control of seismic damage.

7.6.5 Economy Analysis

In previous discussion, the advantage of the cassette structure in structural performances are obvious. However, these performances are not the only factors that are considered in the practical application. In addition to structural performance, economic performance is another important factor that should be considered. Therefore, economic analysis is conducted. The material usage and construction cost are based on Chinese code and quota, and the factor of building industrialization is considered.

7.6.5.1 Chinese Quota

The Chinese quota used in this study is listed in Table 7.17 [8]. This quota specifies the essential cost of labor force, material usage and other necessary processes. As building industrialization is the priority in China, all elements are manufactured in factories, and further transportation and assembling are needed. These extra costs are also included.

Table 7.17 Chinese quota for beam and column

Element	Name of cost	Unit	Cost per unit (CNY)
Beam	Concrete	m^3	497
	Template	m^2	10,354
Column	Concrete	m^3	508
	Template	m^2	10,314

7.6.5.2 Economic Analysis

The consumption of concrete and steel in both two structures is nearly the same. The concrete usage of the cassette structure and the frame structure are 2472.4 m^3 and 2422.9 m^3, respectively, and the steel usages are 276.8 t and 282.1 t, respectively. Thus, the material cost for both structures are almost the same.

As these structure elements are made in specialized factories, careful element division is needed. The frame structure is dissembled into various single beams and columns, but the cassette structure is dissembled into four units shown in Fig. 7.37. Unlike the frame structure, elements in the cassette structure are quite similar, only a few kinds of templates are needed. The number of elements used in a typical floor for both structures is shown in Table 7.18 and Fig. 7.37. Considering the construction speed, three sets of templates are needed for the A unit, two sets for the D unit and all columns, and one set for all other elements in both structures. After counting seam and fabrication loss, the templates needed for one typical floor of the cassette structure are 110.75 m^2 and 152.6 m^2 for the frame structure. As two different types of typical floors are designed for both structures, the usage of templates is doubled. The entire template cost for the frame structure is 1.58 million CNY, while for the cassette structure, it is 1.147 million CNY, saving approximately 37% of the cost.

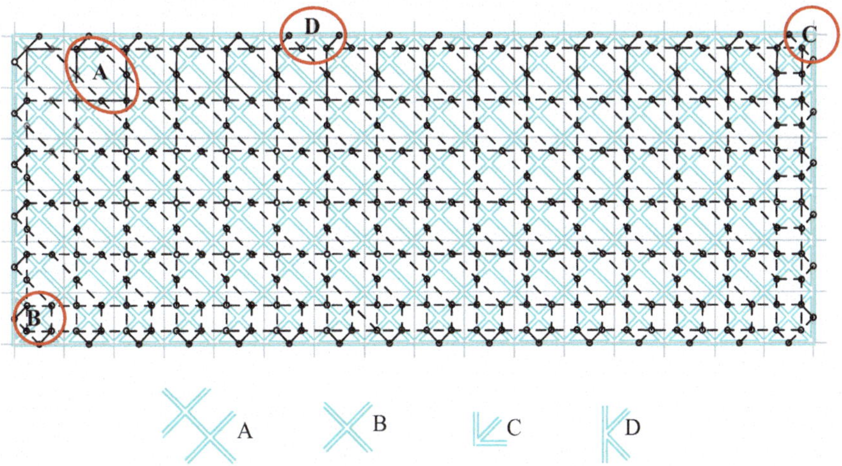

Fig. 7.37 Separated elements in the cassette structure

Table 7.18 Template usage of the two structures (unit: mm)

	Cross-section/Unite	Maximum length	Amount of element	Area of template
Frame	950 × 300	8700	20	1.97×10^7
Structure	650 × 300	8700	28	1.43×10^7
	750 × 350	8700	15	1.66×10^7
	900 × 300	7400	18	1.61×10^7
	400 × 200	9075	7	9.23×10^7
	550 × 250	9200	3	127×10^7
	1000 × 800	5400	32	1.67×10^7
Cassette Structure	A		75	1.22×10^7
	B		21	6.1×10^6
	C		4	4.25×10^6
	D		40	5×10^6
	800 × 700 Column		44	1.56×10^7

After cast and maintenance, individual transportation is needed for the beams and columns of the frame structure, as the weight may exceed the limit of trucks. For the cassette structure, as the elements are comparatively small, two elements can be transported together. Therefore, the entire transportation cost is similar, but the costs for the frame structure will be slightly higher.

When considering the assembly of these two structures, the numerous nodes in the cassette structure is challenging. Take the first floor of cassette structure as an example; there are 121 nodes in the frame structure while there are 436 nodes in the cassette structure, nearly 4 times of those in the frame structure. However, the size of the node in the cassette structure is only 1/3 of that in frame structure is, and all of them are in the middle of beams (shown in Fig. 7.37). Therefore, only 1/5 of time cost is needed when assembling a node of cassette structure in practice.

Through current quota, assembling one frame node costs one worker a day, while the same worker can assemble five nodes of the cassette structure. Considering the entire structure, approximately 22 workdays can be saved in the construction of the 47 m height building.

The entire costs of these two structures are shown in Fig. 7.38. It is obvious that compared with cassette structure, the template cost is much larger in the frame structure, while the other costs are nearly the same. Considering all the procedures, approximately 0.43 million CNY (approximately 7% of the total cost) and approximately 22 workdays can be saved when the frame structure is redesigned to a cassette structure. In addition, the structure can achieve much better performance under seismic and other loads. Therefore, it is a good option to replace the frame structure with the cassette structure when designing some mid-rise buildings.

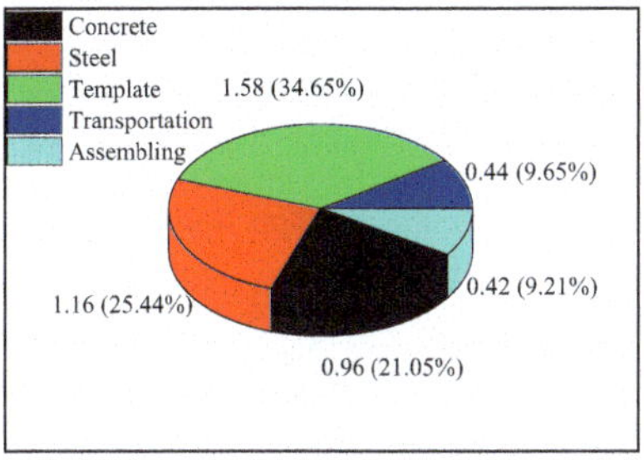

(a) Entire cost of the frame structure

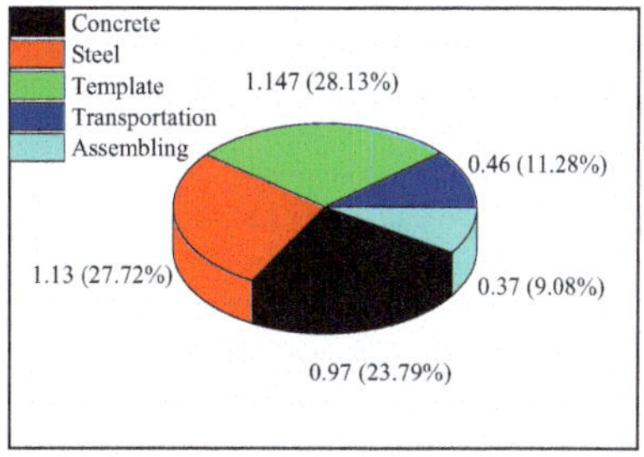

(b) Entire cost of the cassette structure

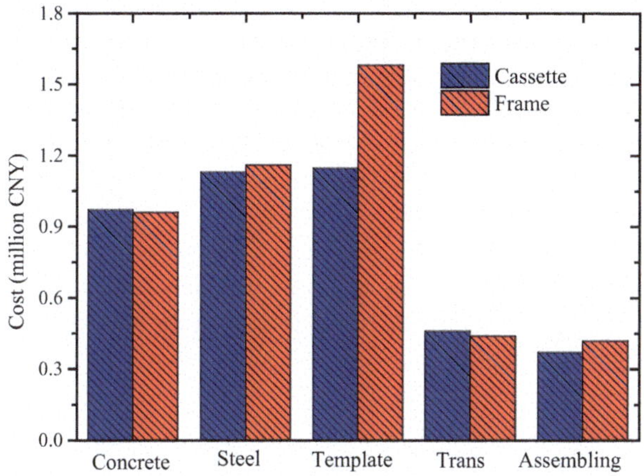

(c) Cost comparison between the two structures

Fig. 7.38 Entire cost of the two structures (unit: million CNY)

7.7 Conclusions

In this book, the pushover analysis, IDA, fragility analysis and economy comparison are used to study the advantage of cassette structure. From all above-mentioned studies, the cassette structure shows an obvious advantage compared to the traditional frame structure, and the cassette structure may be a good substitution of frame structure in some cases.

References

1. Ma K, Huang Y, Jianchun X, Zhu L (1995) The research and application of concrete open-web grid and open-web sandwich slab. Spec Struct 1(3):28–36

2. Ma K-J, Guofu G, Huagang Z, Yaqing L, Haoming H, Sanke Y, Tian J, Guangeng Z, Zhihua C, Wenling Tian, Weihong Niao and Wanjiang Yu (2009) Research and application on space grid frame construction used in multi story and tall buildings with energy saving and large bays. Spetial Struct 15(3): 66–86
3. Chen Z-P, Wu G, Feng D-C, Ma K-J (2022) Experimental study of a novel open-web sandwich slab and modified design procedure. Mag Concr Res (2022) 74(1):22–41
4. Lan Hu (2017) Theoretical analysis and experimental study on semi assembled composite open-web sandwich plate outsourcing U type steel plate in Civil Engineering. Hunan University
5. Fan J-J, Wu G, Xu A-L, Feng D-C, Chen Z-P (2020) Experimental study on the seismic performance of novel precast reinforced concrete grid moment-resisting frames. Struct Concr 3:1–16
6. Park YJ, Ang AHS (1985) Mechanistic seismic damage model for reinforced concrete. J Struct Eng 111(3):722–739
7. Chen Z-P, Wu G, Feng D-C, Ma K-J (2018) Numerical study of the static and dynamic characteristics of reinforced concrete cassette structures for high-rise buildings. Struct Des Tall Spec Build 28:e1574
8. Chen Z-P, Feng D-C, Wu G, Ma KJ (2020) Seismic and economic performance of a mid-rise cassette structure. Adv Struct Eng 23(16):3541–3554

Chapter 8
Modularized Suspended Building Structure

Abstract This chapter introduces modularized suspended building structures, which can be categorized as mega-sub control configurations with precast/prefabricated components. Intrinsic seismic attenuation mechanisms of passive-control suspended buildings are illustrated mainly from the perspective of a large-mass-ratio TMD-structure system. The structural features of prefabricated modular structures are summarized before modularization of the secondary structure within the suspended building is proposed to induce a harmless pattern of secondary-structure-inter-story drift. Shake-table experiments are also conducted. The results of those investigations show that the modularized suspended building has a satisfactory multi-mode tuning effect induced by the aforementioned inter-story drift pattern, and thus an enhanced seismic control performance.

Keywords Suspended building · Prefabricated module · Passive control · Large-mass-ratio TMD · Numerical optimization · Shake-table experiment

8.1 Introduction

Suspended building structures provide structural and architectural benefits, including smaller sections of vertical components, visual transparency, reduced weights, and a large column-free space in the first story. Since 1960s, the application of suspended buildings with precast/prefabricated components has gradually become popular. In this research, highly integrated and prefabricated 3D box-like modules are adopted as the suspended part, in order to overcome the inherent inter-suspended-story-drift-induced damage of passive-control suspended buildings. The target is to further reduce seismic response of suspended buildings. In this section, background introduction of modular structure and suspended building are provided respectively; the benefits of strategically combining the aforementioned two are also discussed.

G. Wu et al., *Novel Precast Concrete Structure Systems*, Springer Tracts in Civil Engineering, https://doi.org/10.1007/978-981-19-6821-1_8

8.1.1 *Prefabricated Modular Building Structures and Corresponding Inherent Features*

Modular building system is a highly prefabricated system, as shown in Fig. 8.1. The box-type 3D modules can be conveniently positioned and interconnected on site. Because nonstructural components, devices, pipelines, and decorations are pre-integrated into the module, time and labor are saved, and no subsequent work is needed [1–4].

Different levels of modular construction applications have long existed in the industry. The temporary structures, the pre-ordered cottage houses, as well as the integrated functioning pod, are lower-level modular structural systems. In the strict version of the definition, modular buildings belong to a higher level of modular construction which is permanent, modularized in a large volumetric proportion, and needs to be designed considering both horizontal and vertical loads [2, 5–8]. Two decades ago they originated in U.K. and Ireland, and then gradually became popular worldwide [2, 6]. Successful applications have also emerged in China recently.

Modular structures are highly prefabricated not only in terms of structure components, but also in terms of non-structural components, decoration, and devices. Essential benefits include [2–4]: (1) High-speed installation and lower demand for labor onsite, as shown in Fig. 8.2; (2) Structural self-sufficiency and convenience in positioning and stacking. In most cases, no temporary braces are needed; (3) Higher quality as demanded by specialized applications, for instance, hospitals; (4) Low demand for construction space/technique, and (5) Lower emission and easier demolition improve the sustainability features.

As a result, an increasing number of applications have taken place in the following scenarios: (1) dormitories/rebuilt classrooms; (2) safer and stronger facilities such as prisons and storm shelters; (3) hotel/hospital/high-end residential buildings, and (4) remote or rural construction.

Fig. 8.1 Modular building structures [2] Photo copyrights belong to Zhuoda Group, China, Shepherd Group, UK, and Mark Lawson

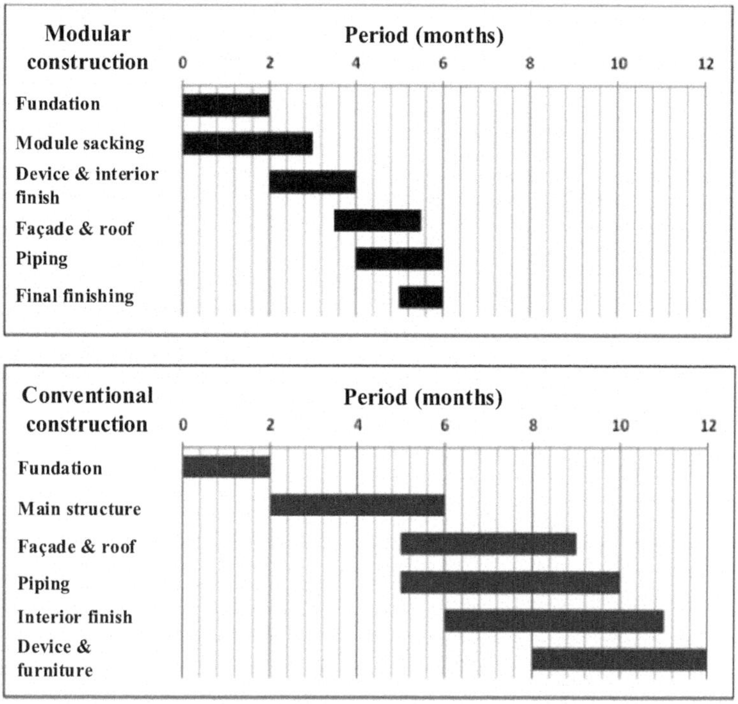

Fig. 8.2 Typical Gantt charts for modular versus conventional building construction [2]

In the early days, manufacturers developed productions with a smaller number of specifications, considering manufacturing and architectural factors. As the market keeps enlarging, structural types and specifications of the products become versatile. Both light (for instance, frame-type) and strong modules (for instance, continuous-spacing-column-type) are commercially available, as shown in Fig. 8.3 [2].

(a)Corner-supported module (b)Continuous-column-supported module

Fig. 8.3 Structures of individual modules [2] (Photo copyrights belong to Kingspan Steel Building Solutions, UK, and Terrapin, UK)

(a) Module structures without additional stability units [2]

(b) Module structures with additional stability units (Photo copyrights belong to AMS Group, China) [9]

Fig. 8.4 Lateral resistance and stability system

When the lateral load is within a certain level, no additional stability units are required. In non-seismic/low-seismic regions, 6-to-8-story buildings [2, 6] can be constructed relying only on the structures of individual modules [7, 10], as shown in Fig. 8.4a. However, modular buildings with additional stability units are much more popular in seismic region and/or high-rise applications [8, 11], RC core-tube being the most popular stability unit, as shown in Fig. 8.4b.

Three major types of load transfer are worthy of investigation: within the same story, to the adjacent stories, and to the foundation. The load transferring path within the same story guarantees cooperation with additional stability units; it has the highest importance in wind/earthquake engineering. The widely-accepted assumption of rigid floor diaphragm is no longer acceptable in modular structures, as the discontinuous diaphragm brings about rocking behavior in the horizontal plane and thus weakens the load transfer. As shown in Fig. 8.5, solutions for discontinuous diaphragm include: (1) sufficient post-poured bands, bolts, welding, shear keys, and/or prestressed friction; (2) 2D truss underneath and connected to each story of modules; (3) partially spaced 2D truss such as corridor units with diagonal braces in the horizontal plane; (4) direct connection between each module and the additional stability units.

In stack-type high-rise modular buildings, the following two limitations are to be solved. (1) In most cases, the additional units withstand a very high proportion of lateral load as they have much higher lateral stiffness, leading to less redundancy of the whole system. If conventional types of viscous dampers are implemented to improve seismic performance, the relatively weak module joints might fail to withstand damper-connection forces. (2) For the stacking-type modular building with multiple stories, the bottom modules sustain larger vertical or axial loads as the total number of stories increase. If the sections of load-bearing components become accordingly larger, a contradiction appears between lightness and standardization of the modules.

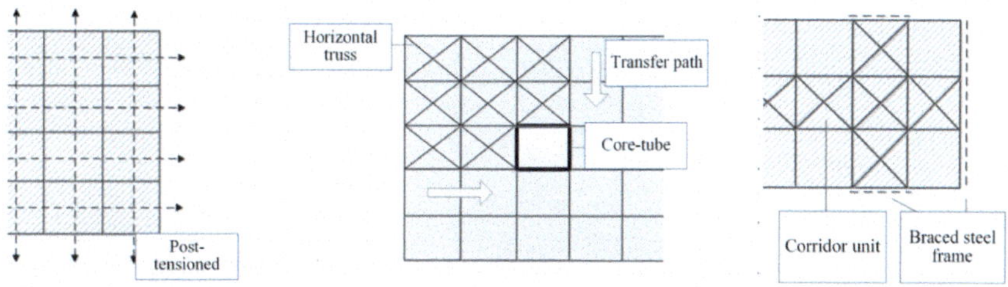

(a) Integrated diaphragm (b) Horizontal truss (c) Corridor units

Fig. 8.5 Load transfer paths within the same story

8.1.2 Passive-Control Suspended Building Structures and Corresponding Inherent Features

Passive-control suspended building structures belong to primary-secondary structural systems, or alternatively, "mega-substructure system". Not all suspended building have passive-control features; not all mega-substructure system is in suspended forms. Generic suspended building is first described in this subsection before the seismic attenuation features are introduced.

The broader definition of suspended structures is "a complex gravity resisting system consisting of main parts, suspended parts, and suspending components, wherein the gravity load is transferred from the suspended parts via the suspending components to the main parts". Cable-stayed bridges, suspended bridges, and stadiums with cable-net-dome structures all belong to this category. Whereas in this section, we focus on the stricter definition, which is "the high-rise suspended building structure, usually in the form of high-rise offices, hotels, or residential buildings".

The main parts of suspended buildings can be single/multiple RC core-tubes [12–14], mega frames [15, 16], arches, etc. Cantilevered transfer stories are often adopted to keep the suspending components vertical. The suspended part provides a major portion of the architecturally functioning space. Multiple connections in horizontal directions may exist between the suspended parts and the main parts [17].

The aforementioned scheme leads to the following benefits. (1) The tensioned vertical components encounter no buckling issues and thus lead to reduced cross-sections. Higher visual transparency and reduced overall weights can be achieved. (2) A large column-free space can be easily formed in the first story to achieve a unique visual expression [18], or to fulfill overtop construction above existing low-rise buildings.

The appealing architectural effect can be achieved if all the parts are well organized and coordinated. The structural type and layout of the main part, the number of transfer stories, and the directions of the suspended components, should be well matched to the architectural design, utilitarian purposes, and overall area. Famous applications include BMW headquarters in Munich, Germany, HSCB headquarters in Hong Kong, China (as shown in Fig. 8.6), Plaza Marquette in Minneapolis, U.S.,

(a) Multiple transfer story (b) Visual transparency (c) Large open space on the first floor

Fig. 8.6 HSCB headquarters (Hong Kong, China) [19] (Photo copyrights belong to EMPORIS)

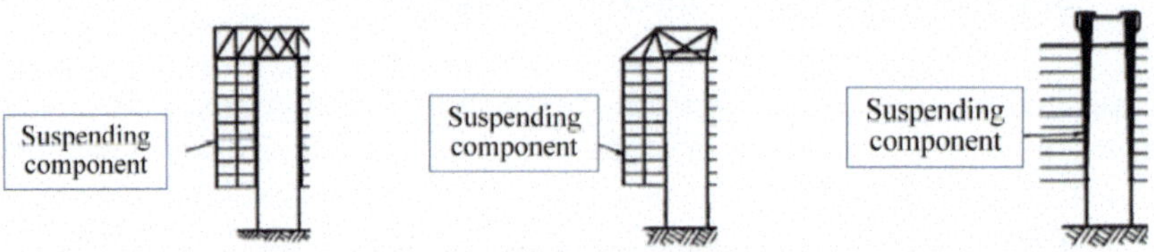

(a) Vertical suspending (b) Inclined suspending (c) Suspending with cantilevering

Fig. 8.7 Suspending types leading to different visual effects [20]

Zhongtian 101 Tower in Guiyang, China, and Guangdong Province Museum in Guangzhou, China. RC core-tube is a popular solution for the primary structure; the suspending plan can be categorized as in Fig. 8.7 [20], achieving different visual effects.

Passive seismic control systems can be formed if relative motion between primary and secondary structures is allowed and the system is supplemented with dampers, as shown in Fig. 8.8. On one hand, the secondary structures act as TMDs if the connection stiffness is set appropriately, reducing the response of the primary structure in a certain range of frequencies. This range of frequencies is often notably larger than that of the ordinary TMD-structure systems, as the mass of a secondary structure is much larger than an ordinary TMD. A larger range of frequencies is beneficial to the robustness against the deviation of structural parameters. On the other hand, as the secondary structure has architectural functions [21, 22], it actually shares some mass from the primary structure. This mass reduction effect is also an attenuation feature. This passive seismic control scheme attenuates both the displacement response of the primary structure and the acceleration response of the secondary structure. High-rise buildings with RC core-tubes can notably benefit from it, due to the fact that ordinary control schemes may be undermined by the following factors: (1) Flexure-type deformation is dominant in RC core-tubes, preventing the dissipation by non-residual-displacement-type ordinary dampers. (2) It is difficult to

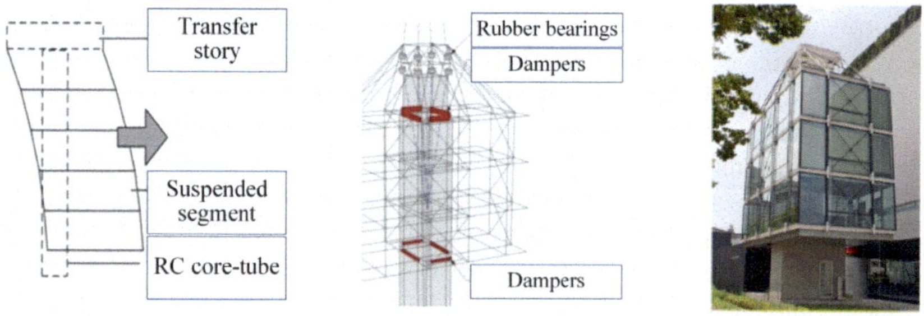

(a) Relative motion (b) Layout of dampers[12] (c) Illustrative building in Japan [12]

Fig. 8.8 Passive-control suspended building diagrams

attenuation displacement and acceleration responses simultaneously. (3) Base isolation might cause overturning issues due to a large total height. The authors believe that passive-control suspended building with a single segment of suspended structure is a prominent solution for the vibration attenuation of core-tube structure about 20-story high.

However, sometimes the seismic control of primary and secondary structures contradicts each other. A large relative motion between the two parts is beneficial to the primary structure [23] but may cause inter-story-drift-related damage in the secondary structure. This issue also undermines the attenuation of acceleration within the secondary structure and may be worsened if a glazed curtain wall is adopted to emphasize the aforementioned architectural merits. Increasing the stiffness of the secondary structure may reduce those unpleasant effects but undermine the attenuation of the primary structure vibration.

8.1.3 Mutually Beneficial Combination of Modular Structure and Suspended Building

As mentioned in Sect. 8.1.1, "module structures + additional stability units" is a popular combination for lateral resistance. Suspended buildings can also be regarded as additional stability units, as the primary structure provides large lateral stiffness while the dampers reduce the force input into the core tube.

The suspended parts in a suspended building are mostly prefabricated, in order to ease the lifting process. From this perspective, the modules fit into suspended buildings naturally. Such a combination is logically feasible.

Such a combination is also mutually beneficial in the following perspectives: (1) A major type of mega-substructure systems develops its flexibility through the secondary structure inter-story drift that occurs entirely inside the story-space and affects non-structural members. On the contrary, the suspended modules let a majority of the drift develop outside the modules, preventing the drift from damaging the non-structural components inside modules. (2) As a result, the inter-story drift limit of the secondary structure can be relaxed, and sufficient relative motion is allowed. Thus, the overall attenuation can be enhanced. (3) The modules are suspended to a chain of rods with pin connections. In this pattern, gravity loads will not accumulate in the modules. Standardization of modules is possible and lighter modules can be adopted.

The aforementioned combination effects lead to the proposal of modularized suspended building structures. The following sections focus on the system-level features of modular structures, as well as attenuation mechanism, optimization, and shake-table test validation of the modularized suspended building structures.

8.2 System-Level Features of Modular Structures

Modular building structures consist of inter-connected prefabricated 3D sub-assemblies and have unique structural features. With stiff and strong inter-connections, modular buildings show similarities to steel MRFs, while differences between them become notable if connections are less than idealistic.

8.2.1 Intrinsic Difference Between Steel Module Groups and Steel Moment-Resisting Frames

To ensure structural stability during storing, transportation, and lifting, individual modules should be structurally integrated and sufficient, even before they are inter-connected onsite and form a larger system. The individual structure and the connecting of individuals show different contributions to the combined structure, leading to the following unique features of modular buildings.

(1) Clustered beam and clustered column. One of the inherent structural features of modular building is the clustered beam and clustered column. A single column in a conventional building is divided into four module columns in a modular building. When these four adjacent modules are positioned and inter-connected, the four module columns behave as a whole unit and then form a "clustered column". The definition of "clustered beam" is mostly the same as the "clustered column", as shown in Fig. 8.9a. They have the potential to raise differences in structural behavior between conventional and modular buildings.
(2) The inter-modular connections either resemble to or are inspired by steel MRF connections. However, as the prefabrication of partition walls separates the whole space into 4 or 8 subspaces, access space to the connections are highly limited [24]. Therefore, inter-module connections are further simplified or highly convenient to lubricate onsite construction, leading to smaller sizes and dense distributions. Bolts, pins, and pre-stressing are usually implemented. As

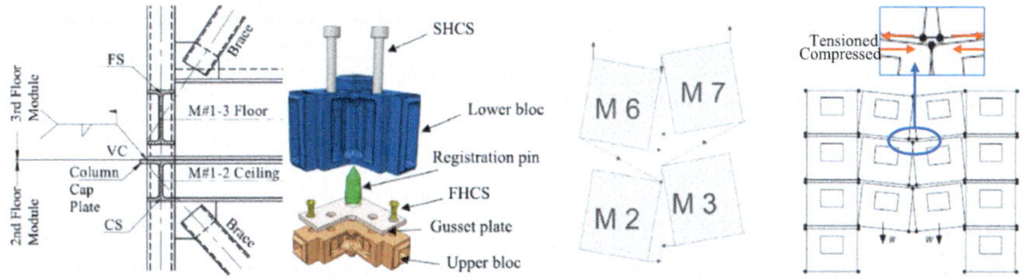

(a) Clustered beam[25] (b) Specialized connection[26] (c) Horizontal rocking[7] (d) Progressive collapse mechanism[2]

Fig. 8.9 Features and challenges of modular structures

shown in Fig. 8.9b, specifications and details are different from those of steel MRFs.

(3) The in-plane stiffness of the floor cassette within each module is very large. However, the horizontal stiffness of the inter-module connection is usually lower than the cassette stiffness, and thus shows semi-rigid features and horizontal rocking mechanisms, as shown in Fig. 8.9c.

8.2.2 Challenges Against Modular Structures

As mentioned previously, for the stacking-type modular building, the bottom modules sustain larger vertical or axial loads as the total number of stories increase. A contradiction appears between the lightness and standardization of the modules. Another challenge of the inter-module connection is that the discrete modules with individual vertical/horizontal diaphragm effects may induce rocking behavior and cause extra demand on the connections. Lastly, in terms of progressive collapse, the discontinuity between upper and lower modules might cause unique collapse mechanisms. Moreover, the horizontal discontinuity may enlarge such an effect as it weakens the horizontal tying, as shown in Fig. 8.9d.

8.2.3 Major Layers of Structural Parts in a Modular Building

Modular buildings consist of four layers of structural parts, namely, connections, individual module structure, path of load transfer, and additional stability units [4]. Each of the four layers can contribute significantly to the overall performance. Among them, connections and individual module structures are mainly introduced in this subsection.

Table 8.1 shows typical connection types. L4 denotes an innovative inter-module connection proposed by Zhihua Chen et al. [24]. In this connection, the plug-in device at the center provides paths for horizontal load transfer, while the blot connections at the modular beams provide paths for vertical load transfer. Access at the corner of a module is specially provided to lubricate fast inter-connecting onsite. The research aims at quasi-static performance under monotonic and cyclic loading schemes, with various combinations of axial compression ratio, height of beam section, and existence of stiffeners. The results show that: (1) Hysteresis curves are plump and adequate energy dissipation is indicated; (2) various types and combinations of failure modes are shown, including local bucking, vertical tearing of beam-column joints, stiffener fracture, gap opening near the plug-in device, and crack at floor-beam joints; (3) the connection remains integrated when aforementioned failure take place, due to a good redundancy; (4) The existence of stiffeners notably improve the load capacity of the connection. The outcome is applicable to multi-story modular buildings without additional stability units.

L6 denotes an innovative type of module-foundation connection proposed by Keum-Sung Park et al. [28]. The end of a "clustered" column at the bottom story is welded at its outer edge to a steel endplate, which is placed at the recess of the foundation with a corrugated pipe. The recess is later filled with mortar to finish the connection. The research conducted a series of quasi-static cyclic loading horizontally at the top of the bottom column, with various depth of the recess, shapes of the endplate, and the existence of studs. The results show that: (1) the depth of the recess is positively contributing to load capacity; (2) the studs help improve the ductility of the connection; (3) shape of the end plate show no obvious effect while thickness of the end plate contributes to load capacity; (4) the major failure mode is a punch of the concrete with a 45-degree crack from the endplate to the corrugated pipe and another 45-degree crack from the top of the corrugated pipe to the surface of the foundation; (5) hysteresis curves are plump and adequate energy dissipation is indicated. As axial load is not considered, the outcome is applicable to low-rise modular buildings.

Table 8.2 shows typical individual module structures. M3 denotes an innovative module structure with double-skin steel panels proposed by Sung-Gul Hong et al. [29]. The panels provide the first line of defense and improve hysteretic energy dissipation as well as redundancy. It is applicable to multi-story modular buildings without additional stability units. The double steel skins are welded to the inner corrugated steel plate to form a panel. A pair of upper and lower modules is inter-connected at the ends of modular columns and the ends of the panels. The experiments include in-plane quasi-static cyclic loading of the panels, and quasi-static cyclic loadings of one and two stories of conventional modules and modules with panels. The results show: (1) the quasi-static features of the double-skin panel resemble those of common steel plates, with the major failure mode being local buckling after the delamination of skin- and inner-plates. Stronger welding helps prevent failure. (2) The major failure mode for modules without panels is plastic hinges at module column ends. (3) In modules with panels, plastic hinges emerge first at the ends of panels instead of column. Thus, the panels provide the first line of defense. Afterward, the delaminating and buckling further develop. The hysteresis curves are plump and adequate energy dissipation is indicated.

Table 8.1 Typical inter-module connections

#	Example figure	Connecting method	#	Example figure	Connecting method
L1	 [25]	Welding	L4	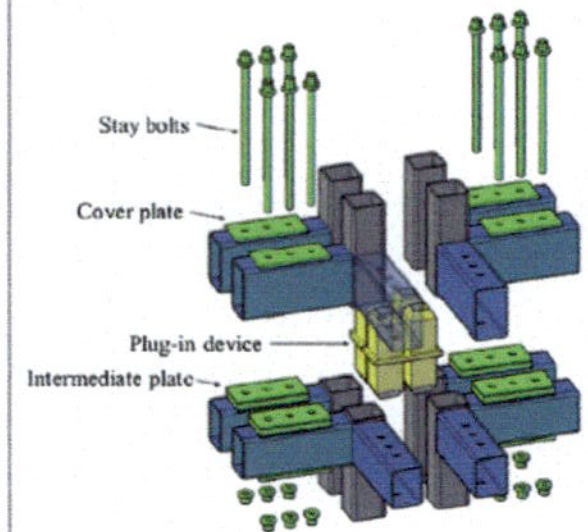 [24]	Blot + pin
L2	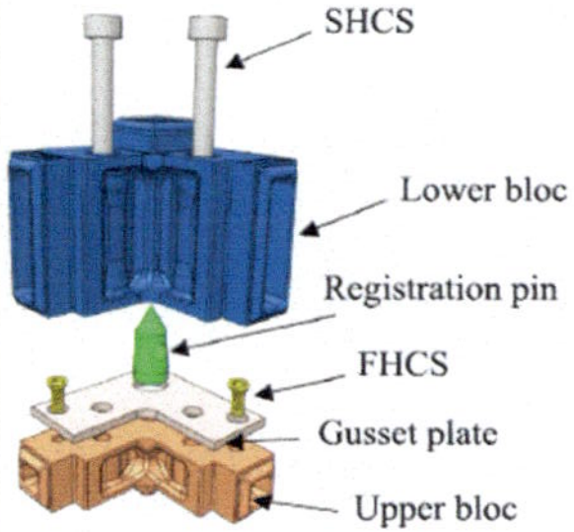 [26]	Blot + pin	L5	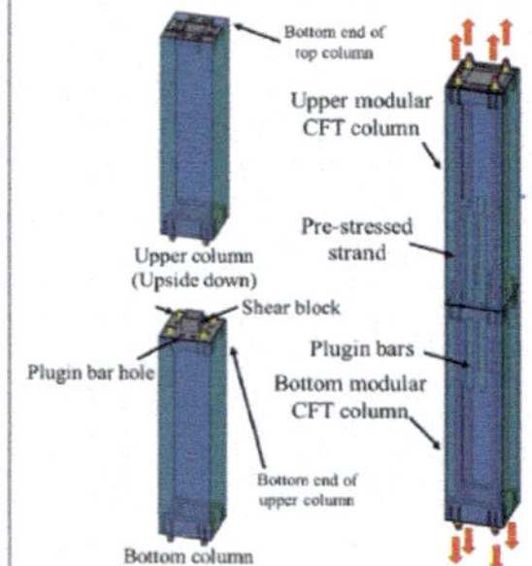 [27]	Positioning block + prestress

(continued)

Table 8.1 (continued)

#	Example figure	Connecting method	#	Example figure	Connecting method
L3	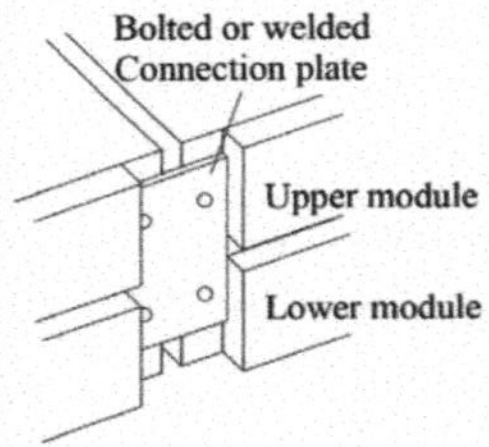 [2]	Blot + connection plate	L6	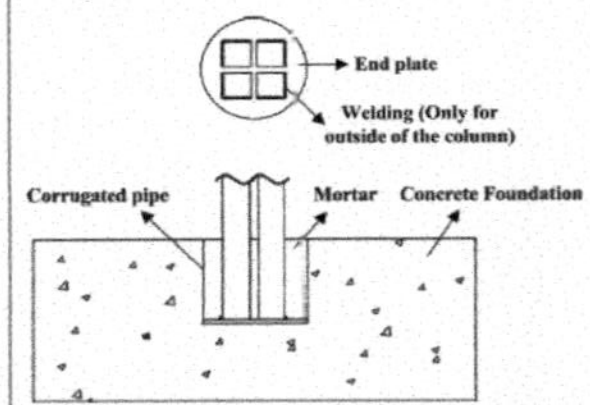 [28]	Post-cast conrete

Table 8.2 Typical individual module structure

#	Example figure	Structure type	#	Example figure	Structure type
M1	 [2]	Steel frame	M3	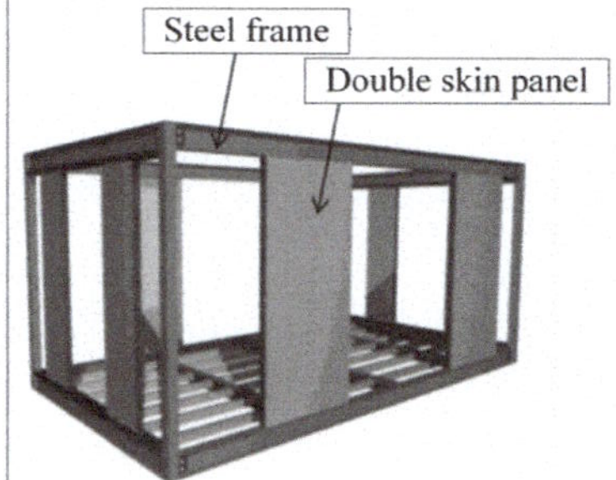 [29]	Steel frame + double skin panel
M2	 [2]	Dense-column steel frame	M4	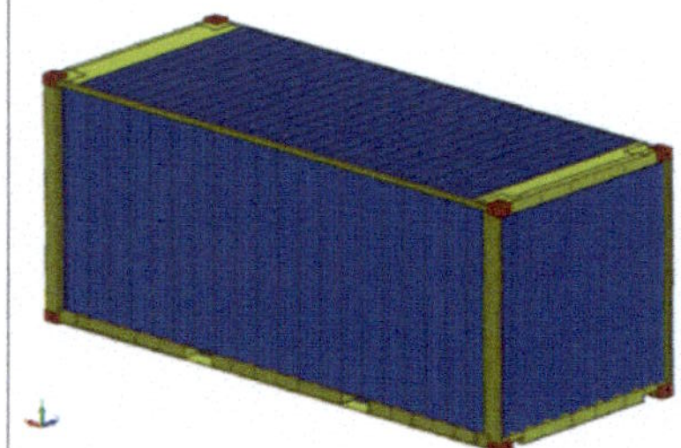 [30]	Corrugated module

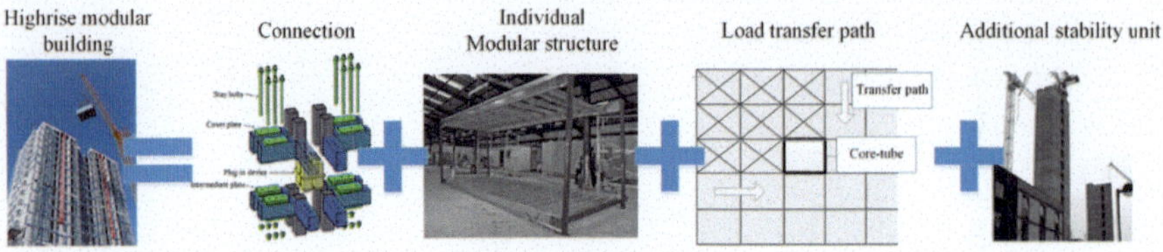

Fig. 8.10 Four layers of structural parts in a modular building

Figure 8.10 presents an example combination of the four layers of structural parts. Such a combination is the most popular one in high-rise residential buildings, as it is efficient structurally and architecturally. During construction, the RC core-tube harnesses slip-form technique while the modules can be stacking and interconnecting at the height of several floors under the slip-form.

8.2.4 Project Examples

The nine-story residential project "Little Hero" in Melbourne, Australia took only 20 days to finish all of its onsite installations [31]. More impressively, it is located in a crowded downtown area with very narrow streets. Such a convincing accelerated process is benefited from the highly prefabricated system: modules plus precast RC cores. For a special architectural design, double-story modules are adopted in certain stories.

In China, multi-story/high-rise applications of modular buildings are gaining popularity in the market. A very successful project is the public housing project in Gangnan Road, Zhenjiang [9]. It is a group of 20-story RC-core-plus-module residential buildings with 18 stories above ground and 2 stories underground. The total height is 56.50 m. Each apartment consists of 2–3 modules. The aforementioned two projects are shown in Fig. 8.11.

(a) "Little Hero", Melbourne, Australia [31]

(b) Public housing project, Gangnan Road, Zhenjiang, China [9]

Fig. 8.11 Project examples

8.3 Seismic Control Mechanisms of Passive-Control Suspended Buildings

From a structural perspective, a suspended building consists of primary and secondary structures. The primary structure is usually a core tube or a mega-frame. With limited total height, fundamental-mode deformation may be dominant in primary structure when subjected to excitations such as wind and earthquake. Based on modal decomposition calculation, thus, the primary structure can be simplified into an SDOF. A single segment of secondary structure with adequate inter-story stiffness and flexible connections to primary structure can also be simplified into an SDOF. Therefore, the complete passive-control structure can be simplified as a two-degree-of-freedom system, with a viscous damper connecting the two degrees of freedom.

The aforementioned 2DOF model of a suspended building has the following intrinsic differences compared to SDOF structure—single TMD system (SDOF-STMD): (1) conventional SDOF-STMD system has a TMD with 1–5% mass of the total system mass, and thus denoted as SDOF-STMD-SMR (SDOF-STMD with Small-Mass-Ratio) in this chapter. Whereas, the secondary-to-primary-structure mass ratio in a suspended building is about 0.5–1.5, which is one order of magnitude larger than that of SDOF-STMD-SMR. Therefore, the 2DOF model of a suspended building is also termed SDOF-STMD-LMR (SDOF-STMD with Large-Mass-Ratio). (2) In other subtypes of passive-control suspended building, especially Type-2 structures in 8.A Appendix, mechanisms could be slightly different from those revealed in SDOF-STMD-LMR.

In this subsection, trends in displacement and acceleration responses of both SDOF-STMD-SMR and SDOF-STMD-LMR are analyzed and compared, with wind and earthquake excitations represented by harmonic ground accelerations and applied forces. The response reduction ratios are also investigated. Those contents are extensions of TMD vibration control theories, as well as a theoretical basis for passive-control suspended buildings.

8.3.1 Numerical Model and Dynamic Equations of the System

SDOF-STMD is a 2DOF system as depicted in Fig. 8.12 wherein m_p, k_p, and c_p denote mass, stiffness, and damping coefficient of the primary structure, while m_s, k_s, and c_s denotes those of the secondary structure. x_p and x_s denote respectively displacement of the primary and secondary structures relative to the ground. The equation of motion is as follows.

$$\boldsymbol{M}\ddot{\boldsymbol{X}} + \boldsymbol{C}\dot{\boldsymbol{X}} + \boldsymbol{K}\boldsymbol{X} = \boldsymbol{F}(t) \tag{8.1}$$

Fig. 8.12 Calculation mode of SDOF-STMD system

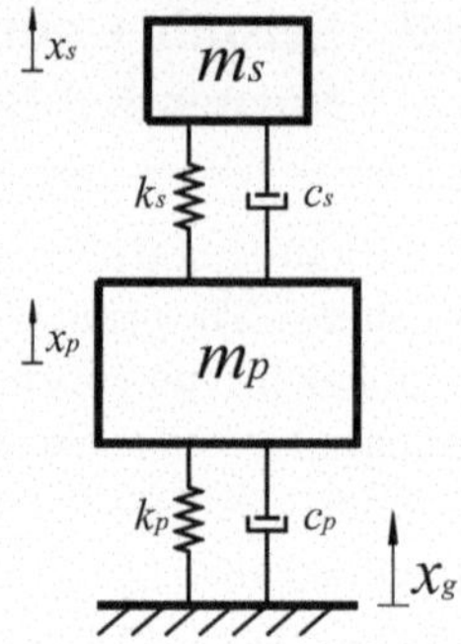

wherein

$$M = \begin{bmatrix} m_p & 0 \\ 0 & m_s \end{bmatrix}, C = \begin{bmatrix} c_p + c_s & -c_s \\ -c_s & c_s \end{bmatrix},$$
$$K = \begin{bmatrix} k_p + k_s & -k_s \\ -k_s & k_s \end{bmatrix}, X = \begin{Bmatrix} x_p \\ x_s \end{Bmatrix}, F(t) = \begin{Bmatrix} F_p \\ F_s \end{Bmatrix} \quad (8.2)$$

For analysis convenience, the frequencies, damping ratios, mass ratios, and frequency ratios of primary and secondary structures are respectively defined in Eqs. (8.3), (8.4), and (8.5). The damping matrix and stiffness matrix in Eq. (8.2) can alternatively be represented as in Eq. (8.6).

$$\omega_p = \sqrt{k_p/m_p},\ \omega_s = \sqrt{k_s/m_s} \quad (8.3)$$

$$\xi_p = c_p/2\sqrt{k_p m_p},\ \xi_s = c_s/2\sqrt{k_s m_s} \quad (8.4)$$

$$\mu = m_s/m_p,\ f = \omega_s/\omega_p \quad (8.5)$$

$$C = \begin{bmatrix} 2m_p\omega_p\xi_p + 2m_s\omega_s\xi_s & -2m_s\omega_s\xi_s \\ -2m_s\omega_s\xi_s & 2m_s\omega_s\xi_s \end{bmatrix}, K = \begin{bmatrix} m_p\omega_p^2 + m_s\omega_s^2 & -m_s\omega_s^2 \\ -m_s\omega_s^2 & m_s\omega_s^2 \end{bmatrix} \quad (8.6)$$

8.3.2 *Dynamic Response Under Harmonic Excitation in Complex Form*

Assuming that excitation is ground acceleration in complex harmonic form, the excitation can be expressed as $\ddot{x}_g = Ge^{i\omega t}$ (G for amplitude, ω for frequency, i for the imaginary unit). If $\lambda = \omega/\omega_p$ denotes the ratio of excitation-to-primary-structure frequency ratio, the force in Eq. (8.1) can be expressed as follows.

$$\boldsymbol{F}(t)/k_p = \frac{1}{k_p}\begin{Bmatrix} F_p \\ F_s \end{Bmatrix} = -\begin{Bmatrix} m_p \\ m_s \end{Bmatrix}\frac{\ddot{x}_g}{k_p} = -\begin{Bmatrix} 1 \\ \mu \end{Bmatrix}\frac{G}{\omega_p^2}e^{i\omega t} \tag{8.7}$$

The following equation represents the displacement of the primary and secondary structures relative to the ground.

$$\boldsymbol{X} = \begin{Bmatrix} x_p \\ x_s \end{Bmatrix} = \begin{Bmatrix} H_{\ddot{x}_g}^{x_p}(\omega) \\ H_{\ddot{x}_g}^{x_s}(\omega) \end{Bmatrix} e^{i\omega t} \tag{8.8}$$

wherein the frequency domain response function $H_{\ddot{x}_g}^{x_p}(\omega)$ has a subscript $\ddot{x}_g$ which represents the acceleration of the ground, and a superscript x_p which denotes displacement of the primary structure. Hereinafter, the frequency domain response function adopts such a symbolic rule.

Substituting Eqs. (8.2) ~ (8.8) into (8.1), we get the displacement response functions as

$$\begin{Bmatrix} H_{\ddot{x}_g}^{x_p}(\omega) \\ H_{\ddot{x}_g}^{x_s}(\omega) \end{Bmatrix} = \frac{G}{\omega_p^2}\begin{Bmatrix} H_{\ddot{x}_g}^{x_p}(\lambda) \\ H_{\ddot{x}_g}^{x_s}(\lambda) \end{Bmatrix} = \frac{G}{\omega_p^2}\begin{Bmatrix} \Pi_{\ddot{x}_g}^{x_p}(\lambda)/\Omega(\lambda) \\ \Pi_{\ddot{x}_g}^{x_s}(\lambda)/\Omega(\lambda) \end{Bmatrix} \tag{8.9}$$

$$\begin{aligned}\Omega(r) = {} & \left[1 + \mu f^2 - \lambda^2 + 2i(\xi_p + \mu f \xi_s)\lambda\right](f^2 - \lambda^2 + 2if\xi_s\lambda) \\ & - \mu(f^2 + 2if\xi_s\lambda)^2\end{aligned} \tag{8.10}$$

$$\Pi_{\ddot{x}_g}^{x_p}(r) = -(1+\mu)f^2 + \lambda^2 - 2i(1+\mu)f\xi_s\lambda \tag{8.11}$$

$$\Pi_{\ddot{x}_g}^{x_s}(r) = -1 - (1+\mu)f^2 - 2i\xi_p\lambda - 2i(1+\mu)f\xi_s\lambda + \lambda^2 \tag{8.12}$$

From Eq. (8.8) ~ (8.9), we get the relative displacement between the primary and secondary structure as

$$x_{s-p} = x_s - x_p = H_{\ddot{x}_g}^{x_{s-p}}(\omega)e^{i\omega t} \tag{8.13}$$

$$H_{\ddot{x}_g}^{x_{s-p}} = G/\omega_p^2 H_{\ddot{x}_g}^{x_{s-p}}(\lambda) = G/\omega_p^2 \Pi_{\ddot{x}_g}^{x_{s-p}}(\lambda)/\Omega(\lambda) \tag{8.14}$$

$$\Pi_{\ddot{x}_g}^{x_{s-p}}(\lambda) = -1 - 2i\xi_p\lambda \tag{8.15}$$

wherein $H_{\ddot{x}_g}^{x_p}(\lambda)$, $H_{\ddot{x}_g}^{x_s}(\lambda)$, $H_{\ddot{x}_g}^{x_{s-p}}(\lambda)$, and $\Omega(\lambda)$ are dimensionless. Furthermore, the following can be derived.

$$\Pi_{\ddot{x}_g}^{\ddot{x}_p}(\lambda) = -2i\xi_p\lambda^3 - (1 + 4f\xi_p\xi_s)\lambda^2 + 2if(f\xi_p + \xi_s)\lambda + f^2 \tag{8.16}$$

$$\Pi_{\ddot{x}_g}^{\ddot{x}_s}(\lambda) = -4f\xi_p\xi_s\lambda^2 + 2if(f\xi_p + \xi_s)\lambda + f^2 \quad (8.17)$$

wherein $H_{\ddot{x}_g}^{\ddot{x}_p}(\omega)$ and $H_{\ddot{x}_g}^{\ddot{x}_s}(\omega)$ are acceleration responses functions of primary and secondary structures. They are dimensionless.

8.3.3 Parametric Analysis in Undamped Primary Structure Cases

Current applications of TMDs to structures usually have mass ratios around 1–5%. The design objective is to control the dynamic responses of the primary structure, alternatively speaking, minimization of the maximum/peak response of the primary structure. The responses of the secondary structure are set as constraints, ensuring the response amplitude of the secondary structure is under a limit when the control is oriented towards the primary structure.

The control objective of the primary structure response can be defined as the following dimensionless quantity.

$$N_{\ddot{x}_g}^{x_p} = \min_{f,\xi_s}\left(\max_{\lambda}\left(\left|\frac{\omega_p^2 H_{\ddot{x}_g}^{x_p}(\omega)}{G}\right|\right)\right) = \min_{f,\xi_s}\left(\max_{\lambda}\left(\left|H_{\ddot{x}_g}^{x_p}(\lambda)\right|\right)\right) \quad (8.18)$$

Common methods to minimize the quantity above can be categorized as either analytical or numerical. For SDOF-STMD with an undamped primary structure, analytical expressions of optimum parameters can be derived via fixed-point theory. Figure 8.13a shows the frequency domain response function curves of SDOF-STMD system with $\mu = 0.05$, $f = 0.94$, and $\xi_p = 0$. Figure 8.13a reveals that the group of curves, with the same mass ratio and frequency ratio but different damping ratios, intersects at two fixed points. Den Hartog and other researchers proposed that the system almost reaches its optimum when the two fixed points have the same amplitude and both are peak points. The procedure to optimize the system is as follows. Firstly, adjust the amplitudes of fixed points until they are the same, and then get the optimum frequency ratio f_{opt}; secondly, assume that the fixed points are respectively the peak point, and the two damping ratios are derived. Then the optimum damping ratio is derived as the average of the two values. The aforementioned procedures form the fixed point method or Den Hartog method. Figure 8.13b shows the frequency domain response function curves of SDOF-STMD system with $\mu = 0.05$, $f = 0.94$, and $\xi_p = 0.05$. It is obviously shown that there does not exist a fixed point for SDOF-STMD system with a damped primary structure and Den Hartog method is not applicable in this situation. Numerical methods are usually adopted instead.

For SDOF-STMD system with undamped primary structure subjected to sinusoidal acceleration of the ground, the optimal parameters and corresponding displacement response amplitude can be derived via the fixed point method as:

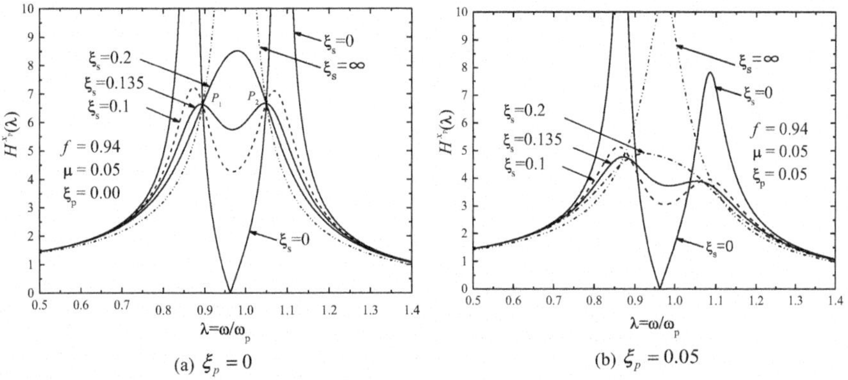

Fig. 8.13 Frequency response curves for primary structure displacement of SDOF-STMD-SMR system

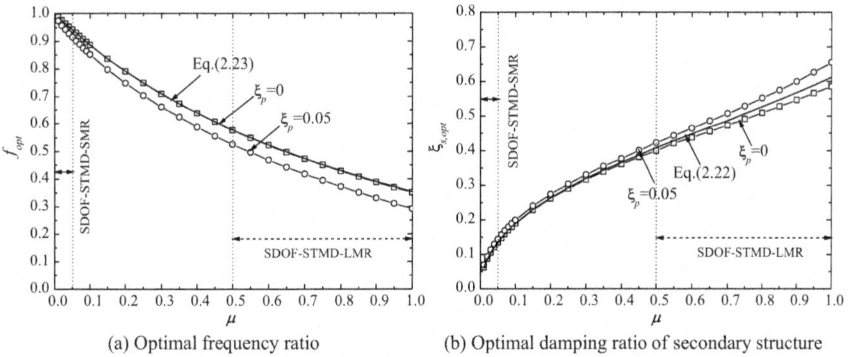

Fig. 8.14 Optimal parameters versus mass ratio of an SDOF-STMD system

$$f_{opt} = \frac{(1-\mu/2)^{1/2}}{1+\mu} \tag{8.19}$$

$$\xi^2_{s,opt} = \frac{3\mu}{8(1+\mu)(1-\mu/2)} \tag{8.20}$$

$$N^{x_p}_{\ddot{x}_g,opt} = \sqrt{\frac{2}{\mu}}(1+\mu) \tag{8.21}$$

Figure 8.14 shows the curves of optimal parameters versus the mass ratio of an SDOF-STMD system, wherein curves marked as $\xi_p = 0$ and $\xi_p = 0.05$ are derived via numerical methods. In Fig. 8.14a, it is revealed that the optimal frequency ratio f_{opt} decreases as the mass ratio increases. If the primary structure is undamped ($\xi_p = 0$), numerical results match analytical results, validating Eq. (8.19). With

$\xi_p = 0.05$, f_{opt} is lower than the previous case. ξ_p has a smaller impact on f_{opt} when the mass ratio is small (0 ~ 0.05), but the impact becomes notable as mass ratio increases. In Fig. 8.14b, it is revealed that the optimal damping ratio $\xi_{s.opt}$ increases as the mass ratio increases. If the primary structure is undamped ($\xi_p = 0$), numerical results match analytical results (Eq. (8.20)) only when the mass ratio is small, due to approximate procedures in the Den Hartog method. With $\xi_p = 0.05$, $\xi_{s.opt}$ is larger than the previous case. ξ_p has a smaller impact on $\xi_{s.opt}$ when the mass ratio is small (0 ~ 0.05), but the impact becomes notable as mass ratio increases.

Based on Fig. 8.14, for SDOF-STMD-LMR systems, classic analytical methods may not be suitable for deriving optimal parameters, because of the notable discrepancy. Further investigation of SDOF-STMD-LMR systems mechanisms is necessary.

8.3.4 Attenuation Indices for Large-Mass-Ratio TMD System

The vibration reduction of SDOF-STMD system subjected to ground acceleration can be quantified as follows.

$$\delta = 1 - \frac{|N_{\text{TMD}}|_{\max}}{|N_p|_{\max}} \tag{8.22}$$

wherein N_{TMD} and N_p denote respectively vibrational response with and without TMD implemented. In complex harmonic terms,

$$N_{TMD} = H_{\ddot{x}_g}^{x_p}(\omega);\quad N_p = H_{\ddot{x}_g,p}^{x_p} = -\frac{G}{\omega_p^2}\frac{1}{1 + 2i\xi_p\lambda - \lambda^2} \tag{8.23}$$

wherein $H_{\ddot{x}_g,p}^{x_p}(\omega)$ denotes frequency domain response function of primary structure displacement without TMD. Substituting Eq. (8.27) into Eq. (8.26), we have the following.

$$\delta = 1 - \left|H_{\ddot{x}_g}^{x_p}(\lambda)\right|_{\max} \cdot 2\xi_p\sqrt{1 - \xi_p^2} \tag{8.24}$$

Figure 8.15 shows the vibration reduction ratio of SDOF-STMD system subjected to ground acceleration δ versus mass ratio μ. Overall δ increases as μ increases, before finally approaching a limit. Damping of primary structure has an impact on vibration reduction ratio δ; a smaller inherent damping of primary structure corresponds to a larger reduction ratio. The following paragraphs further illustrate the vibration control effect of a suspended building, which is a protocol of SDOF-STMD-LMR systems.

As shown in Fig. 8.16, the illustration is via comparing the forming process of SDOF-STMD-SMR and SDOF-STMD-LMR. For the former, as the TMD with a

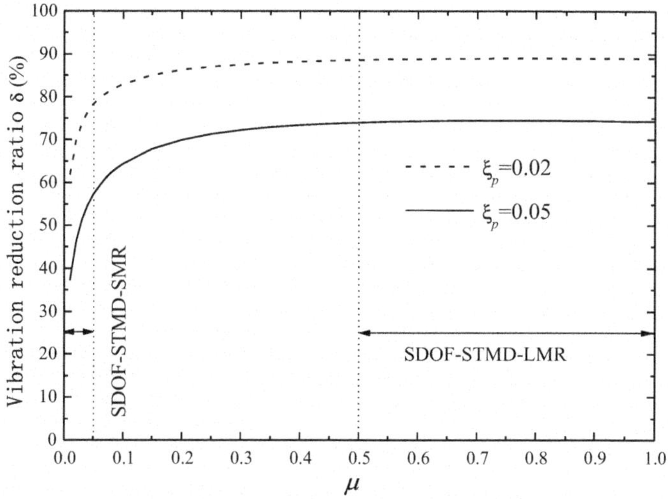

Fig. 8.15 Vibration reduction ratio versus mass ratio

small mass is supplemented, i.e., additional attached, the complete reduction effect is quantified by δ in Eq. (8.22). However, for SDOF-STMD-LMR, the large mass of TMD is of architectural function, and it can be regarded as part of the total mass separated from the primary structure. As a result, the SDOF-STMD-LMR has two steps in its forming: firstly, isolating partial mass from primary structure and thus reducing the inertial forces; secondly, connecting the partial mass to the primary structure with well-designed/optimized stiffness and damping coefficient. The first step induces a mass-reduction effect while the second step induces a vibration-reduction effect. Alternatively speaking, SDOF-STMD-LMR vibration control involves both mass-reduction effect and vibration-reduction effect. Mass-reduction effect can be quantified as follows.

$$\zeta = 1 - \frac{|N_p|_{\max}}{|N_a|_{\max}} \tag{8.25}$$

wherein N_p and N_a denote the dynamic response of an SDOF with respectively the primary structure mass and the total mass. Assuming the damping ratio for the aforementioned SDOF is kept unchanged, we have

$$\zeta = 1 - \frac{\omega_a^2}{\omega_p^2} \tag{8.26}$$

wherein ω_p and ω_a denote the free vibration frequencies respectively of the two SDOF systems. Further assuming the isolation of partial mass cases no change in stiffness, we have

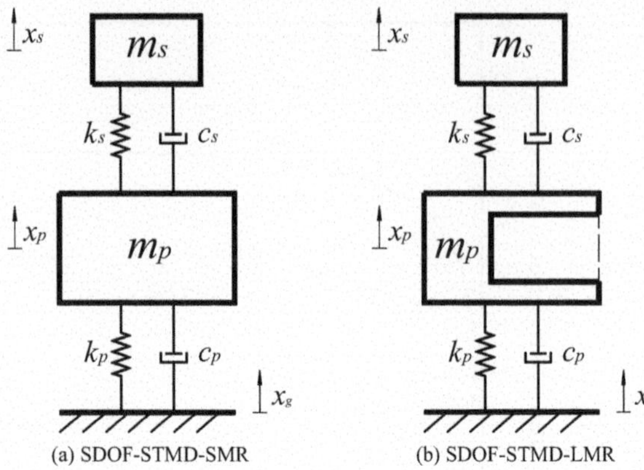

Fig. 8.16 The forming of SDOF-STMD-SMR and SDOF-STMD-LMR

$$\zeta = \frac{1}{1 + \mu} \tag{8.27}$$

Therefore, SDOF-STMD-LMR vibration control effect can be defined as follows.

$$\tau = 1 - \frac{|N_{\text{TMD}}|_{\max}}{|N_a|_{\max}} \tag{8.28}$$

From Eqs. (8.22) (8.25) and (8.28), the following can be derived.

$$1 - \tau = (1 - \zeta)(1 - \delta) \tag{8.29}$$

A positive vibration control requires $\tau > 0$ and the following is to satisfy.

$$\delta > -\zeta/(1 - \zeta) \tag{8.30}$$

Based on the aforementioned equations, it can be deduced that SDOF-STMD-LMR vibration control effect can take place even if the vibration reduction ratio δ is accidentally negative. Thus the control effect and robustness are both further enhanced by the mass-reduction effect.

In this subsection, seismic excitation is simplified into a complex sinusoidal acceleration of the ground, and the SDOF-STMD systems are subjected to such ground accelerations. Optimization based on fixed point theory, parametric study, and comparative study of SDOF-STMD-SMR and SDOF-STMD-LMR (especially the mechanisms) are conducted. Following conclusions are reached.

(1) To minimize displacement of the primary structure, optimal frequency ratio and optimal damping ratio of the secondary structure exist for SDOF-STMD systems subjected to a complex sinusoidal acceleration of the ground. For SDOF-STMD-LMR, classic analytical equations without considering primary structure damping may cause notable inaccuracy. Deriving the optimal values via numerical methods is recommended.

(2) For SDOF-STMD-LMR systems subjected to a complex sinusoidal acceleration of the ground, vibration control consists of both mass-reduction effect and vibration-reduction effect, achieving a more effective and robust control compared to SDOF-STMD-SMR systems.

8.4 Modularization of Secondary Structure in Suspended Buildings

8.4.1 Structural and Utilitarian Schemes

In order to fulfill the mutually beneficial combination of modular structure and suspended building as discussed in Sect. 8.1.3, modularization of the suspended segment is proposed by Ye and Wu [32]. In other words, the suspended segment consists of "suspending components and discretely suspended modules" instead of "suspending components, suspended floor systems, and non-structural components". The primary structure, which is an RC core tube in this subsection, can be regarded as an additional stability unit from the perspective of modular buildings. Moreover, the connections between the modules and the additional stability unit are flexible and able to achieve energy dissipation.

8.4.1.1 Overall Scheme

Figure 8.17 shows the illustrative figure of the proposed modularized suspended building [32, 33]. The primary structure is a core tube with a steel truss on the top; the secondary structure is a segment of suspended discrete modules. The modules are vertically series-connected, by a chain of steel tubes with pin connections, at their column-ceiling beam joints. Mechanical springs are installed in the inter-story gaps, thus providing lateral inter-story stiffness. Prefabricated floor cassettes with fuse-type connections are adopted as a path to transfer wind loads from the modules to the core tube; the connections are expected to break automatically during the earthquake, allowing floor cassettes to slide and relative motion to occur. The damage within the floor cassettes is rather limited and they can still serve as personnel passages during an emergency.

In this innovative structural system, the relative displacement between the primary and secondary structures is of large amplitude. Flexible pipelines and connectors are necessary to avoid piping damage during earthquakes. They are commercially available and widely adopted in base-isolated structures. Prefabricated modules are easy to combine and divide, and thus some of the modules can be cantilevered from the core tube but not suspended, in order to facilitate the piping. In other words, the kitchen/bathroom modules which are implemented with pressured piping are not allowed to sway, but the living room/bedroom modules which are implemented with non-pressured piping are allowed to.

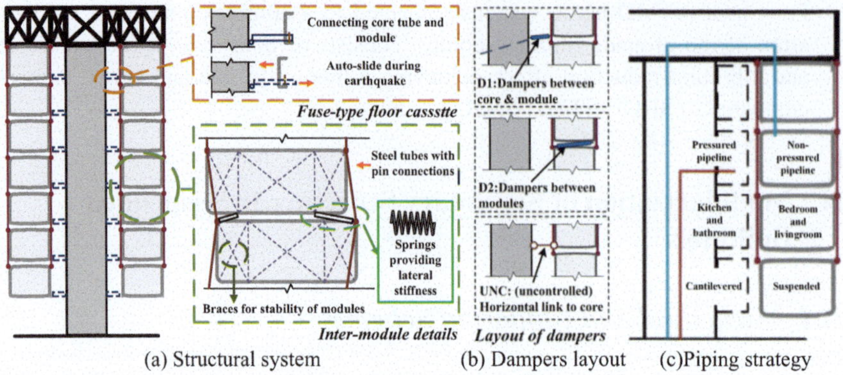

Fig. 8.17 The illustrative diagram of the proposed modularized suspended building [32]

The key details of the Chinese National Patent ZL201610596905.9 are shown in Fig. 8.18. Besides the aforementioned details, a limiting buffer device 7 is arranged between the vertically adjacent modules 3 and helps avoid the modules from a vertical collision when the sway is large. The A-shaped rod members 8 play the role of a leverage, which aims at reducing the elongation of the steel wire rope to avoid premature rupture. The ceiling beam is implemented with a stiffening rib 12 at the intersection with the module columns. The hanging rod member 4 is a steel tube with a hollow circular section. Its ends are connected to universal joints 13, which are connected to the web of a T-shaped steel plate 14 through a bolt. Such a T-shaped steel plate is an extension of the ceiling beam which plays the role of allowing the hanging rod member 4 to swing in a certain range without collision. A prefabricated floor slab 15 is arranged between each module 3 and the core tube, so as to retain the access of personnel while the module and the core tube structure move relatively. Channel steel 20 is pre-buried in the prefabricated floor slab 15 after being connected to a steel plate 19 which is subsequently welded to the beam-column node of the module on site. The other end of the prefabricated floor slab 15 is placed on a groove of a core tube floor slab 17, with each of the two sides of a contact surface being a pre-buried contact steel plate 21. The vertical end face of the groove is provided with a pre-buried anti-collision steel plate 22. All the aforementioned pre-buried members have pre-buried anchor bars 23.

8.4.2 Protection Effects for Non-Structural Components

If a flexible suspended segment is directly connected to the primary structure, the relative motion is developed through deformation within the suspended segment. However, non-structural members within multi-story suspended segments are susceptible to inter-story drift. For example, racking-type deformation can easily

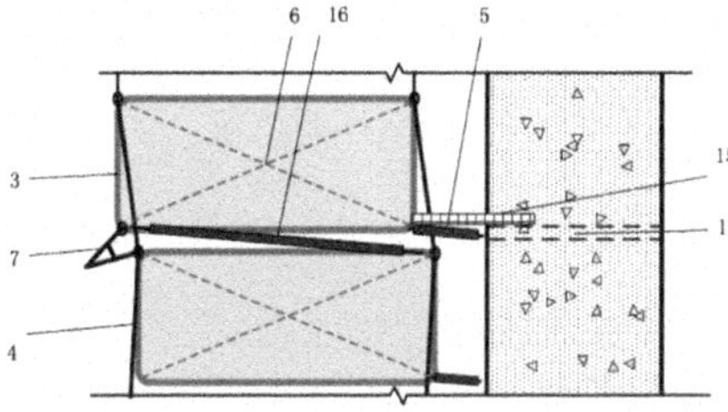

(a) Discretely suspended modules

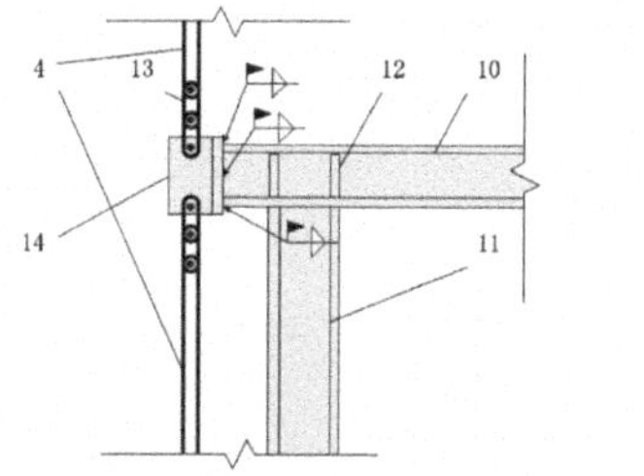

(b) Connection between suspended rods and modules

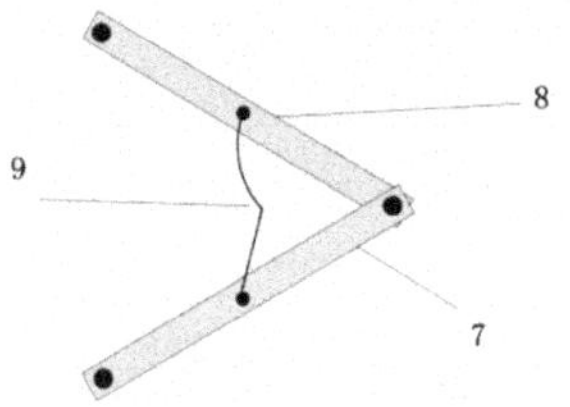

(c) "A-shaped" buffering device

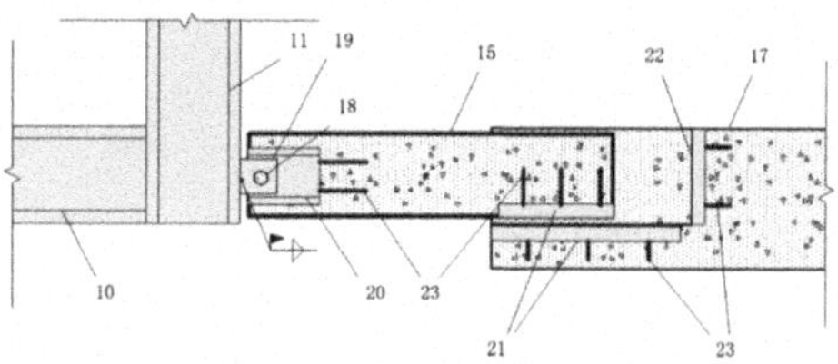

(d) Prefabricated sliding floor slab

Fig. 8.18 Key details of the proposed system (3—module, 4—hanging rod member, 5—damper, 6—diagonal supporting member, 7—limiting buffer device, 8—rod member, 9—steel wire rope, 10—ceiling beam, 11—module column, 12—stiffening rib, 13—universal joint, 14—T-shaped connecting steel plate, 15—prefabricated floor slab, 16—module inter-story damper, 17—core tube floor slab, 18—high-strength bolt, 19—connecting steel plate, 20—pre-buried channel steel, 21—pre-buried contact steel plate, 22—pre-buried anti-collision steel plate, 23—pre-buried anchor bar)

cause the brittle failure of the curtain wall at 1% drift, as shown in Fig. 8.19a. As a result, an inter-story drift limit is therefore set far lower than the optimum in terms of seismic control. This issue affects both displacements of the primary structure and acceleration within the secondary structure.

The mechanism of the protection effect provided by suspended discrete modules is shown in Fig. 8.19b. With connections between modules being flexible and modules themselves being relatively stiff, a minor part of drift, which is defined as harmful drift and denoted as a_1, occurs inside modules, whereas a major part of drift, which is defined as harmless drift and denoted as b_1, occurs in the connections. However, in the original type-2 solutions, the entire drift, which is denoted as a_0, is harmful, as it occurs inside modules and affects nonstructural members. Thus, the system can reduce harmful drift when total drift is kept constant, or extend the drift limit when the allowable harmful drift is kept constant.

If 0.4 m free space is provided between the upper and lower modules, the total story height being 4 m, the device occupation 0.3 m, and the free space 0.1 m, then the allowable inter-story drift can be as large as 22%, as shown in Fig. 8.19c. In applications, the specifications of inter-module connections including mechanical springs and dampers pose different demands for the minimum vertical space, possibly leading to an allowable inter-story drift of less than the aforementioned values.

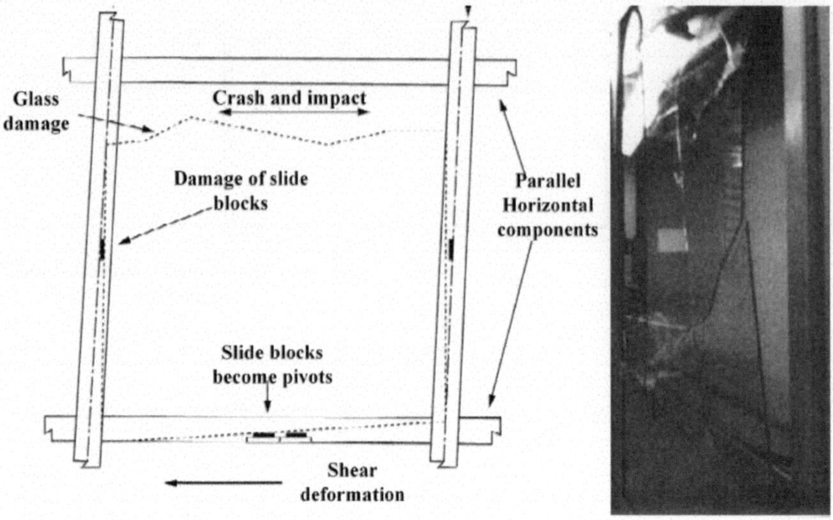

(a) Racking-induced damage of curtain wall [34]

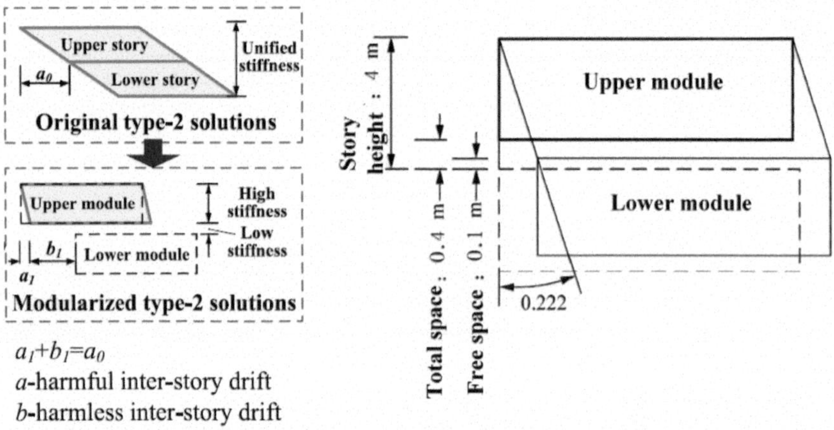

(b) Protection effect [32]

(c)Allowable inter-module drift

Fig. 8.19 The protection effect of the modularized secondary structure

8.4.3 Simplified Inter-Story Relation of Modularized Secondary Structures

In conventional frame structures or conventional suspended buildings with a frame-type suspended segment, the inter-story stiffness depends not only on structural components but also on non-structural components, such as in-fill walls, facades, and pipelines, etc. Such a complex inter-story relationship may bring about the following impacts:

(1) Horizontal and vertical irregularities. Although the structural components can be well designed to avoid any irregularity, the non-structural components can not. Usually, the utilitarian consideration causes a certain level of irregularities. For instance, the soft-story effect caused by an open first story and highly infilled upper stories, and the torsional effect caused by non-symmetric infill walls at the first story, can serve as ideal examples of irregularities.
(2) Short-column effect. Openings on the infill walls for windows and doors can cause a variance of effective length among the columns within the same story. Thus, the inner force can be much deviated from the original design and brittle failure during earthquake becomes possible.
(3) Difficulties in designing the modal characteristics. Even though the mass distribution is more or less controllable, the stiffness feature may be much perturbed by non-structural components, leading to unclear modal characteristics of the structural system. Thus, the optimization of the dampers and their layout could be based on an inaccurate model.

Contrarily, simple inter-story relations of modularized secondary structures are beneficial and cause no aforementioned problems. Since the modules are suspended in a discrete manner, the inter-story stiffness is provided by the inter-module springs and the gravity-induced pendulum stiffness, both of which are easy to determine. Accurately determining inter-story stiffness is a prerequisite for further optimization of the system.

8.5 Mechanic Characteristics of Suspended Building

8.5.1 Mechanic Characteristics of Primary Structure

Suspended building structures belong to primary-secondary passive-control structural systems and the responses of the primary structure are among the prioritized objectives of the seismic control. Mechanic characteristics of the primary structure include the following:

(1) High axial loads and notable P-delta effect. Since the gravity loads of the secondary structures are first delivered upwards, all the vertical loads go through the primary structures before being delivered to the foundation, leading to high axial loads of the primary structure. P-delta effect can be notable if a single core-tube is adopted as the primary structure, posing a greater challenge to the overturning issue.
(2) Lower robustness in the vertical direction. Due to the aforementioned pattern of vertical load delivery, it is recommended that higher redundancy is designed for individual components. Progressive collapse should be avoided at its origin.
(3) The very bottom section is the critical section. Usually, there are multiple connections between the primary and secondary structures and thus the vertical

distribution of the inner forces of the primary structure may confront abrupt changes. However, in most cases, maximum inner forces occur at the very bottom section. Although the phase angles of multiple inputs from the secondary structures to the primary structure may be diverse at certain frequency ranges, the accumulation effect prevails when the structure is subjected to earthquake excitation with various frequency components.

(4) Notable moment at the top stories of the primary structure. Usually, a core tube with flexural-type deformation is adopted as the primary structure. Thus, the rotary acceleration of the transfer truss on top is coupled with the rigid-body inertia of the secondary structure, inducing notable moment input to the top stories of the primary structure.

Although flexible connections between the primary and the secondary structure are widely adopted, such flexibility occurs only in the vertical direction. It is obviously difficult to implement vertical flexible connections when the suspended part is of large mass. Alternatively, Y. Nakamura developed a double-layered rubber bearings matrix to form multiple modes of relative motion in order to decouple the aforementioned undesirable coupling effect and further reduce the horizontal-excitation-induced vertical responses of the secondary structures [12]. However, similar devices are very delicate and not easy to implement when the suspended segment is massive. Thus it is not widely applicable.

Overall, the control objective should be set as the maximum displacement of force response among all the stories of the primary structure, especially when a single core tube is adopted as the primary structure. The primary structure responses are much more important than those of the secondary structures considering the primary structure is the last defense line with low robustness and high axial loads.

8.5.2 Layout Patterns of Dampers

The layout pattern of dampers is a major factor for the attenuation effect of the scheme. If a single core tube is adopted as the primary structure, a layout of dampers within the primary structure can be regarded as inefficient due to the flexural-type deformation of the core tube. Recommended layouts of the dampers are shown in Fig. 8.20.

Layout A and B are the most common layouts which are able to harness the relative motions between the two parts of the system, regardless of the configurations of the secondary structures. Thus, with Layout A or B, higher inter-story stiffness within the secondary structure can be set to prevent excessive deformation and non-structural damage. Type-1 primary-secondary passive-control structural systems (Appendix) serve as examples in this regard. In an extreme case, even if the secondary structure and its interface to the primary structure are both very stiff, dissipation takes place as long as the core-tube bends.

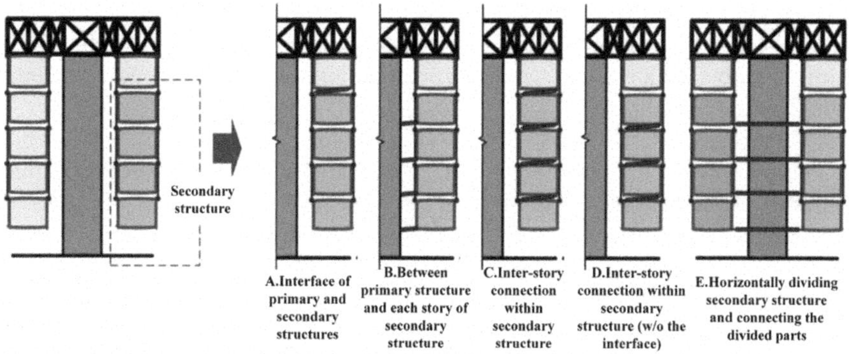

Fig. 8.20 Recommended layout patterns of dampers

Layout C and D are uncommon layouts in applications. The dissipation of the inter-story dampers depends on the amplitude of the inter-story drift which causes non-structural damage within the secondary structures if the secondary structures are of conventional type. Thus, the modularization of the secondary structures is the prerequisite of the Layout C and D. The unique feature of Layout C and D is that they better harness the higher-mode deformation of the secondary structures. The driving force input to the primary structure is reduced due to fewer connections to the primary structure. Layout D removes the dampers in the interface between the two parts and further reduces such driving forces when compared with Layout C.

Type E layout requires a division of the secondary structures within the same story and it harnesses the different features of the substructures and the consequent relative motion among these substructures. Analysis results show that Type E has a similar attenuation effect to that of Type D but less demand for dampers. However, extra fuse-type floor cassettes are needed among the substructures.

8.5.3 The Handicap of Secondary Structure Inter-Story Drift Limits

The flexibility of the secondary structure is the prerequisite of tuning and dissipation. While deformation within the secondary structure is a major way to fulfill the flexibility, such a deformation is the origin of non-structural fragilities which is sensitive to the inter-story drift. Limits are often set on the inter-story drift to protect the non-structural components, leading to restrained relative motion between the two parts of the system, and thus, a handicap to the attenuation effect.

Relative motion between the two parts of the system generally includes the following three types. Firstly, inherent relative motion occurs even if the secondary structures are rigid and are rigidly connected to the primary structure. The flexural-type deformation of the primary structure causes rotation of the secondary structure and forms the relative motion. Secondly, relative motion can occur via the deformation in the interface between the two parts of the system (type-1 as in Appendix). Thirdly, relative motion can occur via the deformation within the secondary structure (type-2 as in Appendix).

If the relative motion of the two parts is developed via the deformation within the secondary structure, multiple modes of the secondary structure take part in dynamic responses. These multiple modes provide a good space for a more delicate design to achieve better passive control. If the inter-story drift within the secondary structure is subjected to a limit, the following undermining effects exist: (1) the amplitude of relative displacement is reduced and so is the dissipation by viscous dampers. If the damping coefficient is increased to compensate for the dissipation, the relative motion is further restrained and the complex mode shapes become undesirable. (2) Certain layouts of dampers (C and D) become not applicable due to the trivial strokes. (3) The modal frequencies of the secondary structure become higher and the tuning relationship between the primary and secondary structures becomes unsatisfactory. (4) A stiffer secondary structure suffers more from the coupling effect between its rigid body motion and the rotation of the transfer truss. The demand on secondary structure accelerations, moments in top stories of the primary structure, and moments in the transfer truss will all be increased.

In short, relaxing the limit on the inter-story drift within the secondary structure and alleviating the aforementioned handicaps are highly beneficial to the passive-control performance of the system. So far, the modularization of the secondary structure is presumably a prerequisite of further improvement of the seismic control performance.

8.6 Genetic Algorithm Optimization of Seismic Control Performance of Modularized Suspended Buildings

The inter-story stiffness of the secondary structures and the damping coefficients of the supplemental dampers are the most important parameters, in terms of the seismic control performance of modularized suspended building. Therefore, calculating the optimal values of the aforementioned parameters using genetic algorithms is necessary for both the fair comparison among structural systems performance and the precise presentation of seismic control mechanisms at the near-optimal level. The MOGA-II algorithm [35], which is a widely-applied genetic algorithm, is adopted in this study to reveal the advantage and the corresponding mechanisms caused by the relaxed inter-story drift limit and the allowed vertical irregularity of the secondary structure parameters which are brought by the modularization.

8.6.1 Setting of the Optimization

Figure 8.21a, b depict the calculation model while the illustration of sub-model categorization is shown in Fig. 8.21c. Ten stories of consecutive elastic beam members represent the core tube and a one-story steel truss is located on the top of the core tube, adding up to the total of 44 m. The secondary structure is one-segment nine-story suspended rigid modules. Interconnections of a pair of upper and lower modules are horizontally flexible but vertically stiff. The bottom modules are connected to the core tube by horizontal links. The secondary structure exhibits only shear-type deformation. Three controlled models were set up for comparison between the type-1 (or the "block-type") and the type-2 (or the "chain-type") strategies of the secondary structures. As shown in Fig. 8.21, (a) the stiff inter-story connection (SI) model represents the type-1 solutions; (b) the flexible inter-story connection-1 (FI-1) and the flexible inter-story connection-2 (FI-2) models represent the modularized type-2 solutions; the difference is that the FI-1 model has dampers connecting each module to the core tube, and the FI-2 model has inter-story dampers. For the FI-2 model, dissipation occurs only inside its secondary structure, because there are no dampers connecting the top two or the bottom two modules, which are the interfaces between primary and secondary structures. (c) An uncontrolled model (denoted as UNC), with no lateral relative motion allowed, is also set.

The concerned responses in the following subsections include: (a) the maximum mean square value of the inter-story drift of primary structure (MMSDP, defined by Eq. 8.41; (b) the maximum mean square value of the moment within primary structure (MMSMP) (c) maximum mean square value of the inter-story drift of secondary structure (MMSDS); and (d) maximum mean square value of the relative displacement between primary and secondary structures (MMSUR).

$$\mathrm{MMSDP} = \max_i \sigma_{\mathrm{dp},i}^2 = \max_i \int_{-\infty}^{\infty} h_{\mathrm{dp},i}(\omega)^* \cdot h_{\mathrm{dp},i}(\omega) S_g(\omega) \mathrm{d}\omega \tag{8.41}$$

wherein $h_{\mathrm{dp},i}(\omega)$ denotes the transfer function of the drift of the ith floor in primary structure, while $S_g(\omega)$ denotes the power spectral density function of the ground motion.

8.6.2 Advantageous Mechanism of Optimized Modularized Secondary Structure with Inter-Story Dampers

Assuming the inter-story drift within the secondary structure is no longer a concern due to modularization, comparison among the type-1(SI models) and type-2 (FI models) suspended buildings is the focus of this section. The reasons are not only that different configurations have different control mechanisms, but also that type-1

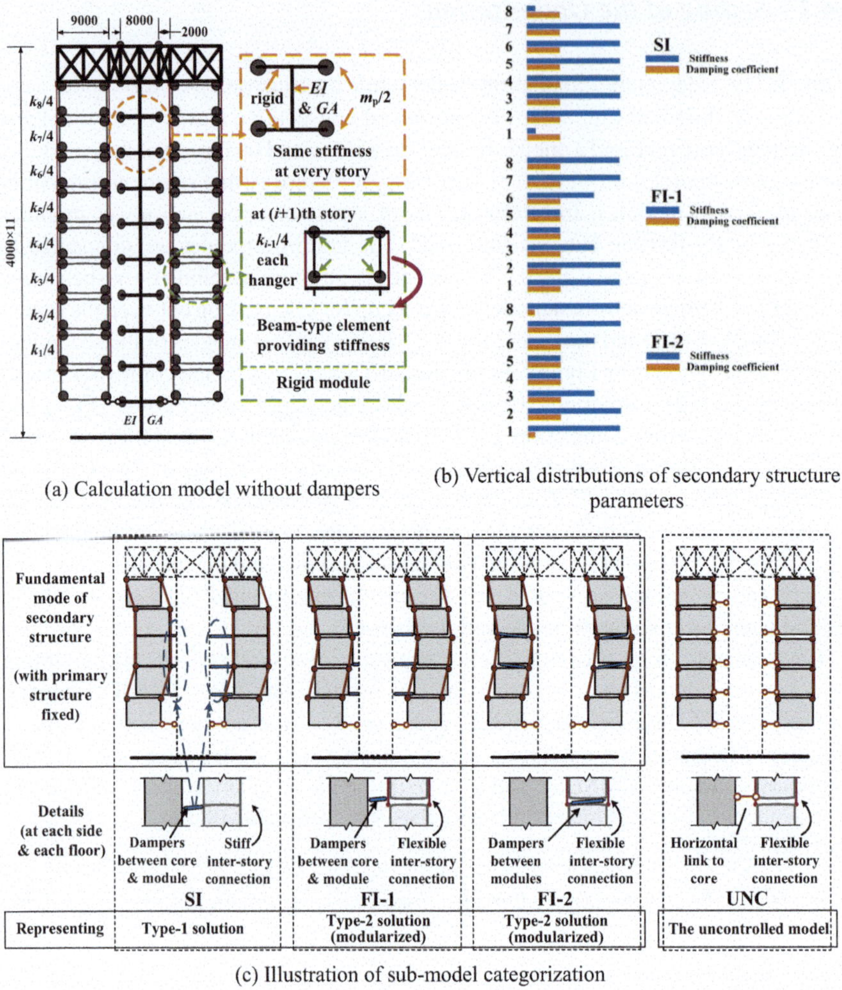

Fig. 8.21 Calculation model and illustration of sub-model categorization [32]

suspended buildings can alternatively be achieved without modularization, representing less features of modularized suspended buildings. Single-objective optimizations aiming at MMSDP have been conducted. The optimal MMSDP values are listed in Table 8.3. It is shown that the FI-2 model has a stable advantage over the SI model in all of the scenarios while the FI-1 model has a similar performance as the SI model.

Advantageous mechanisms of the FI-2 model are discussed from the perspectives of both the transfer functions and the complex modes. The MMSDP transfer functions are shown in Fig. 8.22a. Firstly, all passive-control models outperform the UNC model in terms of the MMSDP transfer function, especially around the fundamental

Table 8.3 Single-objective optimal MMSDP ($\times 10^{-7}$ m^2)

Mass ratio R_m	Soft soil			Firm soil		
	SI	FI-1	FI-2	SI	FI-1	FI-2
0.5	2.8	2.3	2.0	2.8	3.1	2.4
1	4.0	3.5	2.5	2.9	3.1	2.4
2	8.5	6.9	4.3	3.6	3.3	2.6
4	12.9	13.9	8.2	4.0	3.8	3.2

frequency of the UNC model (0.85 Hz), where the original peaks become troughs and where dominating first-mode response of the primary structure takes place. Secondly, different passive-control models show different mechanisms which can be revealed in the low-frequency segments of the transfer functions and the shapes of major complex modes. The low-frequency segments of the MMSDP transfer functions are shown in Fig. 8.22b, c. FI-2 has similar transfer function curves for the two soil conditions. However, for SI and FI-1 models, optimized transfer function curves have deeper troughs but higher peaks under the soft soil condition, compared to those for firm soil condition. In order to evade the narrow-band excitation, the troughs always correspond to the peak of the PSD function. The FI-1 model has multiple peaks and does not have a sole trough in most cases. The FI-2 model has a single wider and lower trough with lower peaks on both sides, when compared with the SI model. The wider trough of FI-2 results from the existence of multiple modes of the secondary structure which allows a tuning between a higher mode of the secondary structure and the first model of the primary structure. Thus, the frequency difference between the first and second peak is larger. The lower peaks and trough of FI-2 results from the dissipation of inter-module dampers excels in higher modes of the secondary structure.

The major complex modes of the optimized models are shown in Fig. 8.23. Herein, the numbers of the included modes are three, six, and three, for the SI, FI-1, and FI-2 models respectively. For the SI and FI-2 models, the two major peaks in Fig. 8.22a correspond to the first two modes. However, for the FI models, the second major peak does not necessarily correspond to the second mode but one with higher-mode response of the secondary structure instead. Thus, the width of the trough is able to become wider, as the result of optimization. Modes around 1 Hz show a high level of complexity, with considerable amplitude of imaginary component and different shapes between real and imaginary components. Because of the higher-mode response of the secondary structure, the inter-story dampers of the FI-2 model over-damp the modes of minor peaks and make the second major peak lower, as indicated by the real part of the eigenvalue.

Key mechanisms that are revealed in the transfer function and complex mode analyses include the following. (1) For the FI models, multiple modes of secondary structure are remained. The optimization actually chooses one of the higher modes of secondary structure, and tunes it to the first mode of the primary structure. Thus, the first two peaks of the transfer function have a larger frequency distance and a wide

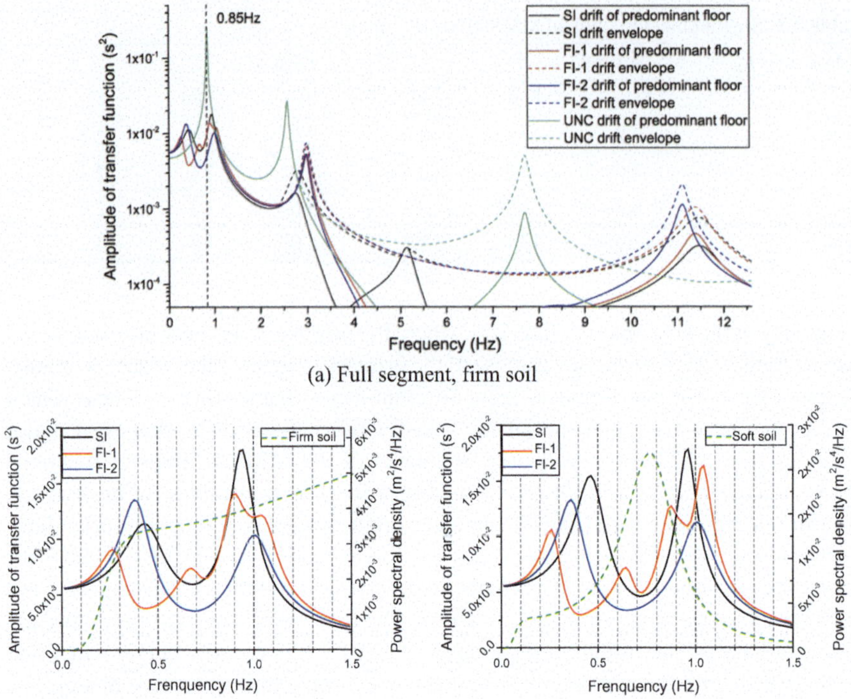

(a) Full segment, firm soil

(b) Low-frequency segment, firm soil

(c) Low-frequency segment, soft soil

Fig. 8.22 Optimized transfer functions of inter-story drift and PSD functions of excitations (Nominal mass ratio = 2)

trough is formed. (2) Inter-module dampers are adopted in the FI-2 model, leading to a better dissipation of the higher-mode energy within the secondary structure. One consequent feature is that the intermediate peaks are damped out, leading to an intact trough between the two major peaks. The other feature is that the mode corresponding to the second peak shows a considerable real part of the eigenvalue, indicating efficiency of the dampers, and it leads to a relatively low second peak.

The aforementioned features and mechanisms are all related to the modularization of the secondary structure; they are revealed not only in single-objective optimizations but also in dual-objective optimizations. The result of multi-objective optimization is a Pareto front whose individual element reaches Pareto optimality. Pareto optimality is a situation where no individual can be better off without making at least one individual or preference criterion worse off. Figure 8.24 shows the MMSDP-R_c Pareto fronts under both soil conditions and corresponding transfer functions when $R_c = 1$. FI-2 model has inter-module dampers whose stroke demands are very low. Thus, it requires a notably larger optimal damping coefficient than the others. As the damping coefficient becomes smaller in the Pareto fronts, the performance of FI-2 drastically worsens. However, as the damping coefficient becomes very small, the

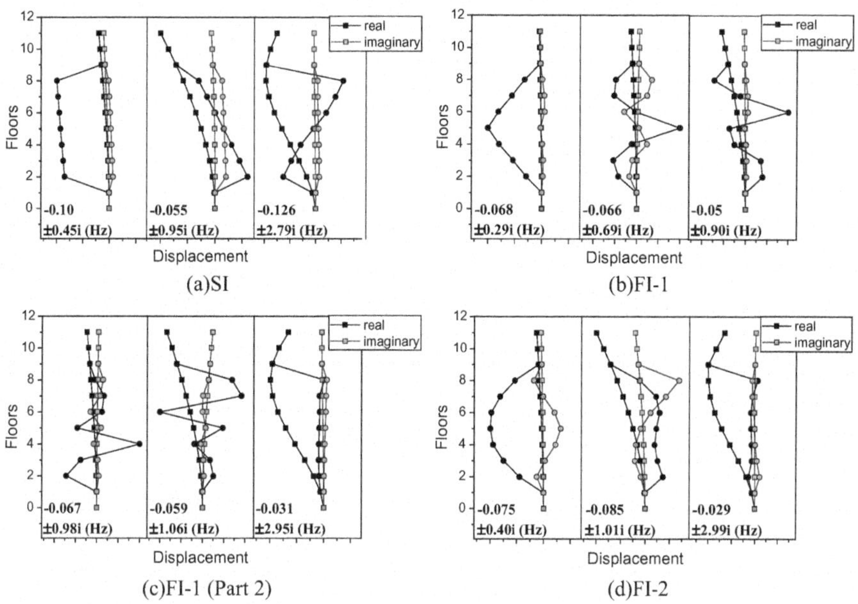

Fig. 8.23 Complex modes of optimized models with notable contribution to low-frequency peaks in transfer functions (firm soil condition, nominal mass ratio = 2)

FI-2 model regains its advantage over other models. Corresponding transfer functions when $R_c = 1$ under firm soil condition are shown in Fig. 8.24b. When $R_c = 1$, the frequency values of the peaks are mostly the same as those of sufficient damping values, only that the peaks are narrower and higher. For the FI-2 model, intermediate peaks previously corresponding to the overdamped intermediate modes are re-emerged. Also for the FI-2 model, the peaks are slightly shifted towards lower frequencies, indicating a tuning between an even-higher mode of the secondary structure and the first mode of the primary structure, in order to compensate for the insufficient dissipation brought by low damping coefficients.

8.6.3 Optimization of Vertical Distributions of Secondary Structure Parameters

Due to the module protection effect and the simplified inter-module relationship mentioned respectively in Sects. 8.4.2 and 8.4.3, the irregular vertical distribution of the secondary structure parameters is allowable in modularized suspended buildings. Therefore, further optimizations have been conducted based on the aforementioned features, in order to attain better modal characteristics and multi-mode tuning effect between the primary and the secondary structures [33]. Two scalars and two

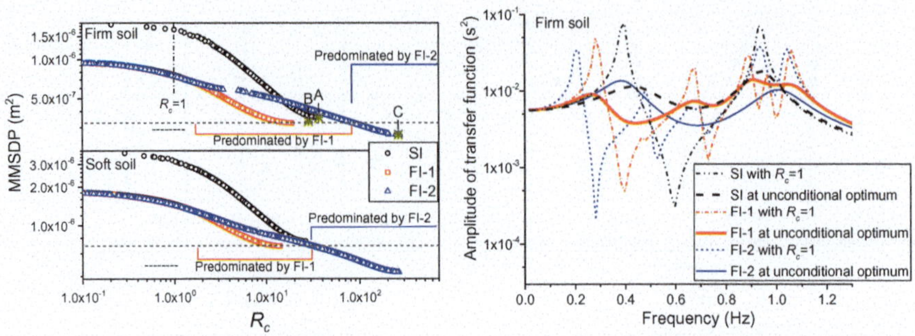

(a) MMSDP-R_c Pareto fronts

(b) corresponding transfer functions when R_c =1 (firm soil condition)

Fig. 8.24 MMSDP-R_c Pareto fronts under both soil conditions and corresponding transfer functions when $R_c = 1$ under firm soil condition ($R_m = 2$). (MMSDP = maximum mean square value of the inter-story drift of primary structure, R_c = damping coefficient scalar.)

vectors are set as the optimizing variables, including the inter-module stiffness scalar, the dampers' damping coefficient scalar, and the corresponding vertical distribution vectors. The main objective of the optimization is the Maximum Mean Square Moment of the Primary Structure (MMSMP). The FI-2 configuration is focused on in this section. We set six different levels of the vertical distribution vectors, as shown in Table 8.4, to account for the effect of different degrees of vertical irregularities, considering the constraint on numbers and available specifications of springs and dampers. In the naming of the models, VD is short for "vertical distribution".

As shown in Fig. 8.25, allowable irregularity levels have a notable influence on the results of single-objective optimizations. As the irregularity level increases, the moment in primary structure decreases, while the secondary structure response depends on the type of constraints on the distribution vectors. The amplitude-type constraints on the distribution vectors are beneficial to the secondary structure response (MMSDS, MMSUR) while the level-type have the opposite effect. The VD-AMP high model can attain the best balance between the primary and the secondary

Table 8.4 Allowable range of elements in distribution vectors

Vertical distribution level		A_k for $r_{k,i}$	A_c for $r_{c,i}$
NVD	(Non-vertically distributed)	Constant*	Constant**
VD-AMP low	(Low allowable amplitude)	[1, 3]	[1, 5]
VD-AMP high	(Higher allowable amplitude)	[1, 15]	[1, 20]
VD-Bind	(Direct binding)	$\{1, +\infty\}$	constant**
VD-LEV	(Scattered levels of allowable values)	{1, 50, 100}	{1, 225, 450}
VD	(Full vertical distribution)	[1, 100]	[1, 450]

* Constant vector $\mathbf{r}_k = \{3, 3, 2, 1, 1, 2, 3, 3\}^T$
**For FI-2 model, constant vector $\mathbf{r}_c = \{0, 1, 1, 1, 1, 1, 1, 0\}^T$.

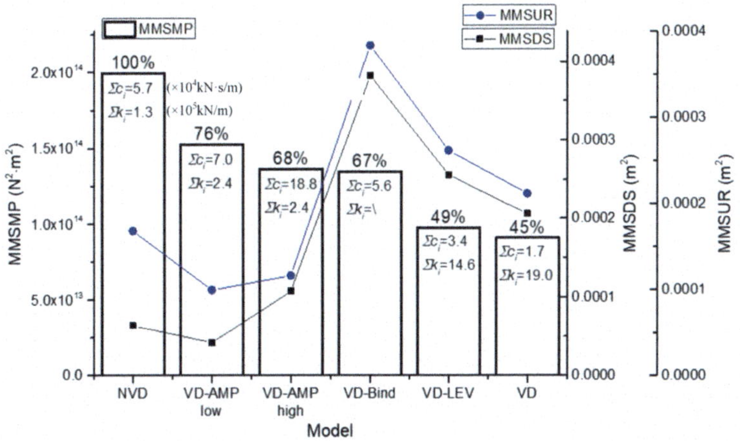

Fig. 8.25 Responses in single-objective optimization of the FI-2 models (nominal mass ratio = 2). (*MMSMP* Maximum mean square moment of primary structure, *MMSDS* Maximum mean square drift of secondary structure, *MMSUR* Maximum mean square relative displacement between the primary and secondary structures)

structure response, while VD-AMP low model has a similar performance but requires much fewer dampers than the VD-AMP high model.

The optimal distribution vectors of the optimizing variables are shown in Fig. 8.26a. The VD models experience drastic changes of the distribution factors only at lower stories. The upper stories have higher stiffness (except for the top) and very low damping coefficients, while high damping appears in lower stories. In terms of stiffness and damping coefficient, the peaks of one parameter tend to correspond to the other's troughs. At the medium levels of irregularities, the aforementioned trends of distribution vectors are weakened. The LEV model exhibits distributions similar to those of the VD model, while the AMP low model better obeys the aforementioned "peak to trough" trend. Thus, the LEV model has a lower primary response, and the AMP low model has lower secondary responses. Although the LEV and VD models show a similar "bound block" in higher stories, the Bind model shows a different pattern of binding due to its pre-set distribution of dampers, indicating the strong interaction between stiffness and dampers.

Frequency-domain transfer functions of the optimized models in single-objective optimizations are shown in Fig. 8.26b. All models with vertical irregularities tend to have lower peaks instead of deeper troughs of the primary structure moments when compared with the NVD model. They show similar trends with different amplitudes. However, only the Bind model fails to form its first trough around 0.7 Hz due to its rigidity and only the VD model manages to form two minor peaks around 3.2 Hz showing the best multi-mode control effect. The three targeted responses in Fig. 8.26b have their peaks emerge at similar frequencies. That results from the fact that each peak is dominated by one major mode, which has considerable responses in both parts of the structure. Compared with the NVD model, the VD and LEV models experience

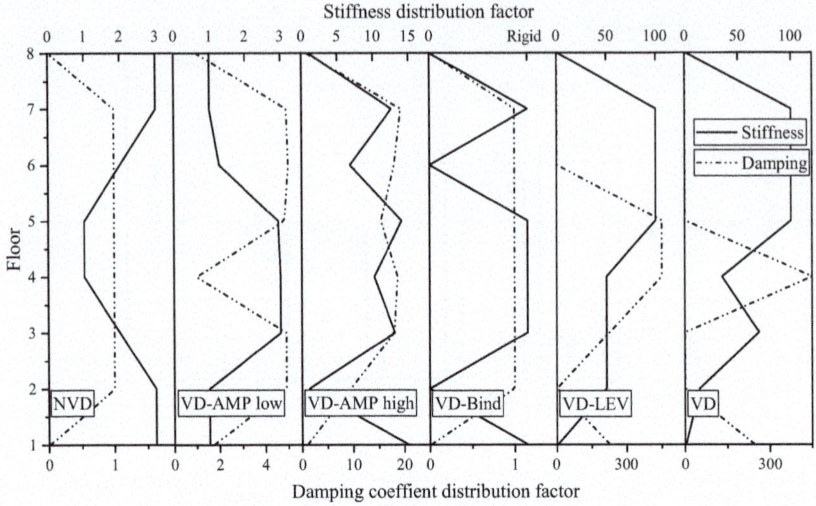

(a) The distribution vectors of the optimizing variables

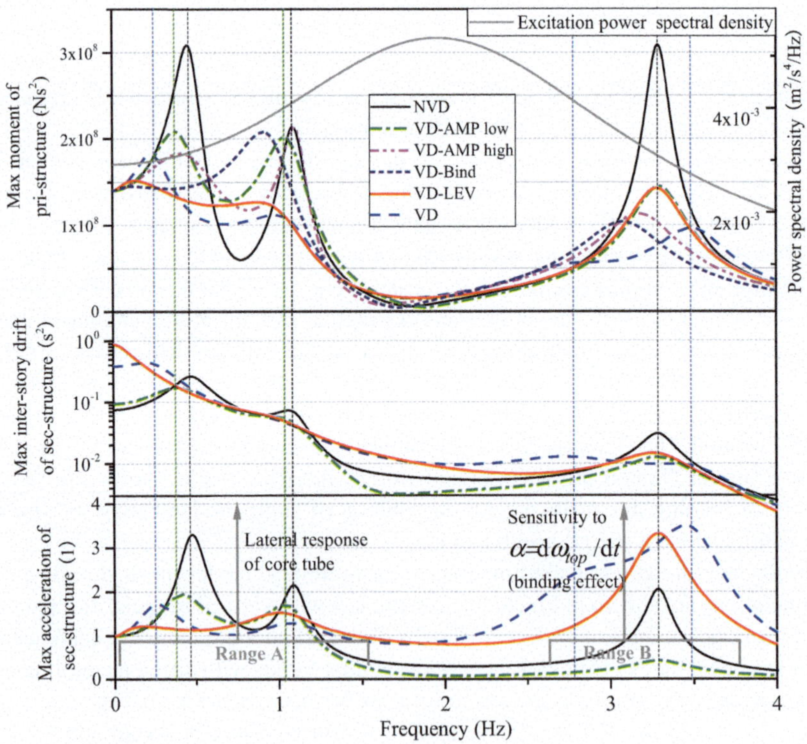

(b) Frequency-domain transfer functions

Fig. 8.26 The distribution vectors of the optimizing variables and transfer functions of different models in single-objective optimizations larger secondary drifts in the low-frequency range and larger secondary accelerations at higher frequencies, while the AMP low model has appealing secondary responses in the entire range.

8.6.4 Time-Domain Performance Verification of Optimized Models

Figure 8.27 shows the time-history responses under the El Centro wave. In terms of the base moment, the passive-control models all show features of lower peak and quicker decay, when compared to the UNC model. The peaks of the passive-control models range from 21 to 33% of that of the UNC model. Among the passive-control models, the VD-LEV model has the lowest peak while the VD model has the lowest RMS value of time-history. Compared with the UNC model, the VD-AMP models show a slight advantage in the base moment, but a notable advantage in the module accelerations. The aforementioned observations in the frequency-domain are thus verified.

The non-stationary dynamic features of the optimized models can also be investigated in a statistical perspective. 400 artificial ground motions, which are non-Gaussian, are generated, using an orthogonal expansion of a basis of random trigonometric functions. The stationary Clough-Penzien type power spectral density function was first adopted to generate a stochastic process, before a shaping function defined how frequency contents evolve over time. The resulting response features under the excitation of these 400 artificial ground motions are shown in Fig. 8.28. As for the moment at the base of the primary structure, the ranking is UNC/NVD/VD-AMP high/VD in terms of peak amplitude. The superiority of the VD model over the NVD model is consistent with the aforementioned findings. Additionally, the quick-decay feature of the passively-controlled models showed in the non-stationary analysis, as earlier peaks and shorter durations of vibration can be observed when compared with the UNC model (whose vibrations last much longer than the excitations). These differences are also shown between the VD model and the UVD model but with smaller margins. Among the cumulative probability curves, the dash-dot lines being far from the others indicates that the decay is quick for the passively-controlled models; however, for the UNC model, the response first evolves to be higher before slightly drops at the 30th second.

8.7 Shake-Table Testing of Modularized Suspended Building Structures

To validate the passive vibration control features of the proposed modularized suspended building, a series of shake-table tests have been conducted on the 1.5 ×

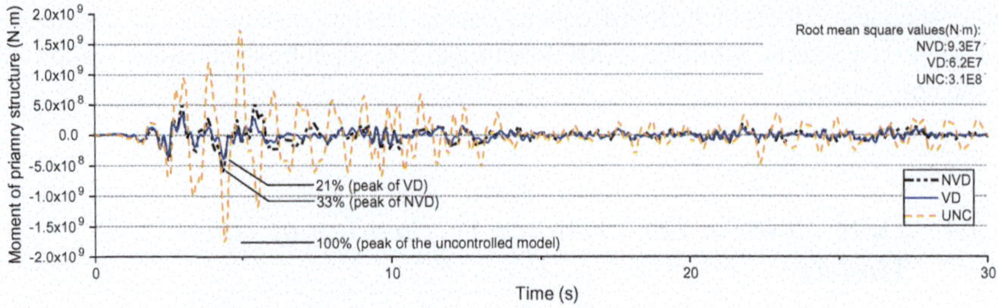

(a) Base moment of the primary structure (with comparison to the UNC model)

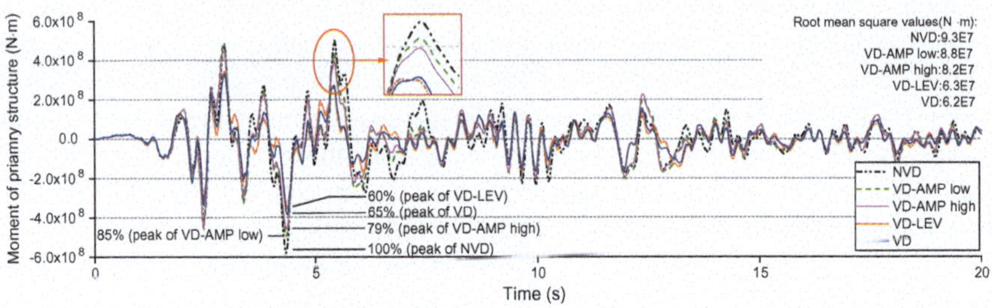

(b) Base moment of the primary structure (without comparison to the UNC model)

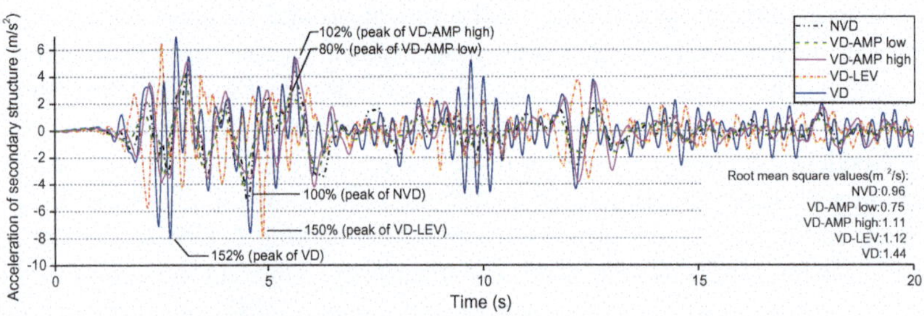

(c) Acceleration of the secondary structure (without comparison to the UNC model)

Fig. 8.27 Time-history responses under the El Centro wave

1.5 m^2 shake table (maximum force: 2 ton, maximum acceleration: 1 g) in the IIUSE lab in Nanjing, China. Results show the satisfactory performance of the structure system [36].

8.7.1 Experiment Set-Up

The 1:15 T-shape steel structure specimen with two suspended segments of modules, as shown in Fig. 8.29, was fixed on the sake table. The total weight of the specimen is 260 kg, among which the 1410 mm-long primary column weights 52 kg, the 1290 mm-wide transfer beam 72 kg, and the suspended modules 94 kg. The sections

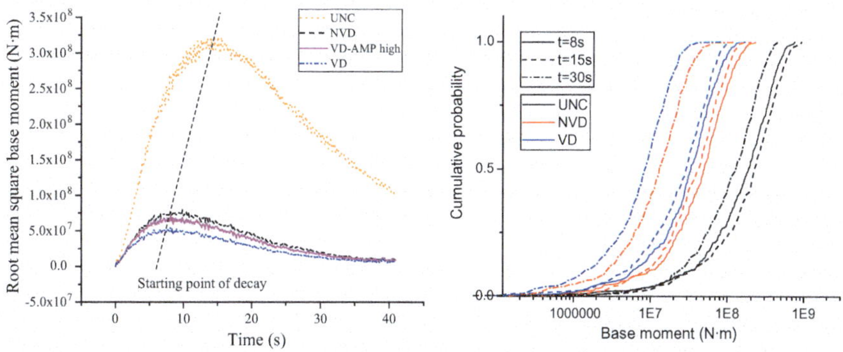

(a) The root mean square curves of the base moment time-histories

(b) Evolutionary cumulative probability curves of the base moment

Fig. 8.28 The root mean square curves and the evolutionary cumulative probability curves of primary-structure base moment

for the primary column and the transfer beam are HN100 × 50 × 5 × 7 (H Section-Narrow Height-Width-Web-Flange) and HW125 × 125 × 6.5 × 9, respectively (according to Chinese code GB/T 11263–2017). The sections are selected according to the desired cross-section *EI* value. To attain the desired flexibility with limited choices of cross-section specifications and similarity ratios, the column section has its flange parallel to and its web perpendicular to the excitation direction. An end plate and a series of stiffeners were used to fabricate high-strength friction grip bolted connections between the column and the transfer beam. Additional stiffeners were welded to the transfer beam to achieve higher stiffness and stability in suspending a chain of modules. A C20A steel is bolted to each end of the transfer beam; two 16 mm rods are inserted through four holes on the C20A steel and corresponding holes in four 6 mm-thick steel strips; the strips are inserted by two rods at its ends and the rods suspend a module via rod bearings and connection pieces bolted on the module. Therefore, the upper and lower strips on the same rod can rotate respectively to allow overall sway; a chain of suspended discrete modules with pin connections is thus formed, in a planar way. For adjustable inter-module stiffness, mechanical springs are pre-stretched before they are pinned to connection pieces on upper and lower modules to avoid any relaxation during vibration.

The similarity ratios are listed in Table 8.5 and the key parameters of the prototype and specimen are listed in Table 8.6. The fundamental frequency is 0.68 Hz (2.73 Hz for the specimen due to a 1:4 similarity ratio) for the UNC configuration (The uncontrolled configuration).

Airpot dampers were adopted in this experiment to suit for the small-scale specimen, to avoid notable temperature-rising and performance deviation caused by a large number of stroke cycles, and to achieve smaller initial resistance which is required by the reduced time scale.

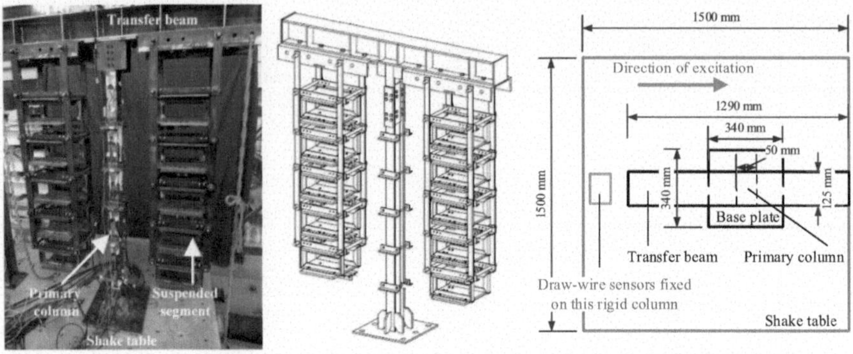

(a) Overall view of the test specimen (b) Layout of shake table

Fig. 8.29 Overall view of the test specimen and layout of the shake table

Table 8.5 Similarity ratios

Dimension	Similarity ratio	Dimension	Similarity ratio
Length	6.67E-02	Stiffness	7.50E-04
Time	2.50E-01	Damping coefficient	1.87E-04
Force	5.00E-05	Strain	1.00E+00
Acceleration	1.07E+00	Stress	1.12E-02
Mass	4.69E-05	Elastic modulus	1.12E-02
$E{\bullet}I$	2.22E-07	$E{\bullet}A$	5.00E-05

Table 8.6 Key parameters in the prototype and specimen

Key parameters	Prototype scale	Specimen scale
Story height (mm)	3000	200
Width of the transfer truss (mm)	1.88E+04	1.25E+03
Story mass (t)	5.00E+02	2.35E-02
Expected fundamental frequency of the UNC model (Hz)	6.75E-01	2.70
Expected damping coefficient of damper (N.s/mm)	9.50E+01	1.72E-02
Expected stroke of damper (mm)	1.50E+02	1.00E+01
Expected inter-story stiffness (N/mm)	9.00E+02	6.80E-01

Five ground motions (referred to as GM.1-GM.5) were chosen, listed in Table 8.7. They have similar intensities, different frequency contents, different far/near-field features, different mechanisms and the same site class (Site Class D).

The vertical distributions of the dampers are all uniform; those for the inter-module stiffness are approximately uniform. We set up different models by varying the damper coefficients and inter-model stiffness, as shown in Fig. 8.30. The "center model" of each group is set by pre-calculation and optimization. However, the

Table 8.7 Input ground excitations

Ground motion	Event	Station	Component
GM.1	ChiChi 1999	CHY101	EW
GM.2	Christchurch 2011	Pages Road Pumping Station	0
GM.3	Hollister 1961	USGS1028	270
GM.4	Imperial Valley 1940 (the El Centro)	USUG0117	180
GM.5	Loma Prieta 1989	CSMIP 47,381	90

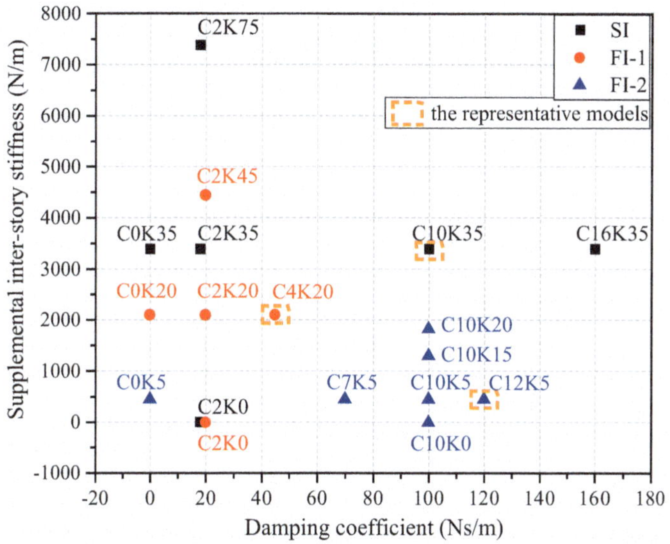

Fig. 8.30 Combinations of model parameters

samples in yellow squares in Fig. 8.30 are those with best experimental performance in each group and thus termed "the representative models".

8.7.2 *Experiment Result and Discussion*

Figure 8.31 shows the key performance ratios corresponding to the five ground motions. Each depicted ratio is the average among all models with the same configuration, compared with the UNC model. Reduced peaks and accelerated decay of the seismic-induced vibration are shown. The ratios for the peak value and the RMS of the top displacement, the peak value and the RMS of the acceleration of the bottom module are respectively around 0.6, 0.5, 0.55 and 0.6. In terms of the RMS value, the smaller ratio of the primary structure response indicates a quicker decaying

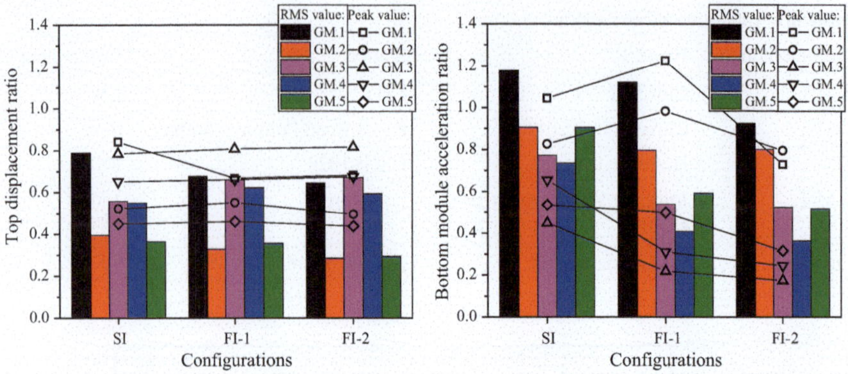

(a) Primary structure top displacement

(b) Secondary structure bottom module acceleration

Fig. 8.31 Overall performance in terms of average performance ratio compared to the UNC model

than that of the secondary structure. The two main factors for the slower decay of secondary structures are listed as follows. First, the secondary structures are isolated absorbers whose vibration may start later and end later than that of the primary structure. Second, the low-frequency contents of ground motions decay slower. Scattering exists among the performance subjected to different ground motions and it is larger in the acceleration indexes. However, even with this scattering, all the performance ratios for the primary structure response are way below one, while those for the secondary structure response are mostly below one. This validates the inherent vibration control feature of the system. However, configuration matters. Although all three configurations have similar levels of performance in terms of the peak value of the top displacement of the primary structure, for all the other indexes the FI-2 configuration shows its advantage.

Figure 8.32a shows the time-history curves of the top displacement of all the configurations when subjected to the 2011 Christchurch ground motion (GM.2). It can be observed that the UNC and the only-primary-structure (OP) configurations have very different dominant vibration frequencies compared to the vibration-control configurations; the UNC configuration has a lower frequency and the OP configuration has a higher one. Though the OP configuration exhibits a lower peak, both the OP and UNC configurations show slower decay when compared to the others because of insufficient dissipation. Whilst the vibration-control configurations have similar peak top displacements, the FI-2 model has the lowest peak by a small margin. On the other hand, notable differences in decaying features are revealed, subjected to this pulse-type excitation. The FI-2 configuration reaches its peak very quickly at 2.9 s, then it starts to decay drastically. This is partially due to the fact that the pulse contains a wide-band frequency content and thus induces both lower-mode and higher-mode deformations for the FI configurations (with flexible suspended segments). The SI model exhibits a very late peak at around 3.6 s, along with several peaks of similar amplitudes. Figure 8.32b shows the inter-story drift at each floor of the FI-2 configuration, revealing a high demand of 3.75% drift between the bottom two stories under the excitation with a 0.12 g PGA, thus proving the need for modularization. It can be observed that the drifts in the higher stories are activated earlier than those of the lower ones, as a result of the module's discretization. Asynchronous responses among stories gradually turn into synchronous ones, as the high-mode responses in the suspended segment are dissipated quicker than the lower-mode responses. Figure 8.32c shows the comparison among displacement time-histories of different models of the FI-2 configuration. Among the models with the same configuration, slightly different peak amplitudes with notably different decaying features can be observed.

The envelopes of the strains of the primary structure and those of the accelerations of the secondary structure are depicted in Fig. 8.33. Overall, for the vibration-control configurations, it is shown that the strains are all remarkably reduced along the height of the primary column and the accelerations of the modules are moderately reduced. The reductions of the strains of the 6th floor are less significant compared to those of the base. For the SI configuration, which has the block-type suspended segments without inter-module drift, the strains on the 6th floor are always larger than those of the FI configurations. This is also in accordance with the observation that it has a K-shaped distribution of the module accelerations, with obviously larger ones in the top and bottom modules, though those for the intermediate modules may be satisfactorily reduced. This indicates that the pattern of the suspended segment motion is mainly rotation combined with sway.

8.7.3 Experimentally Validated Numerical Modeling and Further Analysis

For the specimens, a numerical modeling strategy based on experiment phenomena is developed considering different friction features of different specimen parts and the Airpot dampers features, in order to create a solid base for further numerical investigation. A numerical model with beam elements, mass elements, and regional Rayleigh damping is combined with the proposed Airpot damper numerical model, before matching to the experiment data. The process of calibration is as follows: (1) adjust the mass distributions while keeping the total mass as measured, until the modal information is matched to the value measured using white noise excitation; then adjust the regional Rayleigh damping until the time-history response of the structure without dampers match the experiment data; (2) propose a physics-based model of Airpot dampers as depicted in Fig. 8.34, and calibrate parameters until the hysteretic curves match to the measured ones; (3) combine the structure model and the damper model, before comparing the simulated time-history curves with the measured ones.

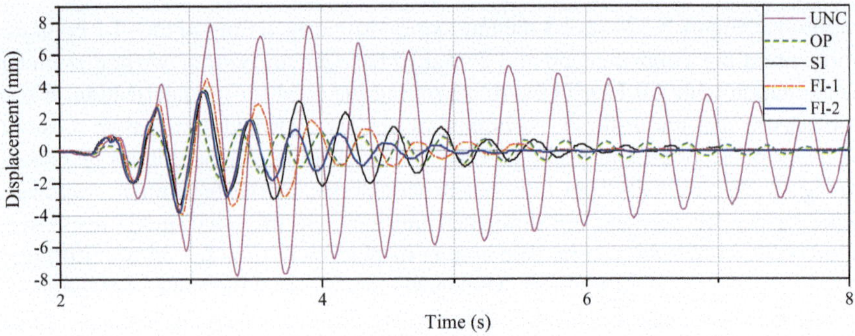

(a) Top displacement

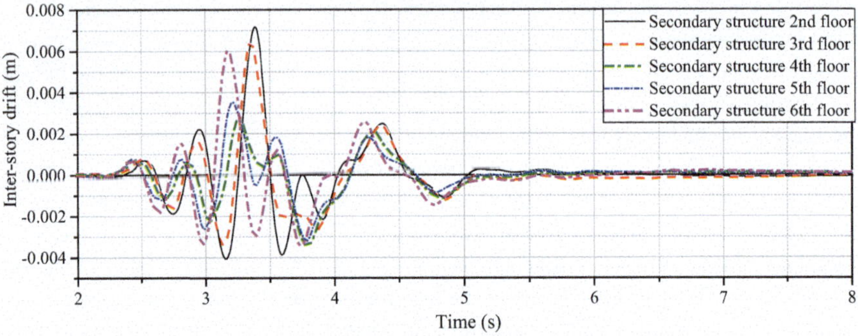

(b) Inter-story drift at each floor (FI-2 configuration)

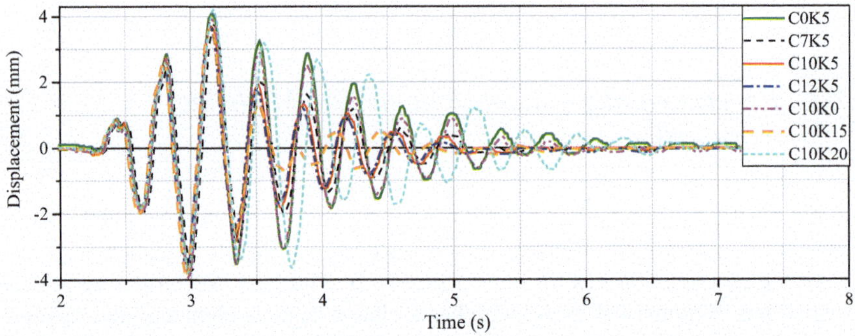

(c) Comparison among models of FI-2 configuration in terms of top displacement

Fig. 8.32 Time-histories performances of the best model in each configuration subjected to the 2011 Christchurch ground motion (GM.2)

To partially capture the physics feature of air, three elements are combined, as depicted in Fig. 8.35. Part A is a Maxwell element with a damping exponent of 0.6 and a high stiffness for the spring, which acts similar to a conventional liquid viscous damper. Part B is hyperviscous-Maxwell with a damping exponent of 1.5 and a low stiffness for the spring, accounting for the air spring effect. When the piston moves slowly, both parts work as viscous dampers and the combined output

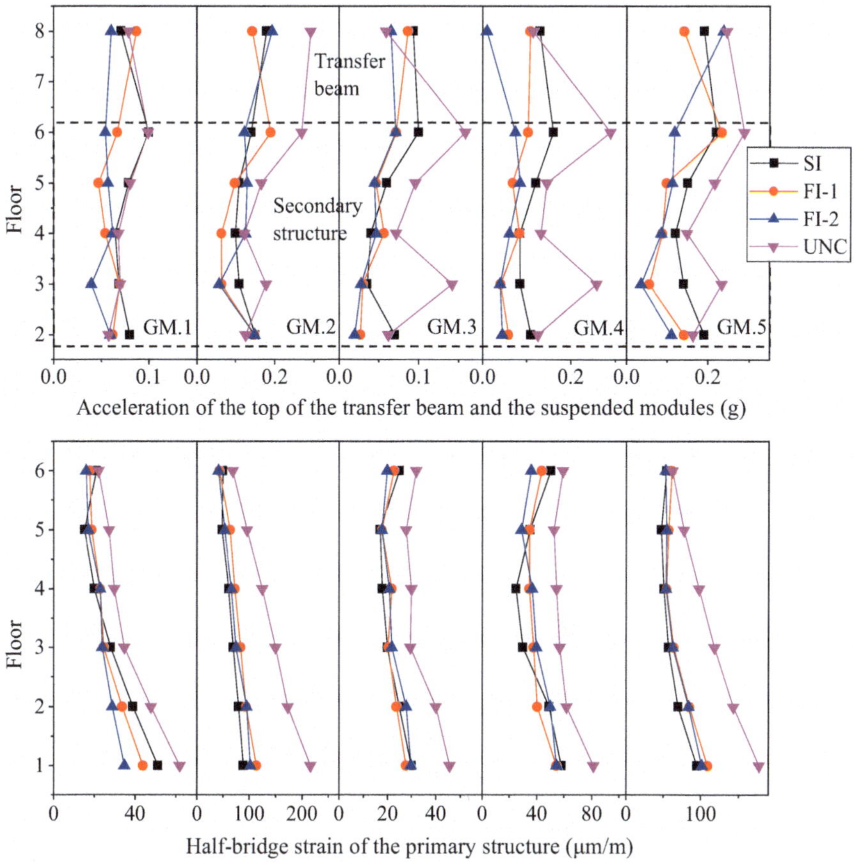

Fig. 8.33 Envelopes of responses under different ground motions

Fig. 8.34 The calibration procedures

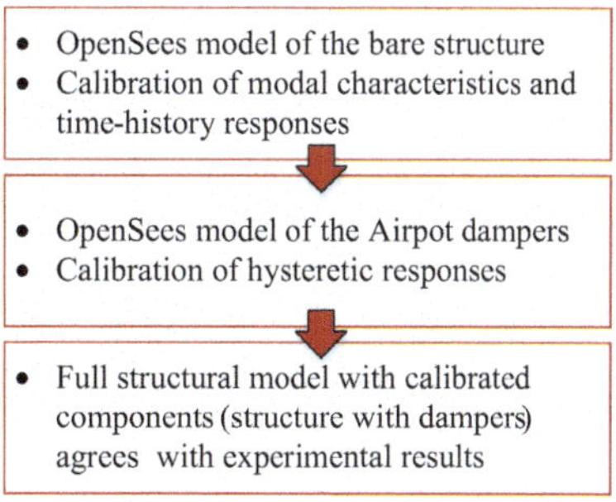

force is roughly linear. When the piston moves rapidly, the spring force in the second Maxwell element is overwhelmed by the viscous force. When the piston moves very quickly, even the first element suffers from air vibration. In this case, the whole damper provides almost no viscous force. The element C is a compression-only element to represent push–pull anisotropy of the air. The full model is calibrated and then subjected to cyclic ground accelerations of 1 Hz and 3 Hz, and the resulting curves approximately match the experiment curves of dampers. After the setup and the calibration of the aforementioned structure model and the Airpot damper model, the two models are combined together to have the full model of the system. The numerically calculated time-history curves match satisfactorily to the experimental ones for all models in all configurations (Fig. 8.36).

The models are respectively assigned with the regional Rayleigh damping (with calibrated values in the primary structure, secondary structure, and the interface [36]) or uniform 1% Rayleigh damping. Then they are subjected to a sequence of the 5 ground motions adopted in the experiment with the interval between ground motions being 2 s. Further numerical modeling is carried out after optimizing the parameters

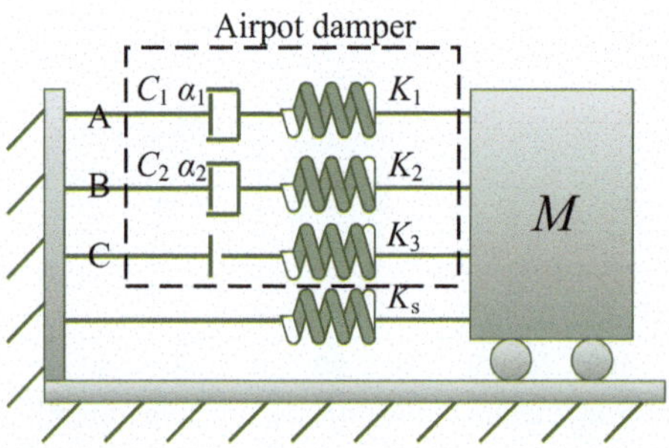

Fig. 8.35 Airpot damper model

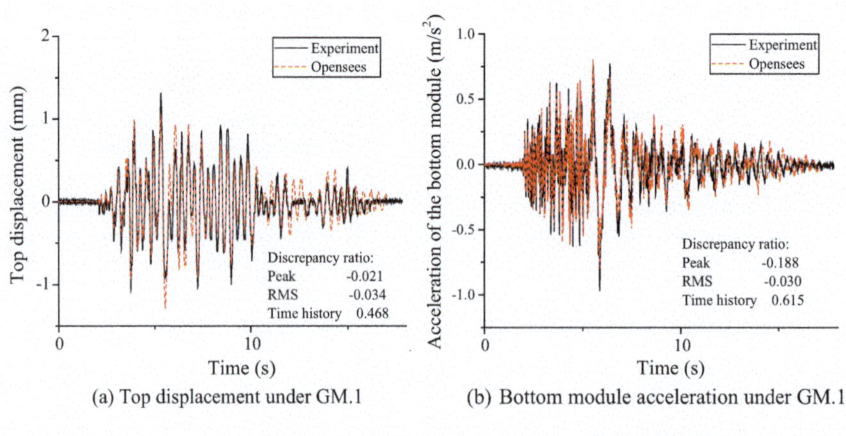

Fig. 8.36 Comparison between experiment and simulated time-histories under GM.1

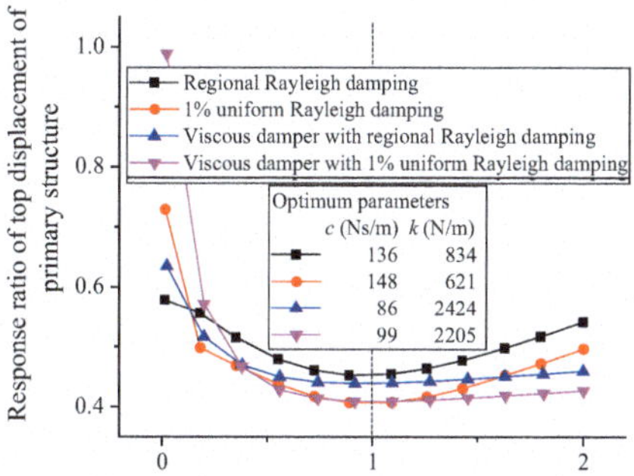

(a) FI-2 varying damping coefficient

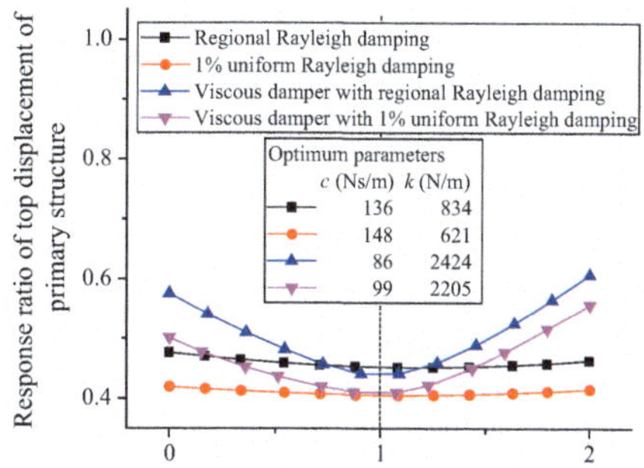

(b) FI-2 varying inter-story stiffness

Fig. 8.37 Parametric analysis of damping coefficient and inter-story stiffness based on OpenSees models at the optimum of each model

for each model using a genetic algorithm, as shown in Fig. 8.37. The results show that models with viscous dampers are rather sensitive to the inter-story stiffness of the secondary structures, while those with airpot dampers are less sensitive. All models are sensitive to damping coefficient except for the one with regional Rayleigh damping and airpot dampers (the experiment-based model), indicating that the overall control is sensitive to the friction level. The setting of the Rayleigh damping notably affects the optimum performance but not the optimum parameters. On the other hand, substituting the Airpot dampers by viscous dampers has a minor influence on the optimal responses but a significant influence on the optimal inter-story stiffness and damping coefficient, as the air spring effect can be compensated by lower inter-story stiffness and higher damping coefficient.

8.8 Conclusions

Passive-control suspended building structures have an appealing architectural effect and structural performance. Although prefabricated components are very common in conventional types of suspended buildings, incorporating the prefabricated 3D modules in an innovative manner can further improve seismic performance. The protection of non-structural components by the suspended module, the relaxed inter-story-drift limit of the suspended segment, and the clear inter-story relationship, are the key reasons for the improved seismic attenuation. From the perspective of modular building, the suspended scheme provides an effective and robust solution of energy dissipation, and avoids vertical accumulation of loads on modular columns. Thus, the scheme helps to achieve the lightness and standardization of prefabricated modules. It has a good potential of application.

The optimized suspended modular building structure attains a wide and deep trough, as well as lowered peaks, of the transfer function, especially near the fundamental frequency. Furthermore, optimum irregular vertical distributions of secondary structure parameters are allowed by this system, leading to efficient multi-mode tuning of the primary and secondary structures. The seismic performance is further improved. Theoretically, the modularization of the suspended segment is not the only method to achieve multi-mode tuning. However, it is a straightforward and feasible one, considering the aforementioned mutual benefits of prefabricated 3D modules and suspended buildings.

Appendix: Categorization of Mega-Substructure Systems

Seismic attenuation strategies of passive-control mega-substructure systems can be categorized based on the pattern of relative motion and the layout of dampers. Since the layout of the dampers is usually optimized based on the pattern of relative motion, categorization in this appendix is mainly based on the pattern of relative motion, as shown in Fig. 8.38.

Type-1: The primary structure is of full-height. The primary and secondary structures are flexibly connected. The inter-story connections within the secondary structure are much stiffer than the primary-secondary structure connections. Relative motions between the two structural regions are developed mainly via the interface.

Type-2: The primary structure is of full-height. The primary and secondary structures are flexibly connected. The inter-story connections within the secondary structure are of the same order of magnitude as those for the primary-secondary structure connections. Relative motions between the two structural regions are developed through the whole secondary structure.

Type-3: The primary structure is of full-height but the secondary structures are of a single story. Secondary structures can be isolated floors or a roof. A large mass ratio of secondary-primary structures may require a large number of isolated floors.

Type-4: The whole structure is vertically segmented via rubber bearings. Each upper segment can be regarded as a secondary structure to the lower segment.

Type-5: The secondary structure consists of little architectural space but a large part of exterior construction, such as flexibly connected curtain walls. The secondary-primary structure mass ratio is usually small; the accelerations of the secondary structures are not to be attenuated; wind load acts only on secondary structures, and wind-induced vibration control is satisfactory.

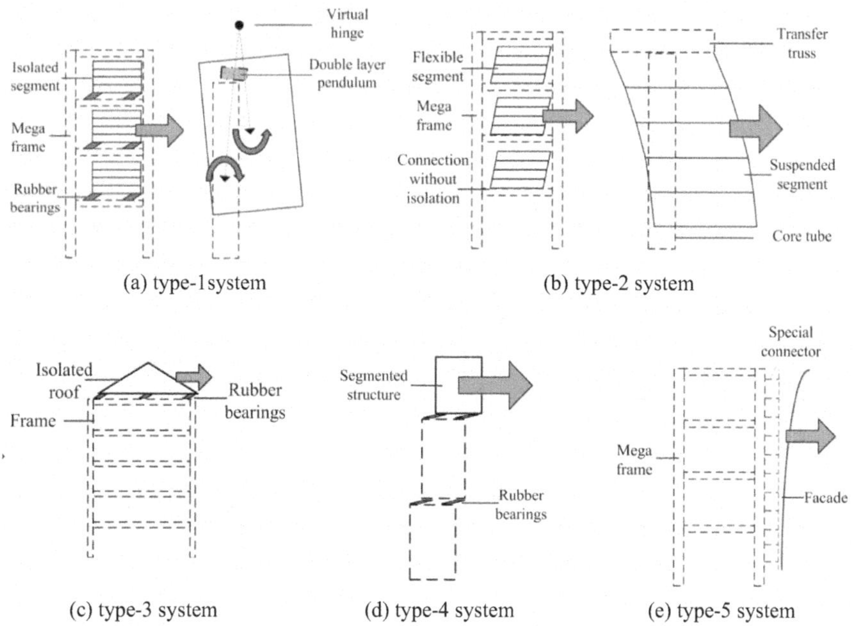

Fig. 8.38 Categorization of mega-substructure systems

References

1. JGJ/T 466–2019. Technical specification for light steel modular building. China Building Industry Press, Beijing, China ([in Chinese])
2. Lawson M, Ogden R, Goodier C, Ogden R, Goodier C (2014) Design in modular construction. CRC Press. https://doi.org/10.1201/b16607
3. Ferdous W, Bai Y, Ngo T, Manalo A, Mendis P (2019) New advancements, challenges and opportunities of multi-storey modular buildings—a state-of-the-art review. Eng Struct 183:883–893. https://doi.org/10.1016/j.engstruct.2019.01.061
4. Ye Z, Giriunas K, Sezen H, Wu G, Feng D-C (2021) State-of-the-art review and investigation of structural stability in multi-story modular buildings. J Build Eng 33:101844. https://doi.org/10.1016/j.jobe.2020.101844
5. Technical specification for modular freight container building, 2013
6. Lawson M, Ray O, Rory B (2012) Application of modular construction in high-rise buildings. J Archit Eng 18:148–154. https://doi.org/10.1061/(ASCE)AE.1943-5568.0000057
7. Fathieh A, Mercan O (2016) Seismic evaluation of modular steel buildings. Eng Struct 122:83–92. https://doi.org/10.1016/j.engstruct.2016.04.054
8. Park H, Ock J (2016) Unit modular in-fill construction method for high-rise buildings. KSCE J Civ Eng 20:1201–1210. https://doi.org/10.1007/s12205-015-0198-2
9. Atlantic Modular System Ltd. (n.d.) http://www.amsgroup.com.cn/en/aboutus.asp. Accessed 21 June 2019
10. Zhang J, Lu Y, Li G (2015) Structural analysis on mechanical properties of 3D steel structure modular buildings. Progr Steel Build Struct 17:57–64 (In Chinese)
11. Zhang P, Zhang X, Liu J, Chen Z (2016) Structural design and analysis of multi-storey steel structure module and steel frame composite structures. Build Struct 46:95–100 (In Chinese)

12. Nakamura Y, Saruta M, Wada A, Takeuchi T, Hikone S, Takahashi T (2011) Development of the core-suspended isolation system. Earthq Eng Struct Dynam 40:429–447. https://doi.org/10.1002/eqe.1036
13. Nakamura Y, Okada K (2021) Effects and dynamic characteristics of the core-suspended isolation system assessed by long-term structural health monitoring. Earthq Engng Struct Dynam 50:2259–2276. https://doi.org/10.1002/eqe.3437
14. Wang C, Lü Z, Tu Y (2011) Dynamic responses of core-tubes with semi-flexible suspension systems linked by viscoelastic dampers under earthquake excitation. Adv Struct Eng 14:801–813
15. Zhang Y, Liang Q (2000) Asynchronous driving principle and its application to vibration control. Earthq Eng Struct Dynam 29:259–270. https://doi.org/10.1002/(SICI)1096-9845(200002)29:2%3c259::AID-EQE904%3e3.0.CO;2-7
16. Wang C, Lu Z (2012) Shaking table experiment of semi-flexible suspended-floor structures, China Civil. Eng J 45:109–117 (In Chinese)
17. Goodno BJ, Gere JM (1976) Earthquake behavior of suspended-floor buildings. J Struct Div 102:973–992
18. Wittfoht H, Cretu G (1987) The 'Bell-Building': a new type of earthquake resistant structure. In: Proceedings of the 5th Canadian conference on earthquake engineering, pp 593–597
19. HSCB headquarters, Wikipedia (2019) https://zh.wikipedia.org/w/index.php?title=%E6%BB%99%E8%B1%90%E7%B8%BD%E8%A1%8C%E5%A4%A7%E5%BB%88&oldid=55089236. Accessed 26 Aug 2019
20. Goodno BJ (1975) Dynamic analysis of suspended-floor highrise building using super-elements. In: Blume JA (ed) Earthquake Engineering Center Technical Report 13. Stanford Digital Repository
21. Wang C (2009) Vibration control mechanism and experimental research of suspended-floor highrise building. Doctoral Thesis. Southeast University, China. (In Chinese)
22. Lan Z, Tian Y, Fang L, Liang S, Wang X (2004) An experimental study on seismic responses of multifunctional vibration-absorption reinforced concrete megaframe structures. Earthq Eng Struct Dynam 33:1–14. https://doi.org/10.1002/eqe.324
23. Feng MQ, Mita A (1995) Vibration control of tall buildings using mega subconfiguration. J Eng Mech 121:1082–1088
24. Chen Z, Liu J, Yu Y (2017) Experimental study on interior connections in modular steel buildings. Eng Struct 147:625–638. https://doi.org/10.1016/j.engstruct.2017.06.002
25. Annan C, Youssef M, El Naggar M (2009) Experimental evaluation of the seismic performance of modular steel-braced frames. Eng Struct 31:1435–1446. https://doi.org/10.1016/j.engstruct.2009.02.024
26. Dhanapal J, Ghaednia H, Das S, Velocci J (2019) Structural performance of state-of-the-art VectorBloc modular connector under axial loads. Eng Struct 183:496–509. https://doi.org/10.1016/j.engstruct.2019.01.023
27. Chen Z, Li H, Chen A, Yu Y, Wang H (2017) Research on pretensioned modular frame test and simulations. Eng Struct 151:774–787. https://doi.org/10.1016/j.engstruct.2017.08.019
28. Park K, Moon J, Lee S, Bae K, Roeder C (2016) Embedded steel column-to-foundation connection for a modular structural system. Eng Struct 110:244–257. https://doi.org/10.1016/j.engstruct.2015.11.034
29. Hong S, Cho B, Chung K, Moon J (2011) Behavior of framed modular building system with double skin steel panels. J Constr Steel Res 67:936–946. https://doi.org/10.1016/j.jcsr.2011.02.002
30. Zha X, Zuo Y (2016) Theoretical and experimental studies on in-plane stiffness of integrated container structure. Adv Mech Eng 8:1687814016637522. https://doi.org/10.1177/1687814016637522
31. Tharaka G, Tuan N, Priyan M, Jose A (2016) Innovative flexible structural system using prefabricated modules. J Archit Eng 22:05016003. https://doi.org/10.1061/(ASCE)AE.1943-5568.0000214

32. Ye Z, Wu G (2017) Optimal lateral aseismic performance analysis of mega-substructure system with modularized secondary structures. Struct Design Tall Spec Build 26:e1387. https://doi.org/10.1002/tal.1387
33. Ye Z, Feng D-C, Wu G (2019) Seismic control of modularized suspended structures with optimal vertical distributions of the secondary structure parameters. Eng Struct 183:160–179. https://doi.org/10.1016/j.engstruct.2018.12.099
34. Behr RA, Belarbi A, Culp JH (1995) Dynamic racking tests of curtain wall glass elements with in-plane and out-of-plane motions. Earthq Eng Struct Dynam 24:1–14. https://doi.org/10.1002/eqe.4290240102
35. Rigoni E, Poles S (2005) NBI and MOGA-II, two complementary algorithms for multi-objective optimizations. In: Dagstuhl Seminar Proceedings, Schloss Dagstuhl-Leibniz-Zentrum für Informatik
36. Ye Z, Wu G, Feng D-C, Shafieezadeh A (2020) Shake table testing and computational investigation of the seismic performance of modularized suspended building systems. Bull Earthq Eng. https://doi.org/10.1007/s10518-020-00902-3

The manufacturer's authorised representative in the EU is Springer Nature Customer Service Centre GmbH, Europaplatz 3, 69115 Heidelberg, Germany. If you have any concerns regarding our products, please contact ProductSafety@springernature.com

Printed and bound by CPI Group (UK) Ltd, Croydon, CR0 4YY
30/09/2026
02225317-0011